Advanced Mass Spectrometry-based Analytical Separation Techniques for Probing the Polar Metabolome

New Developments in Mass Spectrometry

Editor-in-chief:
Frank Sobott, *University of Leeds, UK*

Series editors:
Juan F. Garcia-Reyes, *Universidad de Jaén, Spain*
Marek Domin, *Boston College, USA*

Titles in the Series:

1: Quantitative Proteomics
2: Ambient Ionization Mass Spectrometry
3: Sector Field Mass Spectrometry for Elemental and Isotopic Analysis
4: Tandem Mass Spectrometry of Lipids: Molecular Analysis of Complex Lipids
5: Proteome Informatics
6: Capillary Electrophoresis–Mass Spectrometry for Metabolomics
7: Lipidomics: Current and Emerging Techniques
8: Processing Metabolomics and Proteomics Data with Open Software: A Practical Guide
9: Advanced Fragmentation Methods in Biomolecular Mass Spectrometry: Probing Primary and Higher-order Structure with Electrons, Photons and Surfaces
10: Advanced Mass Spectrometry-based Analytical Separation Techniques for Probing the Polar Metabolome

How to obtain future titles on publication:
A standing order plan is available for this series. A standing order will bring delivery of each new volume immediately on publication.

For further information please contact:
Book Sales Department, Royal Society of Chemistry, Thomas Graham House, Science Park, Milton Road, Cambridge, CB4 0WF, UK
Telephone: +44 (0)1223 420066, Fax: +44 (0)1223 420247
Email: booksales@rsc.org
Visit our website at www.rsc.org/books

Advanced Mass Spectrometry-based Analytical Separation Techniques for Probing the Polar Metabolome

Edited by

Rawi Ramautar
Leiden University, The Netherlands
Email: r.ramautar@lacdr.leidenuniv.nl

ROYAL SOCIETY
OF **CHEMISTRY**

New Developments in Mass Spectrometry No. 10

Print ISBN: 978-1-83916-163-6
PDF ISBN: 978-1-83916-352-4
EPUB ISBN: 978-1-83916-353-1
Print ISSN: 2045-7545
Electronic ISSN: 2045-7553

A catalogue record for this book is available from the British Library

The Royal Society of Chemistry is a charity, registered in England and Wales, Number 207890, and a company incorporated in England by Royal Charter (Registered No. RC000524), registered office: Burlington House, Piccadilly, London W1J 0BA, UK, Telephone: +44 (0) 20 7437 8656.

For further information see our web site at www.rsc.org

Printed in the United Kingdom by CPI Group (UK) Ltd, Croydon, CR0 4YY, UK

Preface

Today, metabolomics is considered a very useful approach for addressing key questions in biology and biomedicine. Although advanced analytical separation techniques are currently used for this purpose, the highly efficient analysis of polar and ionizable metabolites in biological samples remains a huge challenge in metabolomics. As these compounds play a crucial role in basic biochemical pathways (*e.g.*, central carbon metabolism, glycolysis and the pentose phosphate pathway), their comprehensive, reliable and efficient determination is of considerable importance. Over recent years, significant progress has been made in the development of mass spectrometry-based analytical tools enabling the sensitive and efficient profiling of polar ionogenic metabolites in various biological samples. The aim of this book is to give the reader a comprehensive overview of these recent methodological developments, especially from the viewpoint of separation scientists.

This book is composed of 12 chapters in which the state-of-the-art analytical technologies for probing the polar metabolome are critically discussed with illustrative examples. The first chapter sets the stage by providing a helicopter overview of the LC–MS methodologies that have strong potential for analyzing the polar metabolome. Then, in separate chapters, the possibilities of reversed-phase LC–MS using chemical labeling strategies, ion-pair reversed-phase LC–MS and HILIC–MS are discussed in more detail for probing the polar metabolome with an emphasis on the latest methodological developments. The potential of ion-mobility time-of-flight MS, as a new player in this territory, is also considered in this context.

As more and more biological and biomedical problems are intrinsically dealing with minute sample amounts, ample attention has been devoted to

New Developments in Mass Spectrometry No. 10
Advanced Mass Spectrometry-based Analytical Separation Techniques
for Probing the Polar Metabolome
Edited by Rawi Ramautar

Published by the Royal Society of Chemistry, www.rsc.org

recently developed microscale sample preparation (solid-phase microextraction) and microscale analytical separation techniques for analyzing polar metabolites in material-restricted samples (capillary electrophoresis–mass spectrometry and nano-LC–MS). Since this book aims to be comprehensive, the main steps and challenges associated with the assessment of the polar metabolome by nuclear magnetic resonance (NMR) spectroscopy are also considered. The utility and complementary character of the various analytical techniques for probing the polar metabolome are demonstrated by using Alzheimer's disease as a case study.

Finally, this book also covers some challenging and special applications. For example, the possibilities of MS-based analytical techniques for metabolic profiling of live single cells are presented, with a particular focus on challenges related to sampling and isolating, in one chapter, while in another chapter challenges associated with the development of LC–MS methods for probing the chiral metabolome, which is virtually uncharted territory, are outlined.

Overall, this book should give a contemporary representation of the main technological developments in analytical techniques for probing the polar metabolome, including important application fields. I hope that this book is useful for both beginners and experts in the field of metabolomics and bioanalysis, and that it will stimulate further research in this area. I highly appreciate the efforts of all contributing authors and would like to thank them for their excellent chapters. This book would not have been possible without their valuable contributions. I would also like to thank the referees for their help in the evaluation of these papers and the editorial staff, especially Katie Morrey, Janet Freshwater and Liv Towers, from the Royal Society of Chemistry for the support during the preparation of this book.

Rawi Ramautar,
Associate Professor and Principal Investigator,
Leiden Academic Centre for Drug Research,
Leiden University, Leiden,
The Netherlands

Editor Biography

Rawi Ramautar obtained his PhD on the development of capillary electrophoresis–mass spectrometry methods for metabolomics from Utrecht University, the Netherlands, in 2010. Intrigued by metabolomics for disease prediction and diagnosis, Rawi switched to the Leiden University Medical Center to broaden his horizons on this topic. In 2013 and 2017, he received the prestigious Veni and Vidi research grants from the Netherlands Organization for Scientific Research for the development of CE–MS approaches for volume-restricted metabolomics. Currently, he is an associate professor at the Leiden University where his group is developing microscale analytical workflows for sample-restricted biomedical problems. Rawi Ramautar was recently selected for the Top 40 under 40 Power List of the Analytical Scientist. He is an editor of Microchemical Journal (Elsevier).

New Developments in Mass Spectrometry No. 10
Advanced Mass Spectrometry-based Analytical Separation Techniques
for Probing the Polar Metabolome
Edited by Rawi Ramautar

Published by the Royal Society of Chemistry, www.rsc.org

Contents

New Developments in Mass Spectrometry No. 10
Advanced Mass Spectrometry-based Analytical Separation Techniques for Probing the Polar Metabolome
Edited by Rawi Ramautar

Published by the Royal Society of Chemistry, www.rsc.org

CHAPTER 1

State-of-the-art in LC–MS Approaches for Probing the Polar Metabolome

ARTEMIS LIOUPI,[a,b,c,†] MARIA MARINAKI,[a,b,c,†] CHRISTINA VIRGILIOU,*[a,b,c] HELEN GIKA,[b,c,d] IAN WILSON[e] AND GEORGIOS THEODORIDIS[a,b,c]

[a] Laboratory of Analytical Chemistry, School of Chemistry, Aristotle University of Thessaloniki, 54124 Thessaloniki, Greece; [b] Biomic AUTh, Center for Interdisciplinary Research and Innovation (CIRI-AUTH), Balkan Center B1.4, 10th km Thessaloniki-Thermi Rd, P.O. Box 8318, GR 57001, Thessaloniki, Greece; [c] FoodOmicsGR Research Infrastructure, AUTh Node, Center for Interdisciplinary Research and Innovation (CIRI-AUTH), Balkan Center B1.4, 10th km Thessaloniki-Thermi Rd, P.O. Box 8318, GR 57001, Thessaloniki, Greece; [d] Department of Medicine, Aristotle University, 54124 Thessaloniki, Greece; [e] Division of Computational and Systems Medicine, Department of Metabolism, Digestion and Reproduction, Imperial College, South Kensington, London, United Kingdom
*Email: cr_virgi@hotmail.com

[†]Authors of equal contribution.

New Developments in Mass Spectrometry No. 10
Advanced Mass Spectrometry-based Analytical Separation Techniques for Probing the Polar Metabolome
Edited by Rawi Ramautar

Published by the Royal Society of Chemistry, www.rsc.org

1.1 Introduction: Metabolomics and the Polar Metabolome

Metabolomics is a rapidly growing research area that focuses on the systematic study of small molecule metabolites found in various types of biological samples, such as tissues, cells, and body fluids.[1,2] These small molecules (molecular masses typically below 1000 amu) have diverse physicochemical properties and are present over a wide range of concentrations, from fmol to mM, rendering global analysis a challenging task. Recent advances in analytical techniques partially address the challenges related to standardizing metabolomics and maximizing the amount of extracted information. Nuclear magnetic resonance (NMR) spectroscopy and MS-based techniques, such as gas chromatography coupled to mass spectrometry (GC–MS), liquid chromatography coupled to MS (LC–MS), and capillary electrophoresis coupled to MS (CE–MS), are widely and effectively utilized for the analysis of microbial, plant, and animal metabolomes.[3–5] Comprehensive two-dimensional (2D) techniques (LC×LC, GC×GC) or tandem-MS (MS/MS) have also been tested in the quest to obtain the most detailed profiles of metabolites in a single analysis. More recently ion mobility-based techniques have also been integrated into metabolic and lipidomic phenotyping protocols to enhance metabolome coverage.[6]

Considering the scope of the analysis, metabolomics studies are widely categorized into two major approaches: targeted and untargeted metabolomics.[7] Targeted methods aim at the identification and quantification of a small subset of pre-selected molecules (from the low tens to some hundreds of analytes).[8,9] Targeted methods are extensively used in clinical and medical applications and find use in the confirmation of biomarkers and hypothesis testing. Analysis can be carried out in a quantitative or semi-quantitative manner with the use of internal standards, surrogate matrices, or surrogate analytes. These approaches exploit the extensive understanding of a wide range of metabolic enzymes, their kinetics, end products, and the known biochemical pathways to which they contribute.[10] Tandem triple quadrupole instruments (TQ) operating in multiple reaction monitoring (MRM) mode have enabled simultaneous targeted analysis of hundreds of metabolites involved in important metabolic pathways.[11] Untargeted metabolomics, on the other hand, focuses on the detection and identification of as many metabolites as possible in a sample, to obtain accurate and comprehensive metabolic fingerprints.[12] A recent review covered the topic of targeted metabolomics.[5] Untargeted metabolomics relies mainly on either ^{1}H NMR spectroscopy, direct infusion (DI) MS, or the hyphenation of high-efficiency chromatographic separations with the high-sensitivity and high-resolution mass spectrometry analysis on time of flight (TOF) or Orbitrap-MS detectors in full scan mode.[13] Ultimately, untargeted experiments offer wide metabolome coverage, whereas targeted experiments provide better quantification, usually through the usage of isotope-labeled internal standards and defined conditions for the mass spectrometer and the chromatographic separation system.[14]

Even after the continuous improvement and refinement of analytical platforms over the past few years, the analytical research community is still struggling with the analysis of highly polar and charged metabolites. A potential solution is resorting to ^{1}H NMR spectroscopy, where the polarity of the metabolite is irrelevant, and only the concentration and the possession of protons to generate a measurable signal are required. ^{1}H NMR spectroscopy requires minimum sample preparation and is especially suited for the analysis detection and characterization of polar metabolites such as sugars, organic acids, alcohols, polyols, and other highly polar compounds that are less tractable to LC–MS.[15,16] Yet, such analysis is often much less sensitive compared to LC and GC–MS-based techniques, although this depends on the nature of the analyte. GC–MS also presents important limitations in the analysis of highly polar compounds, due to their thermolability and low volatility.[17] Therefore, a derivatization step is required before analysis. CE–MS is highly effective for the analysis of polar and charged metabolites and can offer important and complementary information, supplementing that obtained by LC–MS regarding the biological composition of samples.[18,19] However, CE currently lacks the robustness required for analyzing large sets of biological samples, and its use in the field of metabolomics is still relatively limited.[20]

Here, we describe state-of-the-art in LC–MS approaches for the analysis of the polar metabolome. There are very strong needs to map the polar metabolome because polar hydrophilic metabolites are important in many basic biochemical pathways. With the ever-growing need to explain altered cellular metabolism in cancer and other diseases, the evaluation of polar central carbon metabolites associated with energy production has been at the forefront of metabolomics.[21] Similarly, amino acid metabolism as a whole or specific pathways, *e.g.*, the tryptophan pathway, is increasingly screened in disease biomarker discovery. Continuous enhancements in LC separations, MS technologies, bioinformatic tool development, and database expansion have significantly improved the LC–MS workflow and offer new perspectives on the analysis of polar compounds.[21]

1.2 LC Technologies: General Introduction

HPLC can resolve relatively complex analyte mixtures that contain analytes with various molecular weights, polarities, and acid–base properties. To accomplish this, different modes of separation, such as reversed-phase chromatography (RPLC), normal-phase liquid chromatography (NPLC), ion-exchange chromatography (IEC), and hydrophilic interaction (HILIC) chromatography have been developed. The selection of the appropriate LC mode relies on the analytes and the sample of interest, as well as the available analytical platforms.[22,23]

The combination of the detection capabilities of a mass spectrometer, especially when linked to an LC separation provides advanced analytical capabilities, such as high sensitivity, increased specificity, and rapid analysis. Hence LC–MS is now the most powerful instrumental platform for the analysis of non-volatile molecules and in particular small molecule metabolites.[24] Important trends are seen in the use of ultra-high performance (UHPLC), multi-dimensional (MD) LC, and technical aspects, such as miniaturization/microfluidics, and new MS interfaces.[23] The establishment of commercially available UHPLC has been used to dramatically increase the coverage of the metabolome and/or throughput in comparison to regular HPLC methods.[25] UHPLC typically uses stationary phases of sub-2 μm particle diameter for packed columns or in the range of 2.6–2.8 μm for fused-core particles and high solvent flow rates and pressures, of up to 19 000 psi. Higher flow rates lead to shorter analytical run times and increased peak capacity. The greater chromatographic resolution of UHPLC compared to HPLC leads to reduced matrix effects and a reduction of mass spectral overlap. Capillary (Cap) LC and nano-LC could also be listed among the current topics of interest in liquid chromatography[26] as they offer increased sensitivity and much lower solvent consumption. New column technologies are continuously being introduced, such as specially treated phases and materials with the aim of improving chromatographic efficiency and versatility, especially for the analysis of very polar molecules. Furthermore, novel materials with micro-chips and, in particular, pillar array columns are gaining interest for in-field or real-time analysis.

1.3 LC Technologies to Probe the Polar Metabolome

Among the several modes of chromatography and the stationary phases available in metabolomics–metabonomics, the most common systems rely on reversed-phase (RPLC) and to a lesser extent on HILIC.[27] RP-based approaches utilize non-polar stationary phases and polar mobile phases and are well suited to the LC of "medium" to non-polar metabolites and are not particularly useful for polar metabolites. Several important classes of metabolites *e.g.*, carbohydrates or amino acids have polar, or polar–ionic, properties and are unretained in RPLC. To profile such metabolites using RPLC methods a derivatization step may be required to render them hydrophobic and amenable to this mode of separation.[28]

An alternative option is to use HILIC, which finds increasing application for the analysis of polar metabolites that are not well retained in RP systems.[29] HILIC has grown in popularity in bioanalysis and the field of metabolomics, as it provides a wide range of different retention mechanisms and column selectivities to probe the polar metabolome.[30] A consequence of the differential metabolome coverage of RP and HILIC profiling has been the requirement to use both methodologies to analyze samples leading to the need for each sample to be analyzed multiple times (including analyzes in both positive and negative ionization modes) if maximum coverage is required.

In recent years several other methods such as ion-pair chromatography (IPC), mixed-mode chromatography (MMC), and IEC have been applied to the analysis of potentially "problematic" polar analytes, which are not well served by either RPLC or HILIC. If there is also a need for improved separation efficiencies and maximum metabolite coverage in a single analysis, multi-dimensional separations such as comprehensive 2D liquid chromatography (LC×LC) can be applied.[31] Another alternative could be normal phase (NP) LC but reproducibility in retention can be a practical problem, and the solubilization of ionic analytes in the non-polar solvents required for sample injection may be challenging. These drawbacks have drastically reduced its use in metabolomics.[32]

1.3.1 Hydrophilic Interaction Liquid Chromatography (HILIC)

HILIC is nowadays the first choice for the separation of polar metabolites, due to the ability it offers for the high-resolution separation of many polar or semi-polar metabolites, and the alternative[33] complementarity to RPLC separation selectivity.[34–36] Analyte separation is based on a combination of partitioning and electrostatic interactions with a polar stationary phase and the use of mobile phases with high ratios of organic solvent. A typical HILIC method requires hydrophilic stationary phases such as bare or chemically modified silica (amino, amide, diol, cyano, or some other ligand) or polar polymers. Different HILIC chemistries[37,38] and materials (usually silica gel-based[39]) have been used, depending on the properties of the analytes of interest. These can generally be grouped as functionalized and non-modified separation media[40] and more specifically as neutral, charged, or zwitterionic bonded phases.[41] The most commonly used modifications of silica are amide,[42,43] aminopropyl,[44] diol,[45] and pentafluorophenylpropyl[46] phases; in the case of zwitterionic phases, ZIC-HILIC[47,48] (sulfoalkylbetaine or phosphorylcholine) and ZIC-pHILIC[49,50] (polymer-based) are the most "traditional" columns for HILIC.

As the mobile phase, polar organic solvents (most frequently acetonitrile, sometimes methanol) are usually selected, containing up to 40% water, rendering this technique appropriate for tandem with electrospray mass spectrometry (ESI–MS).[51] Buffer salt concentration (ammonium formate or ammonium acetate) and pH adjustment (value of 3–8) must also be taken into consideration for the development of effective separations. Separations are performed either by isocratic[52] or by gradient elution[53,54] with a high percentage of acetonitrile (98–60%) followed by an increase in the proportion of the aqueous phase up to 40%. Flow rates in UHPLC–HILIC, depending upon column diameter, vary from 0.1 to 0.8 mL min^{-1}. In LC–MS analysis the high portion of organic solvent in the mobile phase provides high MS sensitivity as it typically improves analyte ionization efficiency in ESI.[55,56]

Overall separation of polar analytes in HILIC can be more straightforward compared to RPLC, as there is no need for analyte derivatization and thus no interferences caused by excess of derivatizing reagent.[57,58]

The mechanism of HILIC separation and the optimization of analyte detection in HILIC–MS have been studied extensively by several researchers aiming to highlight differences between HILIC stationary phases and study the effects of mobile phase on analyte separation and detection.[59–61] For instance, Sampsonidis *et al.*[59] compared three columns (an amide, a diol, and a zwitterionic phase) and studied analytical parameters, such as the number of detected compounds, the chromatographic resolution and peak width obtained using the same HILIC methods on the three columns. The scope of this study was to mathematically compare the results obtained from the analysis of several groups of compounds by the use of different combinations of separation parameters.

Recent studies have shown that HILIC is suitable for both untargeted[62–73] profiling and (qualitative or quantitative) targeted[59–61,74–83] approaches. HILIC–TOF–MS has been widely applied in various application areas from disease biomarker discovery[65,68,72] to plant phytochemistry and food chemistry. For example, in a pilot study aimed at the study of chronic kidney disease, an untargeted HILIC–TOF–MS method, with a 30 minute run time, in both positive and negative ESI modes, was used to detect discriminating metabolites in patients.[68] Kok *et al.*[65] used HPLC/MS (–ESI) on an XBridge amide column to study the effects of antibiotic-treatment on the urinary profiles of rats. Using principal component analysis (PCA), twenty-nine polar compounds (such as aconitic acid, sulfates of ferulic acid, and benzene-diol) were found to be differentiated in the samples. In other research, aimed to identify metabolic differences between healthy controls, patients with Mild Cognitive Impairment and those with Alzheimer's disease, HILIC–QTOF–MS profiling detected 6751 ions in plasma samples.[72] Multivariate statistical analysis using OPLS–DA (Figure 1.1) showed clear group separation and indicated effects on the biochemical pathways of polyamines and arginine.

An early example of the potential and power of targeted HILIC–MS/MS in the metabolic profiling of multiple types of biological matrices included the study of urine, serum, amniotic fluid, and cell culture medium. This method used a targeted HILIC–MRM method on a HILIC Amide column that provided semi-quantification of *ca.* 100 polar metabolites (including sugars, amino acids, organic acids, and amines). When applied to the study of *in vitro* embryo growth a range of metabolites, with roles in central biosynthetic routes and embryonic growth, were found to be significant.[76,84,85] Another study described the monitoring of 263 water-soluble metabolites in order to detect biomarkers of PCB exposure in maternal and umbilical cord serum samples.[33] Both studies were based on the use of amide–HILIC columns and acetonitrile–water gradient elution systems, applying the polarity switching mode in ESI. Furthermore, two recent studies[81,83] used similar methods to determine polar compounds, such as amines and amino acids.

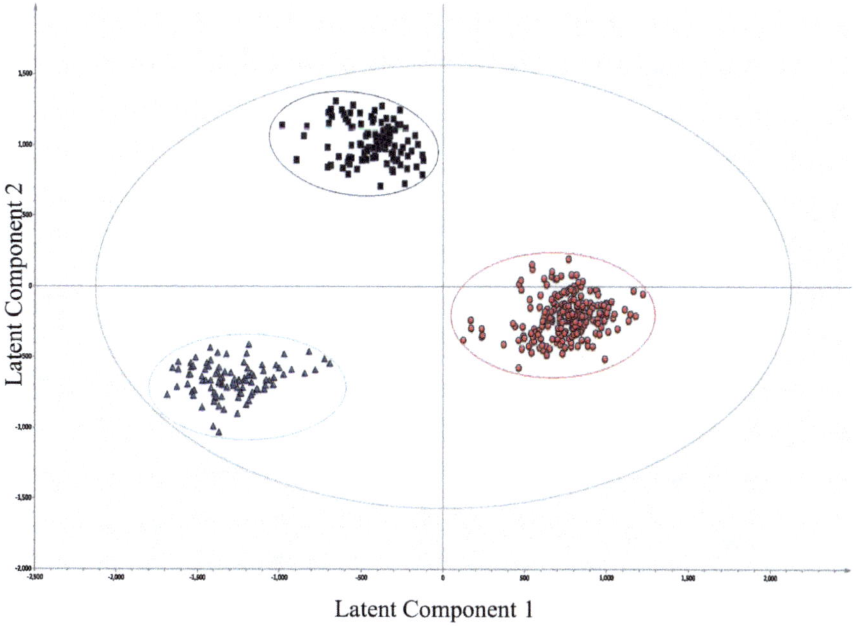

Figure 1.1 Multivariate model generated from data obtained from HILIC–TOF–MS analysis of human plasma from controls and Alzheimer's patients (MCI and MCI_AD). OPLS-DA scores plot shows controls (red circles), MCI (black squares), and MCI_AD (blue triangles).
Reproduced from ref. 72, https://doi.org/10.1371/journal.pone.0119452, under the terms of the CC BY 4.0 license, https://creativecommons.org/licenses/by/4.0/.

HILIC is suitable not only for the analysis of mammalian samples but also for the analysis of food for vitamins, organic acids, and amino acids, *etc.* For example, 23 amino acids were detected in a study of cucumber leaf tissue extracts by a dual eluent linear gradient over 15 minutes.[86] The study showed that exposure to copper-containing nanopesticides resulted in higher concentrations of these amino acids compared to controls.

Gika *et al.*[87] developed a quantitative method for the detection of 135 polar metabolites in grape extracts, in order to classify grape samples according to their skin color. The concentration of some sugars was found to be higher in red varieties in contrast to the concentration of metabolites such as guanidine, mannitol, and glutaric acid, which was higher in green varieties. Representative chromatograms from this analysis are shown in Figure 1.2. Subsequently, Arapitsas *et al.*[88] investigated the effects of storage conditions on red wine in an untargeted LC–MS mode, based on the above targeted method. A combination of chromatographic analysis with optimized sample preparation was key to the development of a robust and sensitive untargeted HILIC–TOF–MS method.

Other examples of development and application of targeted HILIC–MS/MS methods include the analysis of royal jelly samples for the detection of

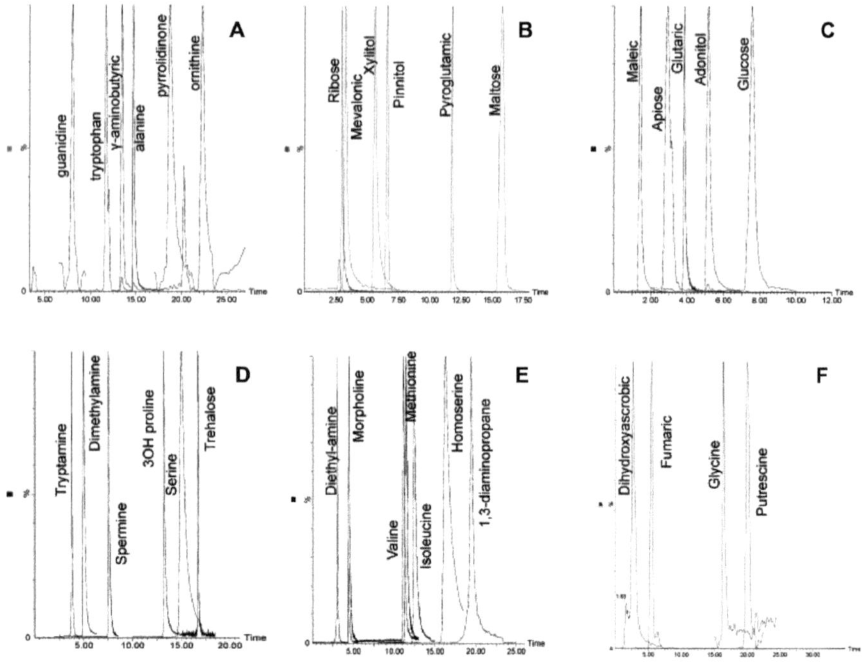

Figure 1.2 UHPLC–MS/MS chromatographic traces for polar analytes analyzed on HILIC Amide column and MRM analysis on UPLC–ESI–QqQ–MS. In the original paper *ca.* 120 analytes were determined. For clarity, 34 different ion traces are depicted in 6 different panes. Metabolite traces from top left to bottom right and in increasing retention time. (A) Guanidine, tryptamine, *g*-aminobutyric acid, alanine, pyrrolidinone, and ornithine. (B) Ribose, mevalonic acid, xylitol, pinitol, pyroglutamic acid, and maltose. (C) Maleic acid, apiose, glutaric acid, adonitol, and glucose. (D) Typtamine, dimethylamine, spermine, 3-hydroxyproline, serine, and trehalose. (E) Diethylamine, morpholine, valine, methionine, homoserine, and 1-3 diaminopropane. (F) Dihydro ascorbic acid, fumaric acid, glycine, and putrescine.
Reproduced from ref. 87 with permission from Elsevier, Copyright 2012.

biomarkers related to adulteration and specifically to sugar supplementation of the diet of bees. Polar metabolites such as amino acids, amines, carbohydrates, and vitamins were differentiated in royal jelly obtained after feeding bees with different types of sugar supplements.[89,90] The combination of HILIC with ion mobility (IM) separation to provide a very rapid HILIC–IM–MS method, suitable for the screening of large numbers of samples, has recently been described with application to mouse urine.[91]

As these examples show, whilst not a complete solution for polar/ionic analytes, HILIC is probably the most widely applied method for the analysis of metabolites that are poorly retained in RPLC. Given its widespread adoption and many successful applications, HILIC-based metabotyping for polar metabolites is likely to remain popular in this role for the foreseeable future. The potential of HILIC in metabolomics studies has been covered by a number of reviews *e.g.*, ref. 92.

1.3.2 Reversed-phase Liquid Chromatography (RPLC) with Derivatization

RPLC remains the dominant mode of LC separation and probably by far the most used separation mode in any application field (including GC, CE, and other chromatographic separations). Improvements in the retention of some polar, ionic, metabolites can be achieved by the addition of additives such as formic acid and salts (ammonium formate, acetate, and carbonate) to the mobile phase to control the pH value and suppress ionization.[93,94]

However, RPLC, as discussed, has limitations for metabolites found in the TCA cycle, glycolysis, amino acid metabolism, *etc.*, where metabolites are either poorly retained, or not retained at all, on the column. A solution to such issues is to modify the polar groups responsible for poor chromatographic properties *via* chemical derivatization. Derivatization may not only improve retention, but can have other advantages such as producing characteristic fragments and superior ionization efficiencies which improve both the sensitivity and selectivity of MS detection.[95,96]

One of the best-known applications of RPLC–MS and indeed HPLC analysis in general following derivatization is the analysis of amino acids.[97] A large number of derivatization reagents have been developed and tested for this purpose, including commercial kits from instrument vendors. An example from the development of a targeted method for the quantification of 20 amino acids in human serum by RP–LC–MS/MS after N-phosphorylation labeling is shown in Figure 1.3.[98]

Sugar phosphate metabolites are very important biomolecules but are not effectively analyzed in HILIC and are practically un-retained in RPLC. Rende *et al.*[99] performed a two-step derivatization procedure in order to render sugar–phosphate metabolites amenable to RPLC. Methoxylamine, propionic acid anhydride, and *N*-methylimidazole were used for the derivatization reaction in the scheme depicted in Figure 1.4. Nineteen compounds were derivatized, and most of them were quantified in Populus leaf and wood extracts by applying a UHPLC–ESI–QqQ–MS method.

The analysis of native carbohydrates and oligosaccharides by MS is technically difficult because of their poor ionization efficiencies. However, by using a combination of derivatization with 3-methyl-1-phenyl-2-pyrazolin-5-one and a pentafluorophenylpropyl stationary phase, which offers retention and selectivity based on a combination of hydrophobic, hydrogen bonding, dipole–dipole, and π–π interactions, a suitable separation can be achieved. This combination was used to provide fingerprints in the screening of oligosaccharidosis in human urine samples from patients with I-Cell disease, mannosidosis, Pompe, galactosialidosis, or fucosidosis.[100]

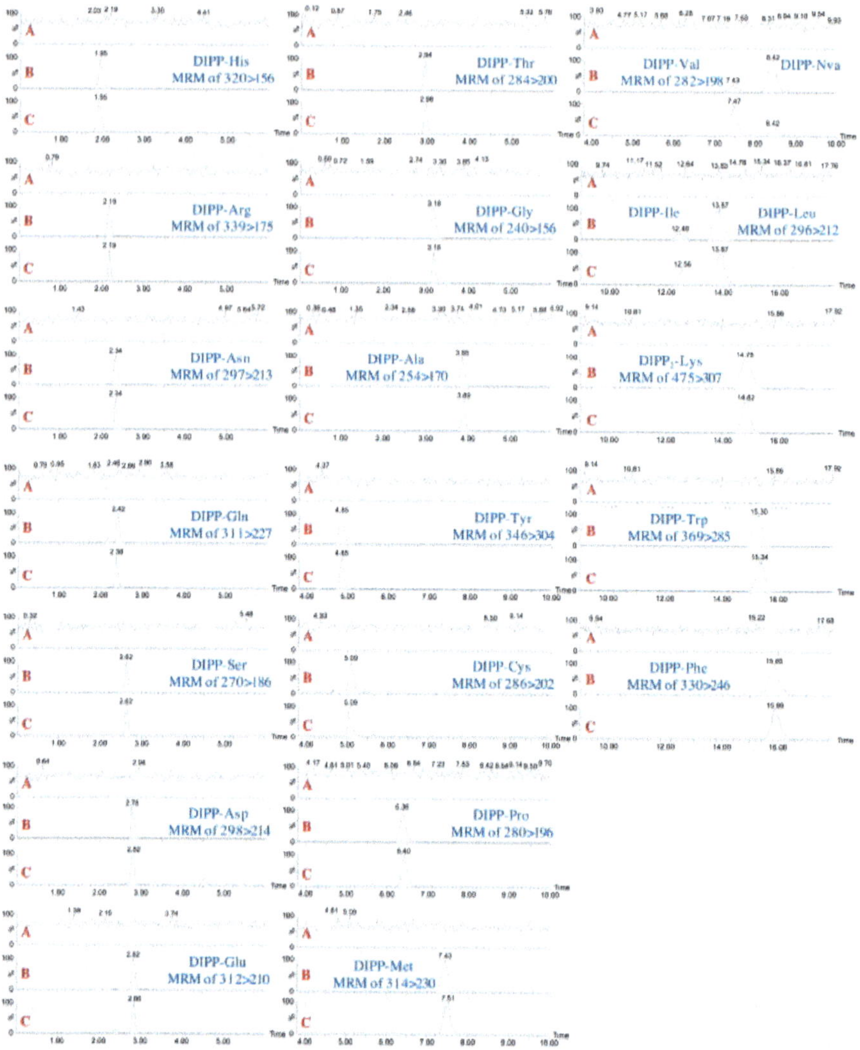

Figure 1.3 LC–MS/MS analysis on MRM mode for twenty-one amino acids separated on an RP-C18 column within 20 min by isocratic elution (0.1% formic acid–acetonitrile, v/v 7 : 3) in positive ESI after labeling with DIPP in blank matrix (A), blank matrix spiked with calibrator (B), and human serum sample (C).
Reproduced from ref. 98 with permission from Elsevier, Copyright 2014.

Dansylation is another approach that can be used in metabolomics studies for the derivatization of metabolites containing amines or phenolic hydroxyls *etc.*,[101] and aliphatic hydroxyl groups.[102] Interestingly the proponents of this derivatization strategy used differential isotope labeling with ^{12}C and ^{13}C-labeled dansylating reagents enabling them to compare the study samples (derivatized with the ^{12}C reagent) to a study pool sample

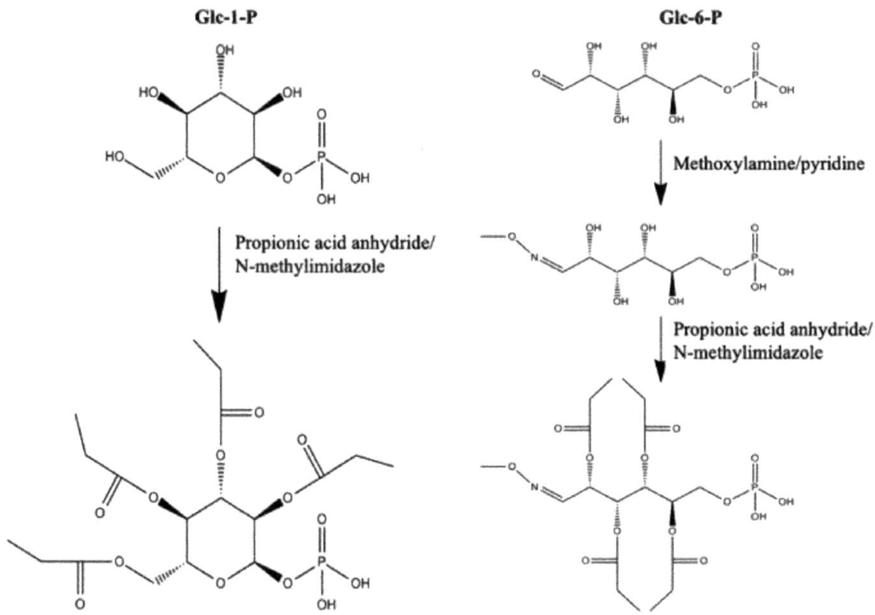

Figure 1.4 Examples of the derivatization of sugar phosphates. Left: propionylation of glucose-1-P (using methylimidazol/propionic anhydrate); right: methoxylation (methoxyamine/pyridine) and subsequent propionylation (using methylimidazol/propionic anhydrate) of glucose-6-P. Reproduced from ref. 99, https://doi.org/10.1186/s13007-019-0514-9, under the terms of the CC BY 4.0 license, https://creativecommons.org/licenses/by/4.0/.

(derivatized with the ^{13}C reagent). The methodology has been utilized in a number of metabolomic investigations including a study on biomarkers of early on-set familial Alzheimer's disease in a TgCRND8 mouse model. Metabolic differences between control and pre-symptomatic AD-prone mice were seen when the mice were 15–17 weeks old and were shown to be more pronounced after 25 weeks. Polar candidate biomarkers identified by this approach included amino acids such as methionine, desaminotyrosine, and taurine as well as N1-acetylspermidine, and 5-hydroxyindoleacetic acid.

Derivatization offers several potential advantages for those researchers who wish to use RPLC for otherwise poorly or unretained metabolites. In the first instance, and most importantly, derivatization can dramatically change hydrophilicity and enable good retention to be achieved on RP-phases. In addition, derivatizing reagents can, depending upon their structure, greatly improve MS signal response and can also provide a common fragment ion that identifies compounds of a similar class. They may also aid in quantification/semi-quantification by reducing differences in ionization properties by "normalizing" them to the derivatizing reagent rather than the analyte. Clearly, the downside of derivatization is the extra steps that it introduces into the analysis.

1.3.3 Ion-pair Chromatography (IPC)

IPC is commonly used for the separation of polar analytes (organic or inorganic) that contain ionizable groups. The separation of these compounds can be performed by the use of a hydrophobic stationary phase (*e.g.*, C8, C18, *etc.*) and an ion-pairing (IP) reagent that is typically added to the mobile phase usually at concentrations between 10–20 mM.[103] Depending upon structure, the IP reagent, which must have an opposite charge to the analyte, interacts with the ionized analyte (ion-pair model) and/or modifies the stationary phase so that it acts as an ion exchange material (dynamic ion-exchange model). As a consequence, retention of the otherwise poorly retained ionized analytes is feasible, and increase in the retention time and the selectivity of the column is observed.[104,105] A problem is that IP reagents typically contaminate the system, and decontamination requires extensive cleaning with concomitant instrument downtime, and thus, a mass spectrometer may need to be devoted only to this type of analysis. Volatile alkylamines and organic acids are usually used as IP reagents when LC–MS is performed, combined with organic solvents such as methanol or propan-2-ol. The mobile phase is at pH 7.0 so as to promote negative electrospray ionization (ESI) and to optimize analyte interactions with the IP reagent.

IPC has found a niche in the analysis of difficult analytes such as sugar phosphates, a wide variety of polar metabolites, including nucleotides, amino acids, organic acids of various types (keto-acids, hydroxyl acids di- and tri-carboxylic acids), coenzyme A derivatives, and metabolites involved in central carbon/energy metabolism. These molecules are very hard to analyze even in HILIC and show poor chromatography or/and ionization efficiency. A large number of studies[106–117] have shown that RP columns used with amine IP reagents provide a very powerful tool for the analysis of such molecules.

In LC–MS metabolomics applications volatile cationic compounds, most often tributylamine (TBA) or hexylamine, are used to form ion pairs with negatively charged analytes. For instance, in a recent study more than a hundred polar, ionic endogenous metabolites were identified in urine, serum, and tissue samples by employing IPC with TBA.[115] A representative extracted ion chromatogram for some of the metabolites determined by this approach is shown in Figure 1.5. Compared to previously developed HILIC and porous graphitic carbon (PGC) separations there was substantial improvement in peak shape and increased sensitivity (~5–10 times) for the vast majority of metabolites when IPC was adopted. In another example, a C8 column was used with detection *via* Orbitrap MS for fatty acid analysis.[118] A water/methanol gradient with TBA as the IP reagent was applied for the effective quantification of long-chain and very-long-chain fatty acid species found in mammalian cells.

Amine-containing structures present in amino acid metabolism and the urea cycle are not adequately retained or resolved by TBA-based IPC methods. In addition, TBA suppresses the ionization efficiency of positively

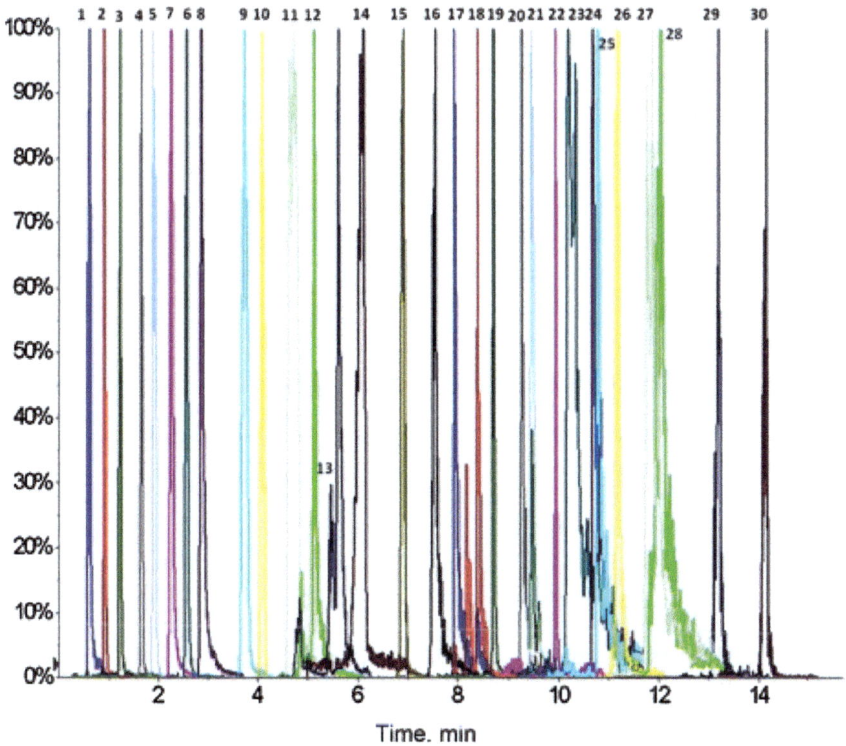

Figure 1.5 Extracted ion chromatogram of 30 endogenous metabolites by IPC–MS/MS analysis. In ascending retention time elution are shown: (1) glutamine, (2) methionine, (3) adenine, (4) thymine, (5) inosine, (6) glutamic acid, (7) phenylalanine, (8) aspartic acid, (9) glucuronic acid, (10) tryptophan, (11) lactic acid, (12) galactose 1P, (13) xylulose 5P, (14) pyruvic acid, (15) NAD, (16) UMP, (17) GMP, (18) AMP, (19) maleic acid, (20) phosphocreatine, (21) malic acid, (22) α-ketoglutarate, (23) G3P, (24) NADP, (25) FBP, (26) isocitric acid, (27) dCTP, (28) ATP, (29) acetyl CoA, and (30) butyryl CoA.

Reproduced from ref. 115 with permission from Elsevier, Copyright 2014.

charged amine molecules.[119] To explore the possibility of coupling IPC with positive ESI a different kind of IP reagent is needed.[106] Volatile perfluorinated acids such as heptafluorobutyric acid (HFBA)[120,121] and perfluoroheptanoic acid (PFHA)[122] have been utilized for the analysis of amino acids and related metabolites. Roy *et al.*[121] proposed a combined targeted and untargeted metabolomics method for studying changes in metabolic pathways associated with cardiometabolic diseases, and IPC was used to separate short-chain acylcarnitines and leucine isomers. Piraud *et al.*[123] developed an IP-RPLC–MS/MS method for the analysis of 76 underivatized amino acids in a single run of 15 min using tridecafluoroheptanoic acid (TDFHA) (0.5 mM) as the IP reagent. In a long-term routine analysis of large numbers of biological samples and after many chromatographic cycles, retention time drift was observed, possibly due to column fouling, but was obviated by extensively flushing the column after every 14–16 chromatographic cycles.

IPC thus represents an elegant chromatographic solution to the problem of retaining ionizable hydrophilic compounds on the types of stationary phase normally employed for reversed-phase chromatography that are often not easily catered for by HILIC approaches. However, this does come at the cost of significant instrument contamination, and this limitation is probably one reason why such methods are not currently more widely applied.

1.3.4 Ion Exchange Chromatography (IEC)

For polar ionic compounds, IEC represents an alternative to the IPC methods described above but, to date, relatively few applications of this approach have been described in metabolomics. Whilst IEC provides an interesting complementary tool for separating charged and highly polar compounds, when coupled with MS, IEC faces challenges, especially in analyte ionization efficiency. IEC typically utilizes high salinity solvents which may be incompatible with MS. Recently IEC–MS has become more attractive because of developments in eluent suppression technology that enable continuous online desalting and conversion of high-salt eluents into MS-compatible pure water.[124] Applications of the technique have included the targeted screening and quantification of metabolites such as carbohydrates, amino acids, organic acids, sugar phosphates, and nucleotides in biological samples.[125–129]

Capillary level flow IEC (capIEC), which uses low flow rates (5–30 $\mu L\ min^{-1}$), can provide enhanced sensitivity and applicability in bioanalysis. CapIEC has been combined with MS/MS for the analysis of 54 metabolites involved in glycolysis, the tricarboxylic acid (TCA) cycle, the pentose-phosphate pathway, and nucleotide metabolism.[130] For robust and reproducible capIEC–MS/MS performance sample preparation is highly important, since the specific technique has a low capacity for ionic species. In another example, a targeted metabolomics approach was developed for the quantification of 28 polar organic acids in quadricep muscles of mice.[131] The metabolic profile of sedentary mice was compared to that of fatigued mice subjected to either a low intensity, long duration (LILD) or high intensity, short duration (HISD) forced treadmill regimen. Significant variations between sedentary and active mice were observed for hippuric acid, malic acid, fumaric acid, and 2-ketoglutaric acid.

Semi-targeted metabolomics has been also proposed for the analysis of "problematic" analytes, particularly di- and triphosphates and organic acids.[132] Many of these compounds are poorly resolved by other chromatographic methods. Specifically, compared to the conventional HILIC method, IEC–MS detected a much higher number (more than double) of key metabolites in samples of the protozoan parasite *Trypanosoma brucei*.

To date, the number of applications of IEC in untargeted metabolomics has been limited. However, an application of CapIEC combined with Orbitrap MS has been applied to the metabolic profiling of head and neck squamous cell carcinoma (HNSCC) cells.[124] Polar anionic metabolites were effectively separated, and the sensitivity was increased up to 100-fold related to RPLC and HILIC methods at either high- or low capillary-flow rates. IEC, thus remains an interesting but as yet under developed methodology, especially as far as untargeted metabolic phenotyping is concerned. However, the technical problems associated with coupling this type of chromatography to MS have now largely been solved opening up interesting possibilities for the future.

1.3.5 Mixed-mode Chromatography (MMC) and Multimode Chromatography

MMC has attracted attention due to the ability to obtain, in a single analysis, metabolome coverage that would normally require analysis on two separate systems. This approach has found widespread applications in the separation of compounds that are not retained or are not well retained/resolved by a single mode of chromatography using conventional RPLC, IEC, or HILIC. More specifically, in MMC, the stationary phase and solutes interact with two or more interaction modes or mechanisms.[133] MMC can be performed by 1. combining two types of stationary phases in one column, 2. packing a single mixed-mode stationary phase into one column, and 3. the connection of two columns with different modes in a type of two-dimensional chromatography (2D-LC).[134] Mixed-mode columns can be categorized into RP/anion-exchange (AX), RP/cation-exchange (CX), HILIC/AX and HILIC/CX bimodal phases, as well as RP/AX/CX and HILIC/AX/CX trimodal materials.[135]

Lately, the combination of IEC and RPLC has been investigated to obtain the retention of both hydrophilic and hydrophobic compounds.[136–138] As an example, a mixed-mode column that combined ionic phases and a C18 phase was used to improve the retention and separation of metabolites covering a wide polarity range.[137] Using this combination over 2100 compounds from several classes, including amines, fatty acids, organic acids, bile acids, cholesterols, carnitines, phospholipids, tri- di- and mono-acylglycerols, nucleosides, *etc.*, were detected in samples obtained from patients with hepatocellular carcinoma (HCC) or non-HCC livers by untargeted UHPLC/QTOF–MS analysis.

As a general note, a mixed-mode column alone, or two columns connected in series, is likely to provide comparable separation to two single-mode columns, while at the same time avoiding solvent waste and consumption of sample and materials. For instance, an in-line two-column IE-RP method has been evaluated for the identification of amino acids, acylcarnitines, and organic acids in human plasma and urine, with an overall time of analysis of 20 min.[138]

HILIC and RP columns have been connected in series for the metabolic phenotyping of biological samples such as mouse plasma.[139] Such column chemistries were effectively combined with the use of a single solvent gradient program. This combination provided mixed-mode separation and improved ESI sensitivity. In particular, for metabolites eluting with apparent capacity factors of >0.7, ion suppression, a common issue in ESI, was significantly reduced. The method enabled the comprehensive analysis of complex mixtures containing a wide range of compound polarities.

MMC was also applied by combining RP, HILIC, and IE chromatographic methods using either a single tri-modal column or two dual-mode columns connected in sequence.[140] Initially, hydrophobic compounds were trapped on a C18-trap in a loop position. Hydrophilic compounds were then separated by a simultaneous gradient for HILIC, anion-exchange, and cation-exchange chromatography. The gradient included a final wash step with high water composition (providing appropriate initial conditions for the RP gradient, used to elute the hydrophobic compounds from the C18-trap).

As these examples show, mixed-mode LC, either based on linked columns, mixed stationary phases, or bespoke mixed-mode phases may represent a very useful alternative to the need to perform multiple analyses using individual stationary phases optimized for one range of polarities or another. To date, it not yet possible to identify which of the several available approaches will come to dominate the MMC methodology, but this is clearly an area of separation science to monitor.

1.3.6 Two-dimensional Liquid Chromatography (2D-LC)

2D-LC has been an area of interest in metabolic profiling for many years as a way of maximizing metabolome coverage by coupling two chromatographic columns of different selectivities/separation mechanisms in a single run, providing increased peak capacity and ideally without compromising either separation.[141,145] In the case of complex matrices, such as those typical of biological samples, this approach offers the possibility of obtaining comprehensive metabolome profiles. 2D-LC can be grouped into two basic categories, heart-cutting (LC–LC),[142] where the selected fractions from the first dimension are transferred to the second, or comprehensive (LC×LC),[143] where the whole sample is analyzed through the two dimensions off-line[144] or on-line[145] depending on how fractions from the first dimension (^{1}D) are transferred to the second (^{2}D). Despite the potential advantages of this technique, the development and optimization of 2D-LC are challenging since the optimal parameters for the first dimension can cause difficulties for the second[146,147] especially with regard to the compatibility of the mobile phases used in the two systems. When a 2D-LC system is coupled to MS, the mobile phases should in addition provide suitable analyte ionization. Sensitivity, robustness, baseline correction, peak alignment, or normalization

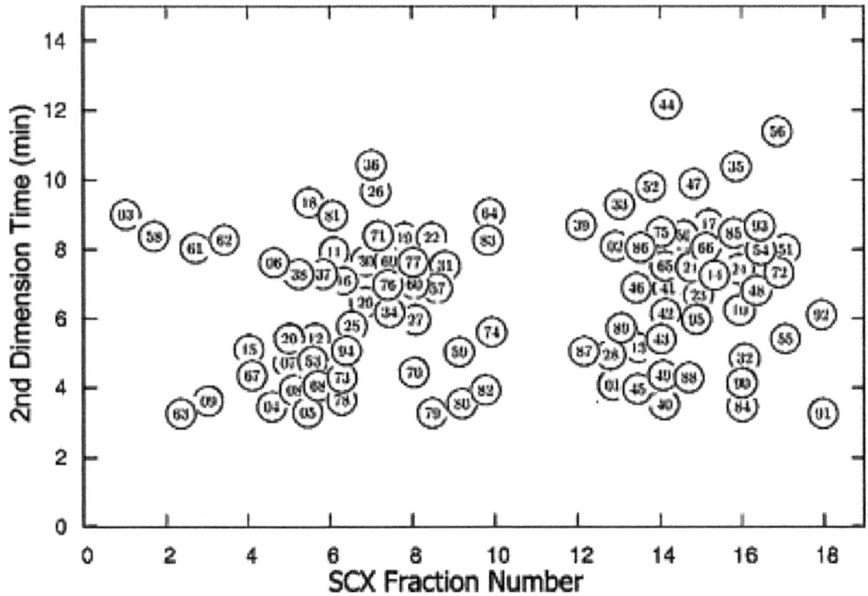

Figure 1.6 The approximate locations of yeast metabolites in a two-dimensional space, based on their relative maximum concentrations detected by a triple quadrupole mass spectrometer in both positive and negative ESI modes. 01-glutamine; 02-4,5-dihydroxy-2,3-pentanedione; 03-ornithine; 04-aspartate; 05-betaine; 06-betaine aldehyde; 07-2-oxoglutatrate; 08-4-hydroxybenzoate; 09-proline; 10-adenosine; 11-iso-leucine; 12-allantoate; 13-5-methylthioadenosine; 14-asparagine; 15-acetyl-CoA; 16-adenine; 17-alanine; 18-glutathione; 19-cytidine; 20-carbamoyl aspartate; 21-cytosine; 22-D-hexose-phosphate; 23-dihexose; 24-citrulline; 25-acetylcarnitine; 26-acetyllysine; 27-D-glyceraldehyde-3-phosphate; 28-folate; 29-carnitine; 30-choline; 31-dimethylglycine; 32-DL-pipecoline; 33-glucosamine; 34-erythrose-4-phosphate 35-flavone; 36-GMP; 37-glucose-6-phosphate; 38-fructose-6-phosphate 39-GABA; 40-epinephrine; 41-*N*-acetyl-L-lysine; 42-*N*-acetyl-glutamine; 43-*N*-acetyl-glutamate; 44-nicotinamide 45-pyridoxine; 46-glutamate; 47-AMP; 48-guanosine; 49-biotin; 50-methionine; 51-phenylalanine; 52-IMP; 53-phenylpropiolate; 54-serine; 55-riboflavin; 56-oxidized glutathione; 57-methylcysteine; 58-methylmalonate; 59-myo-inositol; 60-*N*-acetylglucosamine-1-phosphate; 61-UDPAG; 62-UDPG; 63-aconitate; 64-NAD; 65-*N*-carbamoyl-L-aspartate; 66-nicotinamide ribotide; 67-malate; 68-phenylpyruvate; 69-sedoheptulose-7-phosphate; 70-succinate; 71-trehalose6-phosphate; 72-typtophan; 73-lactate; 74-citrate; 75-tyrosine; 76-DHAP; 77-glycerol-3-phosphate; 78-pyruvate; 79-isocitrate; 80-pantothenate; 81-Sadenosyl-L-homocystein; 82-oxaloacetate; 83-cAMP; 84-S-ribosyl-homocystein; 85-glucose-1-phosphate; 86-threonine; 87-xanthine; 88-nicotinate; 89-inosine; 90-hypoxanthine; 91-fructose-1,6-bis phosphate; 92-gluconate; 93-valine; 94-UTP; 95-xanthosine.
Reproduced from ref. 150 with permission from Elsevier, Copyright 2010.

are some of the bottlenecks of such methods; hence although 2D-LC is a promising technique, it is cumbersome, technically complex, as well as often slow and, as a result, has not yet gained widespread acceptance.[148,149]

In an example of the potential of 2D-LC in metabolic profiling, Fairchild *et al.*[150] created a two-dimensional strong cation exchange (SCX)/HILIC LC system and were able to detect 141 water-soluble metabolites in *Escherichia coli* and *Saccharomyces cerevisiae* cell extracts. In the first dimension the SCX column was used, followed by a diol column in the second dimension. By using both positive and negative ESI, 92 metabolites were detected in *E. coli* and 95 in *S. cerevisiae*, including amino acids, sugars, and vitamins. The system indicated good orthogonality between the two dimensions and grouping of structurally similar metabolites, as shown in Figure 1.6. In one of the first 2D-LC metabolomics applications, two-dimensional HILIC separations were reported with a TSKgel Amide-80 column at the first dimension and a short PolyHydroxyethyl A column at the second dimension. The system was coupled to a QTOF-MS analyzer for the analysis of hydrophilic compounds in complex samples.[151]

1.4 Conclusions

The analysis of the polar, hydrophilic, complement of metabolites in biological samples is not an easy undertaking. Typically, polar metabolites are not well served when analysis by conventional analytical separation approaches, that perform better in the middle of the molecular scales (analyte polarity or molecular weight), rather than at the extremes, is attempted. The truth is, however, that the biochemical interest often lies at the edges of the polarity scale: key metabolic intermediates include polar metabolites such as tricarboxylic acid intermediates, or sugar phosphates and nucleotide mono/di, and tri-phosphates, *etc.* In the analysis of such molecules, chromatographic separation is necessary for the additional reason that many polar metabolites co-exist as isobaric species: *e.g.*, glucose 1-P, glucose 6-P, fructose 1-P, fructose 6-P, galactose 1-P, galactose 6-P mannose 1-P, mannose 6-P, and so forth, to provide only one example. MS analysis alone is severely limited in this respect, and efficient chromatography is necessary, although ion mobility may also be useful in this context.

Whilst the problems of analyzing these polar, hydrophilic compounds by LC–MS cannot be said to have been solved, there have been numerous positive developments in this area. The field is rapidly changing, *via* the development of novel, or improved, separation media such as *e.g.*, new mixed-mode phases, that are designed specifically to serve this purpose.

Abbreviations

AX	Anion-exchange
Cap	Capillary
CX	Capillary electrophoresis
ESI	Electrospray ionization
GC	Gas chromatography
HCC	Hepatocellular carcinoma

HILIC	Hydrophilic interaction liquid chromatography
HNSCC	Head and neck squamous cell carcinoma
HRMS	High-resolution mass spectrometry
IEC	Ion-exchange chromatography
IM	Ion mobility
LC	Liquid chromatography
MD	Multidimensional
MRM	Multiple reaction monitoring
MS	Mass spectrometry
MS/MS	Tandem mass spectrometry
NMR	Nuclear magnetic resonance
OPLS-DA	Orthogonal Partial Least Squares Discriminant Analysis
NPLC	Normal-phase liquid chromatography
PGC	Porous graphitic carbon column
Q-TOF	Quadrupole time of flight
RP	Reversed-phase
RPLC	Reversed-phase liquid chromatography
SCX	Strong cation exchange
SPE	Solid phase extraction
TBA	Tributylamine
TFA	Trifluoroacetic acid
TCA	Tricarboxylic acid
TQ	Triple quadrupole
UHPLC	Ultrahigh performance liquid chromatography
2D	Two-dimensional

References

1. Y. Wang, S. Liu, Y. Hu, P. Li and J.-B. Wan, *RSC Adv.*, 2015, **5**, 78728–78737.
2. H. G. Gika, G. A. Theodoridis, R. S. Plumb and I. D. Wilson, *J. Pharm. Biomed. Anal.*, 2014, **87**, 12–25.
3. W. B. Dunn and D. I. Ellis, *TrAC, Trends Anal. Chem.*, 2005, **24**, 285–294.
4. H. Gika, G. Theodoridis, P. Rainville, R. S. Plumb and I. D. Wilson, in *Liquid Chromatography*, ed. S. Fanali, P. R. Haddad, C. F. Poole and M.-L. Riekkola, Elsevier, 2nd edn, 2017, pp. 245–265.
5. G. Theodoridis, H. G. Gika and I. D. Wilson, *TrAC, Trends Anal. Chem.*, 2008, **27**, 251–260.
6. X. Zhang, K. Quinn, C. Cruickshank-Quinn, R. Reisdorph and N. Reisdorph, *Curr. Opin. Chem. Biol.*, 2018, **42**, 60–66.
7. O. Begou, H. G. Gika, I. D. Wilson and G. Theodoridis, *Analyst*, 2017, **142**, 3079–3100.
8. D. S. Wishart, *Physiol. Rev.*, 2019, **99**, 1819–1875.
9. G. A. Theodoridis, H. G. Gika, E. J. Want and I. D. Wilson, *Anal. Chim. Acta*, 2012, **711**, 7–16.

10. L. D. Roberts, A. L. Souza, R. E. Gerszten and C. B. Clish, *Curr. Protoc. Mol. Biol.*, 2012, **98**, 30.2.1–30.2.24.
11. J. Ivanisevic and A. Thomas, *Methods Mol. Biol.*, 2018, **1730**, 3–28.
12. K. Ortmayr, T. J. Causon, S. Hann and G. Koellensperger, *TrAC, Trends Anal. Chem.*, 2016, **82**, 358–366.
13. X. Liu, L. Zhou, X. Shi and G. Xu, *TrAC, Trends Anal. Chem.*, 2019, **121**, 115665.
14. N. Vinayavekhin and A. Saghatelian, *Curr. Protoc. Mol. Biol.*, 2010, **90**, 30.1.1–30.1.24.
15. A.-H. Emwas, R. Roy, R. T. McKay, L. Tenori, E. Saccenti, G. A. N. Gowda, D. Raftery, F. Alahmari, L. Jaremko, M. Jaremko and D. S. Wishart, *Metabolites*, 2019, **9**, 123.
16. F. Fathi, A. Brun, K. H. Rott, P. Falco Cobra, M. Tonelli, H. R. Eghbalnia, E. Caviedes-Vidal, W. H. Karasov and J. L. Markley, *Metabolites*, 2017, **7**, 61.
17. Y. Iwasaki, T. Sawada, K. Hatayama, A. Ohyagi, Y. Tsukuda, K. Namekawa, R. Ito, K. Saito and H. Nakazawa, *Metabolites*, 2012, **2**, 496–515.
18. R. Vázquez-Fresno, S. Tulipani, O. Khymenets, M. Urpi-Sarda, M. Garcia-Aloy, M. Rabassa, M. Boto-Ordoñez, M. Rotches-Ribalta, R. Llorach and C. Andres-Lacueva, in *Polyphenols in Human Health and Disease*, ed. R. R. Watson, V. R. Preedy and S. Zibadi, Academic Press, San Diego, 2014, pp. 1025–1044.
19. M. C. Gulersonmez, S. Lock, T. Hankemeier and R. Ramautar, *Electrophoresis*, 2016, **37**, 1007–1014.
20. J. M. Büscher, D. Czernik, J. C. Ewald, U. Sauer and N. Zamboni, *Anal. Chem.*, 2009, **81**, 2135–2143.
21. H. Gallart-Ayala, I. Konz, F. Mehl, T. Teav, A. Oikonomidi, G. Peyratout, V. van der Velpen, J. Popp and J. Ivanisevic, *Anal. Chim. Acta*, 2018, **1037**, 327–337.
22. R. Ramautar and G. J. de Jong, *Bioanalysis*, 2014, **6**, 1011–1026.
23. O. Núñez, H. Gallart-Ayala, C. P. B. Martins, P. Lucci and R. Busquets, *J. Chromatogr. B*, 2013, **927**, 3–21.
24. M. Beccaria and D. Cabooter, *Analyst*, 2020, **145**, 1129–1157.
25. J.-L. Ren, A.-H. Zhang, L. Kong and X.-J. Wang, *RSC Adv.*, 2018, **8**, 22335–22350.
26. E. Vasconcelos Soares Maciel, A. L. de Toffoli, E. Sobieski, C. E. Domingues Nazário and F. M. Lanças, *Anal. Chim. Acta*, 2020, **1103**, 11–31.
27. M. Witting and S. Böcker, *J. Sep. Sci.*, 2020, **43**, 1746–1754.
28. S. Wernisch and S. Pennathur, *Anal. Bioanal. Chem.*, 2016, **408**, 6079–6091.
29. G. A. Theodoridis, H. G. Gika, R. Plumb and I. D. Wilson, in *Proteomic and Metabolomic Approaches to Biomarker Discovery*, ed. H. J. Issaq and T. D. Veenstra, Academic Press, Boston, 2nd edn, 2020, pp. 149–169.
30. N. Sillner, A. Walker, E.-M. Harrieder, P. Schmitt-Kopplin and M. Witting, *J. Chromatogr. B*, 2019, **1109**, 142–148.

31. P. Miggiels, B. Wouters, G. J. P. van Westen, A.-C. Dubbelman and T. Hankemeier, *TrAC, Trends Anal. Chem.*, 2019, **120**, 115323.
32. R. A. Sonnenberg, S. Naz, L. Cougnaud and D. Vuckovic, *J. Chromatogr. A*, 2019, **1608**, 460419.
33. A. Eguchi, K. Sakurai, M. Watanabe and C. Mori, *Environ. Int.*, 2017, **102**, 157–164.
34. D. C. Müller, C. Degen, G. Scherer, G. Jahreis, R. Niessner and M. Scherer, *J. Chromatogr. B: Anal. Technol. Biomed. Life Sci.*, 2014, **966**, 117–126.
35. S. Naz, Á. A. Calderón, A. García, J. Gallafrio, R. T. Mestre, E. G. González, C. M. de Cabo, M. C. M. Delgado, J. Á. L. Balanza, A. V. C. Simionato, N. N. Vaeza, C. Barbas and F. J. Rupérez, *Electrophoresis*, 2015, **36**, 2303–2313.
36. J. E. Dalziel, K. Fraser, W. Young, C. M. McKenzie, S. A. Bassett and N. C. Roy, *Am. J. Physiol.: Gastrointest. Liver Physiol.*, 2017, **313**, G62–G72.
37. K. Inoue, C. Tanada, T. Sakamoto, H. Tsutsui, T. Akiba, J. Z. Min, K. Todoroki, Y. Yamano and T. Toyo'Oka, *Food Chem.*, 2015, **181**, 318–324.
38. S. Tufi, M. Lamoree, J. de Boer and P. Leonards, *J. Chromatogr. A*, 2015, **1395**, 79–87.
39. F. Fei, G. C. diCenzo, D. M. E. Bowdish, B. E. McCarry and T. M. Finan, *Metabolomics*, 2016, **12**, 1–14.
40. S. Bawazeer, O. B. Sutcliffe, M. R. Euerby, S. Bawazeer and D. G. Watson, *J. Chromatogr. A*, 2012, **1263**, 61–67.
41. R. Zhang, D. G. Watson, L. Wang, G. D. Westrop, G. H. Coombs and T. Zhang, *J. Chromatogr. A*, 2014, **1362**, 168–179.
42. M. Navarro-Reig, J. Jaumot, B. Piña, E. Moyano, M. T. Galceran and R. Tauler, *Metallomics*, 2017, **9**, 660–675.
43. J. Pekkinen, N. Rosa-Sibakov, V. Micard, P. Keski-Rahkonen, M. Lehtonen, K. Poutanen, H. Mykkänen and K. Hanhineva, *Mol. Nutr. Food Res.*, 2015, **59**, 1550–1562.
44. S. H. Narath, S. I. Mautner, E. Svehlikova, B. Schultes, T. R. Pieber, F. M. Sinner, E. Gander, G. Lihiseller, M. G. Schimek, H. Sourij and C. Magnes, *PLoS One*, 2016, **11**, e0161425.
45. F. Gevi, L. Zolla, S. Gabriele and A. M. Persico, *Mol. Autism*, 2016, **7**, 1–11.
46. S. Yang, M. Sadilek and M. E. Lidstrom, *J. Chromatogr. A*, 2010, **1217**, 7401–7410.
47. J. J. J. van der Hooft, L. Ridder, M. P. Barrett and K. E. V. Burgess, *Front. Bioeng. Biotechnol.*, 2015, **3**, 26.
48. M. A. Kamleh, Y. Hobani, J. A. T. Dow and D. G. Watson, *FEBS Lett.*, 2008, **582**, 2916–2922.
49. A. Teleki, A. Sánchez-Kopper and R. Takors, *Anal. Biochem.*, 2015, **475**, 4–13.
50. M. Cao, K. Fraser, J. Huege, T. Featonby, S. Rasmussen and C. Jones, *Metabolomics*, 2015, **11**, 696–706.

51. D.-Q. Tang, L. Zou, X.-X. Yin and C. N. Ong, *Mass Spectrom. Rev.*, 2016, **35**, 574–600.
52. O. D'Apolito, D. Garofalo, M. Gelzo, D. Paris, D. Melck, R. Calemma, F. Izzo, G. Palmieri, G. Castello, A. Motta and G. Corso, *Metabolomics*, 2014, **10**, 1026–1035.
53. T. Zhang and D. G. Watson, *J. Chromatogr. B*, 2016, **1022**, 199–205.
54. M. G. M. Kok, C. Nix, G. Nys and M. Fillet, *Talanta*, 2019, **197**, 49–58.
55. A. M. Ali, M. Burleigh, E. Daskalaki, T. Zhang, C. Easton and D. G. Watson, *Metabolites*, 2016, **6**, 7.
56. R. T'Kindt, R. A. Scheltema, A. Jankevics, K. Brunker, S. Rijal, J. Dujardin, R. Breitling, D. G. Watson, G. H. Coombs and S. Decuypere, *PLoS Neglected Trop. Dis.*, 2010, **4**, e904.
57. K. Spagou, I. D. Wilson, P. Masson, G. Theodoridis, N. Raikos, M. Coen, E. Holmes, J. C. Lindon, R. S. Plumb, J. K. Nicholson and E. J. Want, *Anal. Chem.*, 2011, **83**, 382–390.
58. A. A. Ebshiana, S. G. Snowden, M. Thambisetty, R. Parsons, A. Hye and C. Legido-Quigley, *PLoS One*, 2015, **10**, e0122883.
59. I. Sampsonidis, M. Witting, W. Koch, C. Virgiliou, H. G. Gika, P. Schmitt-Kopplin and G. A. Theodoridis, *J. Chromatogr. A*, 2015, **1406**, 145–155.
60. B. A. Rappold and R. P. Grant, *J. Sep. Sci.*, 2011, **34**, 3527–3537.
61. S. Schiesel, M. Lämmerhofer and W. Lindner, *Anal. Bioanal. Chem.*, 2010, **396**, 1655–1679.
62. F. Fei, D. M. E. Bowdish and B. E. McCarry, *Anal. Bioanal. Chem.*, 2014, **406**, 3723–3733.
63. Z. Zhu, C. Chen, Y. Zhu, E. Shang, M. Zhao, S. Guo, J. Guo, D. Qian, Z. Tang, H. Yan and J. Duan, *Molecules*, 2019, **24**, 460.
64. D. K. Trivedi and R. K. Iles, *Biomed. Chromatogr.*, 2015, **29**, 240–245.
65. M. G. M. Kok, J. R. Swann, I. D. Wilson, G. W. Somsen and G. J. de Jong, *J. Pharm. Biomed. Anal.*, 2014, **92**, 98–104.
66. T. Pluskal and M. Yanagida, *Cold Spring Harb. Protoc.*, 2016, **2016**, 1081–1086.
67. F. Fei, M. L. Mendonca, B. E. McCarry, D. M. E. Bowdish and M. G. Surette, *Metabolomics*, 2016, **12**, 1–13.
68. J. Boelaert, F. Lynen, G. Glorieux, E. Schepers, N. Neirynck and R. Vanholder, *Anal. Bioanal. Chem.*, 2017, **409**, 2201–2211.
69. S. Nie, Y. Zhao, X. Qiu, W. Wang, Y. Yao, M. Yi and D. Wang, *Evidence-Based Complementary and Altern. Med.*, 2019, **2019**, e3817879.
70. Y. Long, X. Dong, Y. Yuan, J. Huang, J. Song, Y. Sun, Z. Lu, L. Yang and W. Yu, *J. Clin. Biochem. Nutr.*, 2015, **57**, 50–59.
71. L. Zhao, Z. Xiong, X. Lu, S. Zheng, F. Wang, L. Ge, G. Su, J. Yang and C. Wu, *PLoS One*, 2015, **10**, e0129146.
72. S. F. Graham, O. P. Chevallier, C. T. Elliott, C. Hölscher, J. Johnston, B. McGuinness, P. G. Kehoe, A. P. Passmore and B. D. Green, *PLoS One*, 2015, **10**, e0119452.

73. L. Li, F. Zhang, J. Zaia and R. J. Linhardt, *Anal. Chem.*, 2012, **84**, 8822–8829.
74. M. C. Mimmi, N. Finato, G. Pizzolato, C. A. Beltrami, F. Fogolari, A. Corazza and G. Esposito, *Anal. Cell. Pathol.*, 2013, **36**, 71–83.
75. L. Gu, P. Hou, R. Zhang, Z. Liu, K. Bi and X. Chen, *J. Chromatogr. B: Anal. Technol. Biomed. Life Sci.*, 2016, **1033–1034**, 157–165.
76. C. Virgiliou, I. Sampsonidis, H. G. Gika, N. Raikos and G. A. Theodoridis, *Electrophoresis*, 2015, **36**, 2215–2225.
77. T. Sun, S. J. Wetzel, M. E. Johnson, B. A. Surlow and J. Patton-Vogt, *J. Chromatogr. B: Anal. Technol. Biomed. Life Sci.*, 2012, **897**, 1–9.
78. H. Tsugawa, T. Kind, R. Nakabayashi, D. Yukihira, W. Tanaka, T. Cajka, K. Saito, O. Fiehn and M. Arita, *Anal. Chem.*, 2016, **88**, 7946–7958.
79. E. Olesti, J. Rodríguez-Morató, A. Gomez-Gomez, J. G. Ramaekers, R. de la Torre and O. J. Pozo, *Talanta*, 2019, **192**, 93–102.
80. M. Wang, R. Yang, J. Dong, T. Zhang, S. Wang, W. Zhou, H. Li, H. Zhao, L. Zhang, S. Wang, C. Zhang and W. Chen, *J. Chromatogr. B: Anal. Technol. Biomed. Life Sci.*, 2016, **1009–1010**, 144–151.
81. O. Deda, C. Virgiliou, A. Orfanidis and H. G. Gika, *Metabolites*, 2019, **9**, 232.
82. J. Chen, W. Hou, B. Han, G. Liu, J. Gong, Y. Li, D. Zhong, Q. Liao and Z. Xie, *Anal. Bioanal. Chem.*, 2016, **408**, 2527–2542.
83. T. Roggensack, B. Merz, N. Dick, A. Bub and R. Krüger, *Rapid Commun. Mass Spectrom.*, 2020, **34**, e8646.
84. C. Virgiliou, H. G. Gika, M. Witting, A. A. Bletsou, A. Athanasiadis, M. Zafrakas, N. S. Thomaidis, N. Raikos, G. Makrydimas and G. A. Theodoridis, *J. Proteome Res.*, 2017, **16**, 898–910.
85. C. Virgiliou, L. Valianou, M. Witting, F. Moritz, C. Fotakis, P. Zoumpoulakis, A. C. Chatziioannou, L. Lazaros, G. Makrydimas, K. Chatzimeletiou, N. Raikos and G. A. Theodoridis, *Bioanalysis*, 2017, **9**, 37–51.
86. Y. Huang, W. Li, A. S. Minakova, T. Anumol and A. A. Keller, *NanoImpact*, 2018, **12**, 9–17.
87. H. G. Gika, G. A. Theodoridis, U. Vrhovsek and F. Mattivi, *J. Chromatogr. A*, 2012, **1259**, 121–127.
88. P. Arapitsas, A. D. Corte, H. Gika, L. Narduzzi, F. Mattivi and G. Theodoridis, *Food Chem.*, 2016, **197**, 1331–1340.
89. C. Virgiliou, D. Kanelis, A. Pina, H. Gika, C. Tananaki, A. Zotou and G. Theodoridis, *J. Chromatogr. A*, 2020, DOI: 10.1016/j.chroma.2019.460783.
90. A. Pina, O. Begou, D. Kanelis, H. Gika, S. Kalogiannis, C. Tananaki, G. Theodoridis and A. Zotou, *J. Chromatogr. A*, 2018, **1531**, 53–63.
91. A. M. King, L. G. Mullin, I. D. Wilson, M. Coen, P. D. Rainville, R. S. Plumb, L. A. Gethings, G. Maker and R. Trengove, *Metabolomics*, 2019, **15**, 17.
92. K. Spagou, H. Tsoukali, N. Raikos, H. Gika, I. D. Wilson and G. Theodoridis, *J. Sep. Sci.*, 2010, **33**, 716–727.

93. J. P. Scarth, A. D. Clarke, P. Teale and C. M. Pearce, *Steroids*, 2010, **75**, 643–652.
94. W. Skeff, C. Recknagel and D. E. Schulz-Bull, *J. Chromatogr. A*, 2016, **1475**, 64–73.
95. F. Pommier and R. Frigola, *J. Chromatogr. B: Anal. Technol. Biomed. Life Sci.*, 2003, **784**, 301–313.
96. X. Li and A. A. Franke, *Anal. Bioanal. Chem.*, 2015, **407**, 3869–3874.
97. E. Sánchez-López, A. L. Crego and M. L. Marina, *J. Chromatogr. A*, 2017, **1490**, 156–165.
98. X. Chen, D. Gao, F. Liu, X. Gao, S. Wang, Y. Zhao, H. Liu and Y. Jiang, *Anal. Chim. Acta*, 2014, **836**, 61–71.
99. U. Rende, T. Niittylä and T. Moritz, *Plant Methods*, 2019, **15**, 127.
100. J. Sowell and T. Wood, *Anal. Chim. Acta*, 2011, **686**, 102–106.
101. J. Peng, K. Guo, J. Xia, J. Zhou, J. Yang, D. Westaway, D. S. Wishart and L. Li, *J. Proteome Res.*, 2014, **13**, 4457–4469.
102. S. Zhao, X. Luo and L. Li, *Anal. Chem.*, 2016, **88**, 10617–10623.
103. S. C. Moldoveanu and V. David, in *Essentials in Modern HPLC Separations*, ed. S. C. Moldoveanu and V. David, Elsevier, 2013, pp. 145–190.
104. C. G. Campos, H. C. T. Veras, J. A. de Aquino Ribeiro, P. P. K. G. Costa, K. P. Araújo, C. M. Rodrigues, J. R. M. de Almeida and P. V. Abdelnur, *J. Am. Soc. Mass Spectrom.*, 2017, **28**, 2646–2657.
105. S. Wang, I. A. Blair and C. Mesaros, in *Advancements of Mass Spectrometry in Biomedical Research*, ed. A. G. Woods and C. C. Darie, Springer International Publishing, Cham, 2019, pp. 635–647.
106. W. Lu, B. D. Bennett and J. D. Rabinowitz, *J. Chromatogr. B: Anal. Technol. Biomed. Life Sci.*, 2008, **871**, 236–242.
107. F. Ji, Y. Shen, L. Tang and Z. Cai, *Talanta*, 2018, **189**, 1–7.
108. F. Michopoulos, *Methods Mol. Biol.*, 2018, **1738**, 83–97.
109. A. M. Fathima, D. Chuang, W. A. Laviña, J. Liao, S. P. Putri and E. Fukusaki, *Biotechnol. Biofuels*, 2018, **11**, 188.
110. P. N. T. An, M. Yamaguchi and E. Fukusaki, *Metabolomics*, 2017, **13**, 29.
111. H. Mitsunaga, L. Meissner, T. Palmen, T. Bamba, J. Büchs and E. Fukusaki, *J. Biosci. Bioeng.*, 2016, **121**, 413–419.
112. W. Lu, M. F. Clasquin, E. Melamud, D. Amador-Noguez, A. A. Caudy and J. D. Rabinowitz, *Anal. Chem.*, 2010, **82**, 3212–3221.
113. Y. Dempo, E. Ohta, Y. Nakayama, T. Bamba and E. Fukusaki, *Metabolites*, 2014, **4**, 499–516.
114. L. Coulier, R. Bas, S. Jespersen, E. Verheij, M. J. van der Werf and T. Hankemeier, *Anal. Chem.*, 2006, **78**, 6573–6582.
115. F. Michopoulos, N. Whalley, G. Theodoridis, I. D. Wilson, T. P. J. Dunkley and S. E. Critchlow, *J. Chromatogr. A*, 2014, **1349**, 60–68.
116. F. J. Nusantara Putra, S. P. Putri and E. Fukusaki, *J. Biosci. Bioeng.*, 2019, **127**, 52–58.
117. J. M. Buescher, S. Moco, U. Sauer and N. Zamboni, *Anal. Chem.*, 2010, **82**, 4403–4412.

118. J. J. Kamphorst, J. Fan, W. Lu, E. White and J. D. Rabinowitz, *Anal. Chem.*, 2011, **83**, 9114–9122.
119. J. K. Meissen, D. A. Pirman, M. Wan, E. Miller, A. Jatkar, R. Miller, R. C. Steenwyk and M. Blatnik, *Anal. Biochem.*, 2016, **508**, 129–137.
120. S. Vo Duy, S. Besteiro, L. Berry, C. Perigaud, F. Bressolle, H. J. Vial and I. Lefebvre-Tournier, *Anal. Chim. Acta*, 2012, **739**, 47–55.
121. C. Roy, P.-Y. Tremblay, J.-F. Bienvenu and P. Ayotte, *J. Chromatogr. B*, 2016, **1027**, 40–49.
122. S. Benito, A. Sánchez, N. Unceta, F. Andrade, L. Aldámiz-Echevarria, M. A. Goicolea and R. J. Barrio, *Anal. Bioanal. Chem.*, 2016, **408**, 747–760.
123. M. Piraud, C. Vianey-Saban, K. Petritis, C. Elfakir, J.-P. Steghens and D. Bouchu, *Rapid Commun. Mass Spectrom.*, 2005, **19**, 1587–1602.
124. J. Wang, T. T. Christison, K. Misuno, L. Lopez, A. F. Huhmer, Y. Huang and S. Hu, *Anal. Chem.*, 2014, **86**, 5116–5124.
125. J. C. van Dam, M. R. Eman, J. Frank, H. C. Lange, G. W. K. van Dedem and S. J. Heijnen, *Anal. Chim. Acta*, 2002, **460**, 209–218.
126. A. M. Esteves, G. Graça, L. Peyriga, I. M. Torcato, N. Borges, J.-C. Portais and H. Santos, *Extremophiles*, 2019, **23**, 101–118.
127. A. Gomez-Gomez, J. Marcos, P. Aguilera, J. To-Figueras and O. J. Pozo, *J. Chromatogr. B: Anal. Technol. Biomed. Life Sci.*, 2017, **1060**, 347–354.
128. A. C. McGinnis, B. S. Cummings and M. G. Bartlett, *Anal. Chim. Acta*, 2013, **799**, 57–67.
129. C. Bruggink, M. Wuhrer, C. A. M. Koeleman, V. Barreto, Y. Liu, C. Pohl, A. Ingendoh, C. H. Hokke and A. M. Deelder, *J. Chromatogr. B: Anal. Technol. Biomed. Life Sci.*, 2005, **829**, 136–143.
130. H. F. N. Kvitvang, K. A. Kristiansen and P. Bruheim, *J. Chromatogr. A*, 2014, **1370**, 70–79.
131. C. Petucci, A. Zelenin, J. A. Culver, M. Gabriel, K. Kirkbride, T. T. Christison and S. J. Gardell, *Anal. Chem.*, 2016, **88**, 11799–11803.
132. K. Burgess, D. Creek, P. Dewsbury, K. Cook and M. P. Barrett, *Rapid Commun. Mass Spectrom.*, 2011, **25**, 3447–3452.
133. K. Zhang and X. Liu, *J. Pharm. Biomed. Anal.*, 2016, **128**, 73–88.
134. Y. Yang and X. Geng, *J. Chromatogr. A*, 2011, **1218**, 8813–8825.
135. K. Zhang and X. Liu, *J. Pharm. Biomed. Anal.*, 2016, **130**, 19–34.
136. D. G. Sitnikov, C. S. Monnin and D. Vuckovic, *Sci. Rep.*, 2016, **6**, 1–11.
137. K. Nagai, B. Uranbileg, Z. Chen, A. Fujioka, T. Yamazaki, Y. Matsumoto, H. Tsukamoto, H. Ikeda, Y. Yatomi, H. Chiba, S.-P. Hui, T. Nakazawa, R. Saito, S. Koshiba, J. Aoki, D. Saigusa and Y. Tomioka, *Rapid Commun. Mass Spectrom.*, 2020, **34**, e8551.
138. A. Le, J. Mak and T. M. Cowan, *J. Chromatogr. B*, 2020, **1143**, 122072.
139. K. R. Chalcraft and B. E. McCarry, *J. Sep. Sci.*, 2013, **36**, 3478–3485.
140. A. A. Ammann and M. J.-F. Suter, *J. Chromatogr. A*, 2016, **1456**, 145–151.
141. M. Iguiniz, E. Corbel, N. Roques and S. Heinisch, *Talanta*, 2019, **195**, 272–280.

142. S. Wang, L. Zhou, Z. Wang, X. Shi and G. Xu, *Anal. Chim. Acta*, 2017, **966**, 34–40.
143. S. Julka, H. Cortes, R. Harfmann, B. Bell, A. Schweizer-Theobaldt, M. Pursch, L. Mondello, S. Maynard and D. West, *Anal. Chem.*, 2009, **81**, 4271–4279.
144. G. Genta-Jouve, J. Croué, L. Weinberg, V. Cocandeau, S. Holderith, N. Bontemps, M. Suzuki and O. P. Thomas, *Phytochem. Lett.*, 2014, **10**, 318–323.
145. J. Ren, M. A. Beckner, K. B. Lynch, H. Chen, Z. Zhu, Y. Yang, A. Chen, Z. Qiao, S. Liu and J. J. Lu, *Talanta*, 2018, **182**, 225–229.
146. M. Sarrut, F. Rouvière and S. Heinisch, *J. Chromatogr. A*, 2017, **1498**, 183–195.
147. R. López-Ruiz, R. Romero-González and A. Garrido Frenich, *TrAC, Trends Anal. Chem.*, 2019, **118**, 170–181.
148. M. Iguiniz, F. Rouvière, E. Corbel, N. Roques and S. Heinisch, *J. Chromatogr. A*, 2018, **1536**, 195–204.
149. D. Li and O. J. Schmitz, *Anal. Bioanal. Chem.*, 2015, **407**, 231–240.
150. J. N. Fairchild, K. Horvath, J. R. Gooding, S. R. Campagna and G. Guiochon, *J. Chromatogr. A*, 2010, **1217**, 8161–8166.
151. Y. Wang, X. Lu and G. Xu, *J. Chromatogr. A*, 2008, **1181**, 51–59.

CHAPTER 2

Chemical Derivatization for Polar Metabolome Analysis

SHUANG ZHAO AND LIANG LI*

Department of Chemistry, University of Alberta, Edmonton, Alberta T6G 2G2, Canada
*Email: liang.li@ualberta.ca

2.1 Introduction

There are diverse polar compounds present in biological systems. Their high solubility in water helps them to participate in many metabolic reactions commonly taking place in the aqueous environment of an organism. Serving as multifunctional elements (*e.g.*, substrate, product, intermediate, cofactor, *etc.*) in metabolic pathways, relatively polar metabolites play fundamental roles in physiological and pathological processes. Therefore, analytical methods to measure these metabolites with high performance (*i.e.*, highly sensitive detection, accurate and precise quantification) are extremely important in biological studies.[1]

Liquid chromatography–mass spectrometry (LC–MS) has become one of the most commonly used platforms for small molecule analysis.[2] It offers several attractive features, including the ability to detect a wide range of different types of molecules, compatibility with various sample types, consumption of small amounts of samples, and operational robustness with possible automation. However, even with the most cutting-edge LC–MS instrumentation, it is still challenging to analyze many polar

New Developments in Mass Spectrometry No. 10
Advanced Mass Spectrometry-based Analytical Separation Techniques for Probing the Polar Metabolome
Edited by Rawi Ramautar

Published by the Royal Society of Chemistry, www.rsc.org

metabolites with high sensitivity and specificity. Because of their hydrophilic properties, polar metabolites cannot maintain strong hydrophobic interactions with the stationary phase of reversed-phase liquid chromatography (RPLC) (*e.g.*, C18 column), resulting in very weak retention on RPLC. To address this issue, two approaches are traditionally used. The first approach is to use hydrophilic interaction liquid chromatography (HILIC) to target the analysis of polar compounds.[3] Although many methods have been reported, HILIC analysis still suffers from relatively low performance of separation, compared to RPLC. The other approach is to use ion-pairing RPLC.[4] By adding ion pairing reagents as counterions, which are usually large ionic molecules with opposite charge to the analytes of interest and the hydrophobic part, the retention on RPLC can be enhanced. However, the ion pairing reagents often introduce severe ion suppression on the electrospray ionization (ESI) process, which decreases detection sensitivity. In addition to the challenge of separating polar metabolites with high efficiency in LC, many polar metabolites (*e.g.*, sugar molecules) are not readily ionized by ESI. They suffer strong matrix interferences in complex biological samples. Some polar metabolites are readily fragmented in the interface region of a mass spectrometer, which may result in the reduction of molecular ion signals, as well as cause confusion in determining the molecular ions.

Alternatively, instead of optimizing the instrumental setup and running conditions, chemical derivatization can be combined with LC–MS analysis to measure polar metabolites in complex biological samples.[1,5] In this approach, metabolites are derivatized with one or several chemical reagents prior to the LC–MS analysis (see Figure 2.1). As the physicochemical properties of the metabolites are altered, improved analysis performance can be achieved, including enhanced retention and separation on RPLC, higher

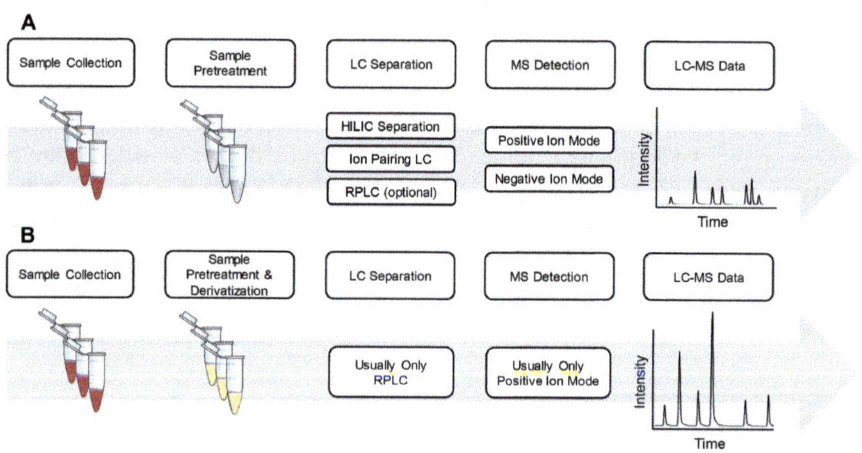

Figure 2.1 Comparison of (A) conventional LC–MS method and (B) chemical derivatization LC–MS method for polar metabolome analysis.

detection sensitivity, and better quantification ability. In this chapter, we present some chemical derivatization methods targeting several important classes of polar metabolites. We then discuss a high-performance chemical isotope labeling (CIL) LC–MS approach[6] as a robust method for high-coverage quantitative analysis of both polar and less-polar metabolites in one instrumental setup.

2.2 Chemical Derivatization LC–MS Analysis of Targeted Polar Metabolites

2.2.1 Amino Acid and Derivative Analysis

In addition to being the basic constituent units of proteins, free amino acids and their analogs and derivatives serve in many biological functions, such as central carbon metabolism, neurotransmitters, *etc.* As a zwitterionic metabolite, the amine group and carboxyl group in this type of molecule greatly increase its hydrophilicity. Derivatization is commonly used in amino acid analysis. There are mainly three types of derivatization reagents: 1) sulfonyl chlorides or acyl chlorides, 2) NHS ester-based reagents, and 3) isothiocyanates. In all three types of reactions, the amine group in amino acids serves as a nucleophile to attack the electrophilic region of a derivatization reagent, forming a stable derivative.

For example, a broadly applicable method of using benzoyl chloride (BzCl) labeling for analyzing neurologically relevant compounds was developed.[7] Many polar metabolites were well separated by a C18 column and detected with greater sensitivity, including amino acids, catecholamines, polyamines, and their metabolites. This approach was successfully applied in studying dopamine regulation in learning and motivation.[8] Multiplex isobaric tags reagents [*e.g.*, isobaric tags for relative and absolute quantification (iTRAQ),[9,10] tandem mass tags (TMT),[11] *N*,*N*-dimethyl leucine (DiLeu)[12,13]] combined with LC–MS/MS have been used in several studies for amino acid analysis. By using multiplex reagents containing different cleavable reporter ions for samples and standards, a wide range of amino acids and amines can be quantified with MS/MS fragmentation in relatively higher sample throughput.

Another highly important area of metabolite analysis is related to the determination of chiral metabolites, such as the analysis of D/L amino acids. Instead of using special chiral chromatography, chiral derivatization reagents that contain another chiral center can be used to label the metabolites, followed by conventional LC separation and MS detection. By forming diastereomers during the derivatization, original enantiomers (*i.e.*, D/L amino acids) can be well separated through achiral chromatography. Toyo'oka and co-workers have developed a number of chiral derivatization reagents for various metabolites.[14,15] For example, by using DMT-(*S* or *R*)-Pro-OSu for chiral amines and DMT-(*S* or *R*)-Apy for chiral carboxylic acids, chiral biomarker candidates were found for Alzheimer's disease.[16]

2.2.2 Nucleotide Analysis

Nucleotides are small molecules consisting of three moieties: a nucleobase, a five-carbon sugar, and a phosphate(s) group. Serving as the basic building blocks of macromolecules such as DNA, nucleotides play essential roles in many biological processes, including DNA/RNA/lipids synthesis, cell signaling, *etc.* However, robust analysis of these molecules by LC–MS is challenging, due to their high polarity. Several derivatization methods combined with RPLC–MS have been reported to improve the detection of nucleotides. Most of the methods use a strategy of blocking the polar and ionic groups in the nucleotides.

A method of using propionyl or benzoyl acid anhydride to esterificate the free hydroxyl groups in AMP, ADP, and ATP was reported.[17] The method could also be used for other types of polar compounds, such as cytokinins, showing the improvement of both ESI response and chromatographic properties on RPLC. Similarly, by using acetone as the derivatization reagent, a simple and low-cost derivatization method was reported for ribonucleoside determination.[18] Improvement of separation and enhancement of sensitivity were achieved. A methylation reagent, (trimethylsilyl)-diazomethane (TMSD), was developed to attach methyl groups to the phosphate groups, primary amines, and secondary amines on nucleotides, leading to much increased hydrophobicity of the metabolites.[19] The derivatized nucleotides could be separated on a C18 column with a common acetonitrile–water elution system. This method also offered a possibility of exhibiting unique ion transition for each metabolite to avoid mutual interferences from each other in MS/MS acquisition, thereby increasing the detection specificity.

2.2.3 Sugar Analysis

Low molecular weight sugars (monosaccharides) and sugar phosphates are key components of many critical energy metabolism pathways, including glycolysis, the pentose phosphate pathway (PPP), tricarboxylic acid (TCA) cycle, *etc.* However, reliable measurement of these compounds by conventional LC–MS is difficult, due to their extremely polar nature. In addition, some sugar phosphates are chemically unstable and easily degraded during sample preparation. Sugars are not ionized by ESI efficiently. To overcome these challenges, derivatization methods of sugars and related compounds followed by LC–MS analysis have been reported.

A two-step derivatization method has been developed, using methoxylamine and propionic acid anhydride to react with the free carbonyl group and hydroxyl groups in the sugar compounds, respectively.[20] After derivatization, the compounds showed good retention behavior on RPLC without the need to use ion pairing reagents, which eliminates the ion suppression effect caused by ion pairing reagents. The detection sensitivity was also increased. The authors successfully measured sugar phosphates in *Populus* leaf extracts. Both reducing and non-reducing sugar phosphates could be analyzed. Similarly, a method using 3-amino-9-ethylcarbazole (3-AEC) to derivatize the sugars and sugar phosphates was reported.[21] The labeled sugars were analyzed on a pentafluorophenyl core–shell LC column. Selective determination of free and phosphorylated reducing sugars could be achieved by this method.

Researchers also used chemical derivatization to improve HILIC separation of sugar compounds. For example, Bawazeer *et al.* found that, if using d_5-aniline to tag the sugars, better separation and peak shape of monosaccharides with a ZIC-HILIC column were observed.[22] With this derivatization reagent, they could perform separation of fructose, glucose, galactose, and mannose with baseline resolution and good peak shape. The approach was validated by applying the method to analyze sugars in urine and brain samples.

2.2.4 Small Organic Acid Analysis

Small carboxylic acids are another important category of molecules that are involved in many essential pathways, including short-chain fatty acids, keto acids, hydroxyl acids, *etc.* Because of the presence of relatively large polar or ionizable groups in small molecules, small organic acids also have weak chromatographic retention behavior on RPLC. Detection of underivatized carboxylic acids is usually conducted in negative ion mode, which suffers from less sensitivity and more interference than positive ion mode detection. Therefore, better methods than the conventional LC–MS approach are needed for small organic acid analysis.

Chemical derivatization can improve the analysis. Many reactions in derivatizing carboxylic acids are based on the condensation reaction with amines and esterification reaction with hydroxyls. The method of using ^{12}C-/^{13}C-3-nitrophenylhydrazine (3-NPH) to analyze carboxylic acids was reported.[23] The labeled metabolites gained significant enhancement of detection in negative ion mode. It was applied in the detection and quantitation of many important categories of acids, including many carboxylic acids in central carbon metabolism, short-chain fatty acids, and bile acids.[24]

Zhao *et al.* developed dansylhydrazine (DnsHz) labeling for untargeted profiling of the carboxyl submetabolome.[25] A labeled-carboxyl-standard library consisting of 193 endogenous human metabolites was constructed for metabolite identification, including various amino acids, short-chain fatty acids, keto acids, polycarboxylic acids, *etc.* With this method, high-performance relative quantification of two comparable groups of urine samples was achieved. Using the similar labeling reaction, Chen *et al.* developed a method with DnsHz labeling to carry out absolute quantification of 25 free fatty acids and 13 acylcarnitines in human plasma samples within 12 min of an LC–MS run.[26]

Among many different types of organic acids, short-chain fatty acids (SCFAs) and their derivatives, as crucial intermediates and elements for metabolism, have attracted great interest. Many reagents have been developed to profile and quantify SCFAs. For example, Girard's reagent T (GT) derivatization can provide a permanent cationic charge to an acid, resulting in a significant increase in detection sensitivity.[27] A fast derivatization reaction was developed with d_0/d_6-DHPP for SCFAs.[28] The authors reported the analysis of 40 SCFA and OH-SCFA species in liver, feces, and serum samples.

2.3 Untargeted Analysis of Polar and Less-polar Metabolites

The examples described in the previous section show that chemical derivatization can significantly improve the analysis of targeted polar metabolites. For untargeted metabolome analysis, the goal is to detect and quantify as many metabolites as possible, including both polar and less-polar metabolites with diverse chemical structures. An ideal method should be able to analyze many compounds with different functional groups (*i.e.*, high metabolic coverage), enable satisfactory retention and separation by LC for a complex matrix (*i.e.*, reducing interference), detect metabolites with wide concentration ranges in samples (*i.e.*, high detection sensitivity), and render reliable quantification (*i.e.*, high accuracy and precision). In the meantime, a compatible data processing method to facilitate data interpretation (*e.g.*, data processing software, databases, *etc.*) is required.

Our laboratory has been working on the development of a high-performance (HP) chemical isotope labeling (CIL) LC–MS platform for high-coverage quantitative metabolome analysis. Chemical derivatization is the central theme of the platform. However, differing from the conventional strategy of chemical derivatization to analyze selected metabolites, HP-CIL is a holistic approach to use chemical derivatization to simultaneously improve the performance of multiple steps in the whole metabolome analysis workflow. For example, a derivatization method used to attach a permanent charge to a metabolite may result in an increase in ionization efficiency. However, the derivatized metabolite is ionic and may not be retained on RPLC, adversely affecting its separation. A preferred method is to rationally design the chemical structure of a derivatization reagent that would allow concomitant enhancement in both ionization and separation efficiency after a metabolite is derivatized. Similarly, a derivatization method can increase the hydrophobicity of polar metabolites; however, we also need to balance the extent of such an increase so that we will not make the less-polar metabolites too hydrophobic to be effectively separated by RPLC. Another consideration is related to the incorporation of isotopic atoms to the reagents. A simple route of synthesizing a reagent would reduce the reagent cost. Robustness of reaction conditions used for chemical derivatization also needs to be carefully weighed. Thus, only a chemical derivatization method that enhances the performance of all different aspects of the whole metabolome analysis workflow can be viewed to have high performance.

Development of the reagents and reaction conditions to perform high-performance metabolome analysis is an active research area. We have developed the 4-channel chemical isotope labeling (CIL) LC–MS technique for metabolomics.[6] We use the term, chemical isotope labeling, to be in line with the metabolic isotope labeling (MIL) approach. MIL uses metabolic reactions in a cell culture containing heavy isotope materials (*e.g.*, $^{15}NH_3$) to introduce heavy isotope atoms into molecules of interest such as metabolites, proteins, *etc.*, while CIL uses chemical reactions to introduce an isotope tag to metabolites. By using four derivatization reactions targeting several functional groups (*i.e.*, amine, phenol, hydroxyl, carboxyl, and carbonyl group), the whole metabolome can be profiled with very high coverage. The rationally designed derivatization reagents ensure the overall improvement in analyte separation and detection. By using a pair of isotope reagents (*i.e.*, $^{12}C_2$-reagent and $^{13}C_2$-reagent) for each derivatization reaction, internal standards can be created.

The workflow of using the CIL LC–MS method to measure both polar and less-polar metabolites in samples is shown in Figure 2.2. Briefly, after general sample pretreatment procedures for biological samples (*e.g.*, protein precipitation for serum/plasma samples, metabolite extraction for tissue/cell samples), derivatization reagents are mixed with samples for chemical labeling. All individual samples are derivatized by $^{12}C_2$-reagents (light labeling), and a reference sample is derivatized by $^{13}C_2$-reagents (heavy labeling). Then the $^{13}C_2$-labeled reference sample is mixed with each $^{12}C_2$-labeled individual sample, followed by LC–MS analysis of the mixture. In the LC–MS data, each metabolite is detected as a peak pair: the light peak originates from a $^{12}C_2$-labeled individual sample and the heavy peak from the $^{13}C_2$-labeled reference sample. This differential isotope labeling approach allows accurate relative quantification. If the reference sample is a standard or a mixture of standards, with known concentration(s), absolute quantification of metabolites in individual samples can also be carried out. With dedicated software, such as IsoMS Pro, peak pair information is extracted, and the chromatographic intensity ratio of a peak pair is calculated to represent the metabolite amount for quantification. Metabolite identification is finally done using a database search approach.

2.3.1 Main Features of CIL LC–MS

Chemical derivatization of the high-performance CIL LC–MS method is based on using different reactions to target four submetabolomes: dansylation reaction for amine/phenol submetabolome,[29] DmPA labeling for carboxyl submetabolome,[30] base-activated dansylation reaction for hydroxyl submetabolome,[31] and dansylhydrazine labeling for carbonyl submetabolome[32] (see Figure 2.3). The derivatization reagents used in these reactions, dansyl chloride, dimethylaminophenacyl (DmPA) bromide, and dansylhydrazine, share several major features for analysis, especially for polar compound analysis.

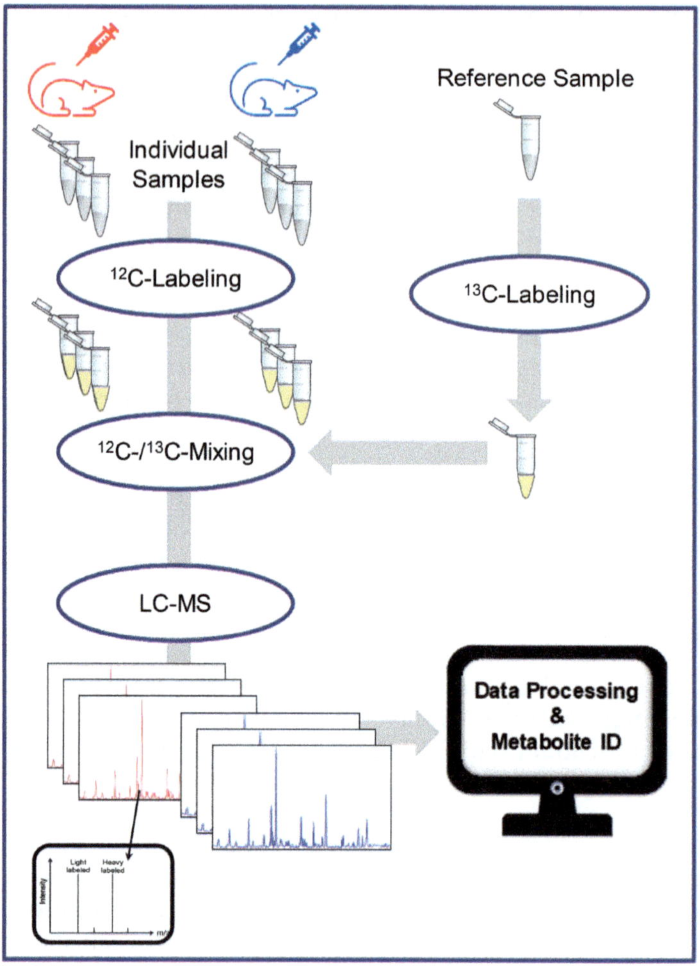

Figure 2.2 General workflow of CIL LC–MS method for metabolome analysis.

First of all, all the reagents contain an aromatic moiety (benzene ring in DmPA and naphthalene ring in the other two reagents), which can significantly increase the hydrophobicity of derivatized metabolites. Thus, only the C18 LC column is required for the 4-channel CIL LC–MS method to separate all labeled compounds, even for the very polar ones, such as amino acids, sugars, *etc.* Compared to the underivatized approach, there is no need for using ion pairing reagents or HILIC column to improve the performance for these compounds. This strategy of using a proper reagent to label polar and less-polar metabolites to alter their chemical and physical properties to an extent that they all can be retained and separated by RPLC decreases the instrumental and operational cost. During the entire metabolome analysis process, there is no need to switch columns and solvents, simplifying the instrument setup and operation.

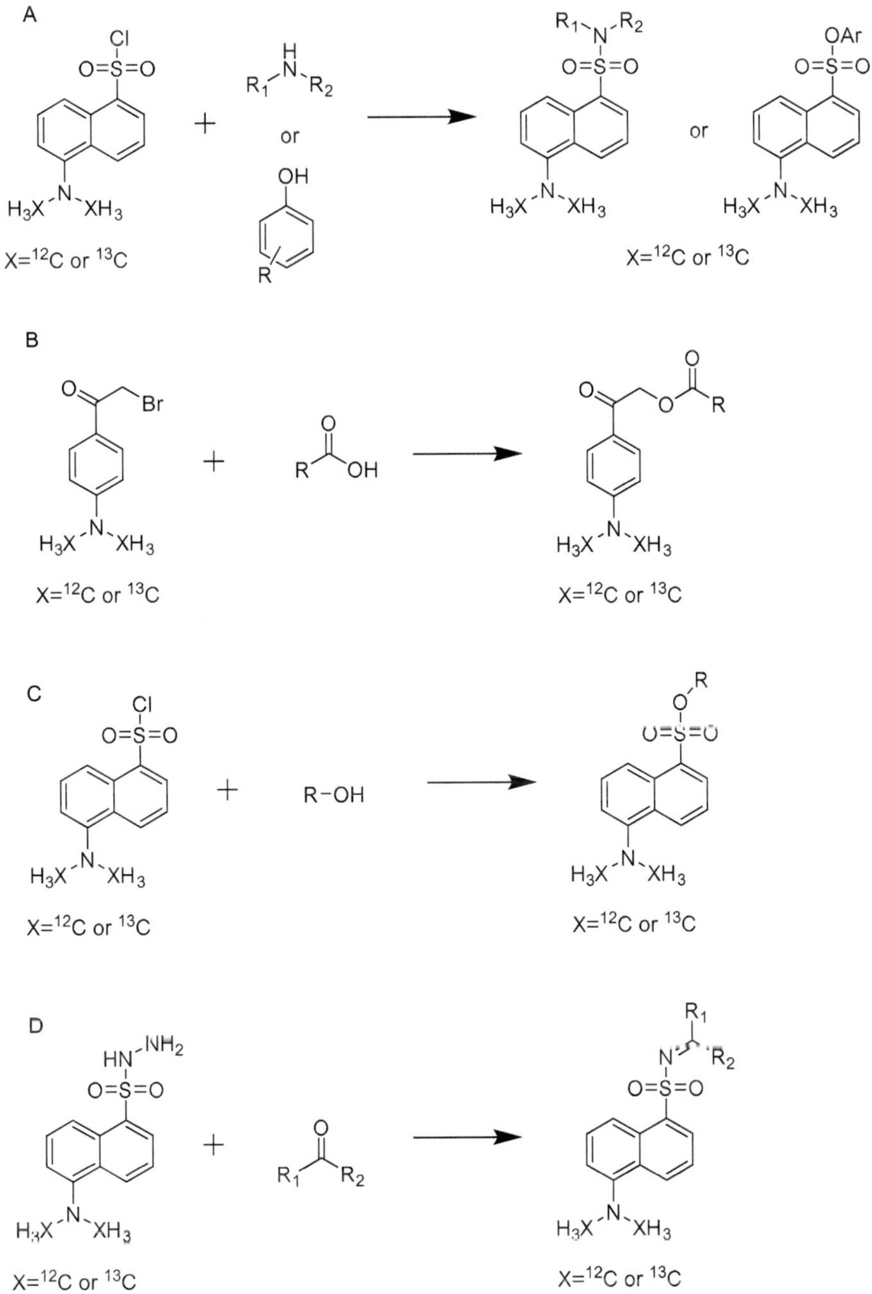

Figure 2.3 Reaction schemes for high-performance CIL LC–MS methods. (A) Dansylation reaction for amine/phenol submetabolome. (B) DmPA labeling for carboxyl submetabolome. (C) Base-activated dansylation reaction for hydroxyl submetabolome. (D) Dansylhydrazine labeling for carbonyl submetabolome.

Secondly, there is a dimethylamino structure located in the labeling reagents. Once introduced into the derivatized metabolite, as an easily ionizable group, it can dramatically increase the detection signal by improving the electrospray ionization process in positive ion mode. In this case, all derivatized metabolite can be analyzed in positive ion mode with much higher detection sensitivity, including those small carboxylic acids that are normally detected in the negative ion mode. Furthermore, for the metabolites with low concentration or low ESI response, they show enhanced signals with better peak shapes, which benefit the metabolite recognition and quantification.

Thirdly, two ^{13}C atoms can be incorporated into the methyl groups in the reagents to generate $^{13}C_2$-reagents for labeling. When the $^{13}C_2$-labeled reference sample is spiked into each individual $^{12}C_2$-labeled sample, internal standards for all labeled metabolites are created. Thus, much improved quantification accuracy and precision can be achieved.

2.3.2 Data Processing in CIL LC–MS

When performing untargeted metabolome analysis, one of the challenging tasks is to extract qualitative and quantitative information of many metabolites from a large batch of LC–MS data. Thanks to generating the heavy-labeled internal standard for every light-labeled metabolite, the metabolite signal picking and quantitative information extraction become more reliable. Because each metabolite is shown as a peak pair (*i.e.*, light peak from $^{12}C_2$-labeling and heavy peak from $^{13}C_2$-labeling) in the MS spectra, an algorithm can be developed to easily differentiate true metabolite signals from the noise and background signals which are shown as singlet peaks. By measuring the peak intensity ratio of light peak and heavy peak, accurate metabolite amount information can be generated. Software, such as IsoMS,[33,34] has been developed to achieve these functions in a batch mode, without needing manual intervention in data processing.

Metabolite identification is essential to obtain qualitative information to understand the biological meaning of the dataset. A metabolome database containing 3-tiered libraries has been constructed for the 4-channel CIL LC–MS method. This database includes information from more than 1500 authentic metabolite standards (1st-tier),[31,35,36] *in-silico*-predicted information for more than 9000 metabolites, covering most of the metabolism pathways (2rd-tier) and a comprehensive evidence-based metabolome library (3rd-tier).[37] By matching the information such as mass and retention time of unknown metabolites from experimental samples with the libraries, rapid and high-confidence metabolite identification can be achieved.

2.3.3 Applications of CIL LC–MS

The CIL LC–MS methods have been applied for metabolomic profiling of various samples, such as urine,[38] blood,[39] sweat,[40] cerebrospinal fluid,[41] feces,[42] tissue,[43] and cells.[44] By using the 4-channel CIL LC–MS method, Zhao *et al.* profiled human plasma and yeast metabolome.[6] A total of 7431 and 5629 metabolite candidates were detected. Among them, 670 and 431 metabolite candidates can be identified by matching the library with high confidence. The metabolites that can be analyzed by this method include many polar metabolites such as amino acids and their derivatives, dipeptides, various carboxylic acids, sugars, *etc.*

In another study, Han *et al.* used a dansylation reaction to measure the amine/phenol-containing metabolites to determine potential metabolite biomarkers associated with Parkinson's disease (PD) and PD with incipient dementia.[45] By analyzing 43 serum samples from PD patients and 42 serum samples from healthy controls, the authors successfully found significantly changed metabolites associated with catecholamine metabolism, tryptophan metabolism, caffeine metabolism, and oxidative stress. By using five small molecules (vanillic acid, 3-hydroxykynurenine, isoleucyl-alanine, 5-acetylamino-6-amino-3-methyluracil, and theophylline) as a biomarker candidate panel, the PD patients and controls can be successfully differentiated. The receiver operating characteristic curve produced an area-under-the-curve value of 0.955 with 87.5% sensitivity and 93.0% specificity. This clearly shows that though only one or two of the four-channel CIL methods were used in some studies, satisfied metabolite coverage can still be achieved with very high detection and quantification performance.

The 4-channel CIL LC–MS method has also been applied to absolute quantification of many targeted compounds. For example, El-Aneed and co-workers performed absolute quantification of potential biomarkers for asthma and chronic obstructive pulmonary disease using dansyl chloride and DmPA bromide.[46,47]

2.4 Concluding Remarks

Chemical derivatization LC–MS can be used to improve the analysis of polar metabolites. Moreover, by rationally designing the chemical labeling reagents, both polar and less-polar metabolites can be derivatized together and detected simultaneously in one LC–MS run. Differential isotope labeling of individual samples and a reference sample (*e.g.*, a pooled sample or standards) can be performed for accurate relative (and absolute) quantification. Using high-performance chemical isotope labeling to target different chemical groups of metabolites or submetabolomes, high-coverage quantitative metabolome analysis can be done.

The limitations of the chemical derivatization approach include 1) the derivatization reaction requires extra time and cost, 2) a small portion of metabolites are not easily derivatized, and 3) labeling efficiencies may not be high for some metabolites. However, much improved analysis performance offered by chemical derivatization often outweighs these limitations. We envisage that this approach will be further advanced in the near future, and many applications in biological studies and biomarker discovery research will emerge.

Acknowledgements

This work was supported by the Natural Sciences and Engineering Research Council of Canada, Canada Research Chairs, Canada Foundation for Innovation, Genome Canada and Alberta Innovates.

References

1. Q. Hu, H. Tang and Y. Wang, *J. Anal. Test.*, 2020, **4**, 140–162.
2. I. Aretz and D. Meierhofer, *Int. J. Mol. Sci.*, 2016, **17**, 14.
3. B. Buszewski and S. Noga, *Anal. Bioanal. Chem.*, 2012, **402**, 231–247.
4. W. Lu, M. F. Clasquin, E. Melamud, D. Amador-Noguez, A. A. Caudy and J. D. Rabinowitz, *Anal. Chem.*, 2010, **82**, 3212–3221.
5. S. Zhao and L. Li, *TrAC, Trends Anal. Chem.*, 2020, **131**, 115988.
6. S. Zhao, H. Li, W. Han, W. Chan and L. Li, *Anal. Chem.*, 2019, **91**, 12108–12115.
7. J. M. T. Wong, P. A. Malec, O. S. Mabrouk, J. Ro, M. Dus and R. T. Kennedy, *J. Chromatogr. A*, 2016, **1446**, 78–90.
8. A. Mohebi, J. R. Pettibone, A. A. Hamid, J.-M. T. Wong, L. T. Vinson, T. Patriarchi, L. Tian, R. T. Kennedy and J. D. Berke, *Nature*, 2019, **570**, 65–70.
9. E. Takach, T. O'Shea and H. L. Liu, *J. Chromatogr. B: Anal. Technol. Biomed. Life Sci.*, 2014, **964**, 180–190.
10. X. Liang, H. Han, X. Zhao, X. Cao, M. Yang, D. Tao, R. Wu and X. Yue, *J. Sci. Food Agric.*, 2018, **98**, 5157–5163.
11. J. P. Murphy, R. A. Everley, J. L. Coloff and S. P. Gygi, *Anal. Chem.*, 2014, **86**, 3585–3593.
12. L. Hao, J. Johnson, C. B. Lietz, A. Buchberger, D. Frost, W. J. Kao and L. Li, *Anal. Chem.*, 2017, **89**, 1138–1146.
13. P. Wei, L. Hao, S. Thomas, A. R. Buchberger, L. Steinke, P. C. Marker, W. A. Ricke and L. Li, *J. Am. Soc. Mass Spectrom.*, 2020, **31**, 1854–1860.
14. T. Toyo'oka, in *Methods in Molecular Biology*, Clifton, N.J., 2019, vol. 2030, pp. 293–306.
15. T. Toyo'oka, in *Methods in Molecular Biology*, 2019, vol. 1985, pp. 57–79.
16. T. Takayama, T. Mochizuki, K. Todoroki, J. Z. Min, H. Mizuno, K. Inoue, H. Akatsu, I. Noge and T. Toyo'oka, *Anal. Chim. Acta*, 2015, **898**, 73–84.
17. A. Nordström, P. Tarkowski, D. Tarkowska, K. Dolezal, C. Astot, G. Sandberg and T. Moritz, *Anal. Chem.*, 2004, **76**, 2869–2877.

18. S. F. Li, Y. B. Jin, Z. Tang, S. H. Lin, H. X. Liu, Y. Y. Jiang and Z. W. Cai, *Anal. Chim. Acta*, 2015, **864**, 30–38.
19. Z. Li, H.-X. Zhang, Y. Li, C. W. K. Lam, C.-Y. Wang, W.-J. Zhang, V. K. W. Wong, S.-S. Pang, M.-C. Yao and W. Zhang, *Anal. Chem.*, 2019, **91**, 1019–1026.
20. U. Rende, T. Niittylä and T. Moritz, *Plant Methods*, 2019, **15**, 127.
21. J. Han, V. Tschernutter, J. C. Yang, T. Eckle and C. H. Borchers, *Anal. Chem.*, 2013, **85**, 5965–5973.
22. S. Bawazeer, A. M. Ali, A. Alhawiti, A. Khalaf, C. Gibson, J. Tusiimire and D. G. Watson, *Talanta*, 2017, **166**, 75–80.
23. J. Han, S. Gagnon, T. Eckle and C. H. Borchers, *Electrophoresis*, 2013, **34**, 2891–2900.
24. J. Han, K. Lin, C. Sequeira and C. H. Borchers, *Anal. Chim. Acta*, 2015, **854**, 86–94.
25. S. Zhao and L. Li, *Anal. Chem.*, 2018, **90**, 13514–13522.
26. G. Chen and Q. Zhang, *Anal. Bioanal. Chem.*, 2020, **412**, 2841–2849.
27. W.-S. Song, H.-G. Park, S.-M. Kim, S.-H. Jo, B.-G. Kim, A. B. Theberge and Y.-G. Kim, *J. Ind. Eng. Chem.*, 2020, **83**, 297–302.
28. J. Wei, L. Xiang, X. Li, Y. Song, C. Yang, F. Ji, A. C. K. Chung, K. Li, Z. Lin and Z. Cai, *Anal. Chim. Acta*, 2020, **1100**, 66–74.
29. K. Guo and L. Li, *Anal. Chem.*, 2009, **81**, 3919–3932.
30. K. Guo and L. Li, *Anal. Chem.*, 2010, **82**, 8789–8793.
31. S. Zhao, X. Luo and L. Li, *Anal. Chem.*, 2016, **88**, 10617–10623.
32. S. Zhao, M. Dawe, K. Guo and L. Li, *Anal. Chem.*, 2017, **89**, 6758–6765.
33. R. Zhou, C.-L. Tseng, T. Huan and L. Li, *Anal. Chem.*, 2014, **86**, 4675–4679.
34. T. Huan and L. Li, *Anal. Chem.*, 2015, **87**, 1306–1313.
35. T. Huan, Y. Wu, C. Tang, G. Lin and L. Li, *Anal. Chem.*, 2015, **87**, 9838–9845.
36. S. Zhao, M. Dawe, K. Guo and L. Li, *Anal. Chem.*, 2017, **89**, 6758–6765.
37. L. Li, R. H. Li, J. J. Zhou, A. Zuniga, A. E. Stanislaus, Y. M. Wu, T. Huan, J. M. Zheng, Y. Shi, D. S. Wishart and G. H. Lin, *Anal. Chem.*, 2013, **85**, 3401–3408.
38. X. Su, N. Wang, D. Chen, Y. Li, Y. Lu, T. Huan, W. Xu, L. Li and L. Li, *Anal. Chim. Acta*, 2016, **903**, 100–109.
39. W. Han and L. Li, in *Clinical Metabolomics: Methods and Protocols*, ed. M. Giera, Springer New York, New York, NY, 2018, pp. 213–225.
40. K. Hooton, W. Han and L. Li, *Anal. Chem.*, 2016, **88**, 7378–7386.
41. K. Guo, F. Bamforth and L. A. Li, *J. Am. Soc. Mass Spectrom.*, 2011, **22**, 339–347.
42. W. Xu, D. Y. Chen, N. Wang, T. Zhang, R. K. Zhou, T. Huan, Y. F. Lu, X. L. Su, Q. Xie, L. Li and L. J. Li, *Anal. Chem.*, 2015, **87**, 829–836.

43. X. Wang, W. Han, J. Yang, D. Westaway and L. Li, *Anal. Chim. Acta*, 2019, **1050**, 95–104.
44. X. Luo and L. Li, *Anal. Chem.*, 2017, **89**, 11664–11671.
45. W. Han, S. Sapkota, R. Camicioli, R. A. Dixon and L. Li, *Mov. Disord.*, 2017, **32**, 1720–1728.
46. M. M. Khamis, D. J. Adamko and A. El-Aneed, *Anal. Chim. Acta*, 2017, **989**, 45–58.
47. H. Awad, K. J. H. Allen, D. J. Adamko and A. El-Aneed, *J. Chromatogr. B: Anal. Technol. Biomed. Life Sci.*, 2019, **1122–1123**, 29–38.

CHAPTER 3

Ion Pair Liquid Chromatography–Mass Spectrometry for Probing the Polar Metabolome

XUN BAO AND JING LI*

Karmanos Cancer Institute, Wayne State University School of Medicine, Detroit, Michigan 48201, United States
*Email: LiJing@wayne.edu

3.1 Introduction

Metabolomics, together with genomics, transcriptomics, and proteomics, are invaluable tools for better understanding system biology. Metabolome refers to the complete set of small molecule metabolites in a cell, tissue, or organism, which represents a biochemical "fingerprint" of the genomic and environmental effects. Metabolomics focuses on comprehensive and quantitative analyses of metabolite concentrations or metabolic changes in a biological system, thus offering a means of deciphering cellular metabolism and metabolic regulation. In contrast to genomics, transcriptomics, and proteomics that deal with the building units (such as nucleotides or amino acids) that share similar chemical structures or biochemical interactions, metabolomics analyzes small molecule metabolites that have diverse chemical structures and physicochemical properties, present at a wide concentration range, and change more rapidly than any genetic material.[1]

New Developments in Mass Spectrometry No. 10
Advanced Mass Spectrometry-based Analytical Separation Techniques for Probing the Polar Metabolome
Edited by Rawi Ramautar

Published by the Royal Society of Chemistry, www.rsc.org

As a result, no single analytical tool is capable of measuring all metabolites. Multiple analytical modalities, including liquid chromatography–mass spectrometry (LC–MS), gas chromatography–mass spectrometry (GC–MS), and nuclear magnetic resonance (NMR) spectroscopy, are required for comprehensive metabolomics profiling. LC–MS offers a versatile metabolomic platform that can analyze a broad range of metabolites with diverse hydrophobicity and polarity. Superior to NMR, LC–MS provides high sensitivity and specificity and a large dynamic range. In addition, advantageous over GC–MS, LC–MS is compatible with most common standard solvents and can measure non-volatile metabolites without the requirement of complicated derivatization reactions.

Effective chromatographic separation is critical to the reliable and reproducible LC–MS analysis. The most common chromatographic separation techniques for LC–MS based metabolomics include reversed-phase liquid chromatography (RP-LC), ion-pair liquid chromatography (IP-LC), and hydrophilic interaction liquid chromatography (HILIC). RP-LC is suitable for the analysis of non-polar compounds because the retention mainly relies on hydrophobic interactions between analytes and stationary phases.[2] HILIC and IP-LC are commonly used for the analysis of hydrophilic and polar metabolites that fail to retain on standard reversed-phase stationary phases.[3,4] The HILIC mode of separation involves multimodal retention mechanisms, including liquid–liquid partitioning of the analyte between the mobile phase and water-enriched solvent layer that is partially immobilized onto the surface of the stationary phase,[3] hydrogen-bonding, hydrophobic and hydrophilic interactions, and ion-exchange interactions.[5] Since the more polar compounds have a stronger interaction with the stationary aqueous layer than the less polar compounds, the HILIC separation mainly relies on differences in polarity and degree of solvation.[3] For the analysis of polar compounds using electrospray ionization mass spectrometry as a detector, HILIC often offers better sensitivity than RP-LC because of the more volatile mobile phase. However, the reproducibility of HILIC may be less optimal than RP-LC or IP-LC, and the peak shape and retention time can be altered by subtle differences in temperature, pH, or solvent additive concentrations. IP-LC is a variant of RP-LC that achieves the retention and separation of charged or ionic compounds on traditional reversed-phase hydrophobic stationary phases by the addition of ion pair reagents into the mobile phase. Through formation of ion pairs with opposite charged ion pair reagents, the retention of charged analytes can be significantly improved on reversed-phase stationary phases.[4] However, IP-LC has its own disadvantages when coupling with mass spectrometry detection, most notably ion-suppression, reduced sensitivity, and system contamination.

Many metabolites involved in the important anabolic and catabolic metabolic pathways are highly polar and often ionized, such as sugar phosphates, nucleotides and nucleosides, carboxylic acids, amino acids, phosphorylated compounds, and co-enzyme A (CoA) derivatives. IP-LC coupled with tandem mass spectrometry (MS/MS) or high-resolution mass spectrometry (MS) is particularly useful for the analysis of these polar and charged metabolites, which provides broad metabolite coverage, versatility, sensitivity, robustness, and capability of separating isomeric metabolites (*e.g.*, isomers of sugar phosphates).[6] In the past decade, a number of IP-LC–MS methods have been published for metabolomics applications, either targeted or untargeted, to decipher core cellular metabolic pathways, including central carbon metabolism, energy transfer, as well as biosynthesis and catabolism of nucleotides and amino acids.

In this chapter, we provide an overview on the basics of IP-LC–MS in terms of retention mechanism, critical parameters, and mass spectrometry detection. In addition, we discuss IP-LC–MS based metabolomics with the focus on sample preparation, chromatographic and mass spectrometry methods, as well as potential pitfalls and key solutions. Overall, this review intends to provide a practical framework and guidance for carrying out robust IP-LC–MS based metabolomics.

3.2 Ion Pair Liquid Chromatography (IP-LC)

3.2.1 Retention Mechanism of IP-LC

Traditional RP-LC is performed on a hydrophobic stationary phase with covalently bonded alkyl chains, where chromatographic retention and separation is dependent on hydrophobic interactions of analytes with immobilized non-polar ligands on the stationary phase (Figure 3.1A).[7] Thus,

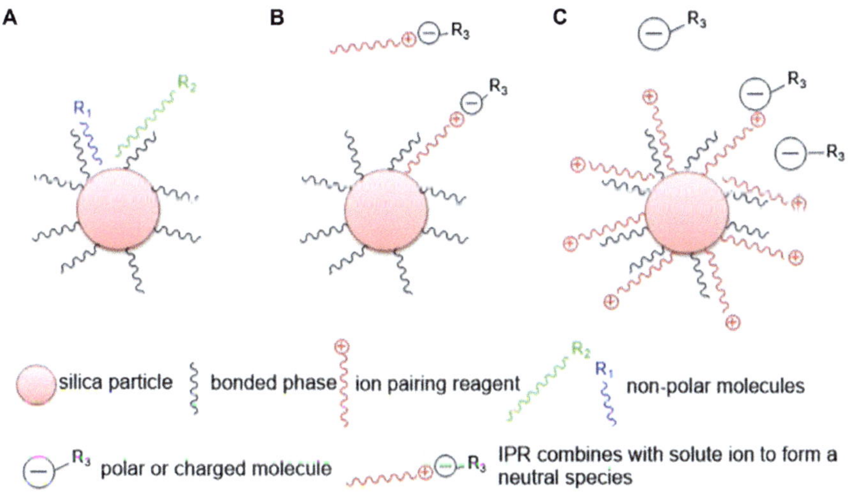

Figure 3.1 Schematic representation of retention mechanisms for: (A) traditional reversed-phase liquid chromatography; (B) ion pair liquid chromatography classic ion pairing model; and (C) ion pair liquid chromatography dynamic ion exchange model.

non-polar or acidic analytes at low mobile phase pH have great retention on a reversed-phase column; in contrast, polar or basic analytes (which remain ionized in the generally acceptable pH range of the column) have poor retention. When ion pair reagents are added in the mobile phase, the retention of charged analytes on reversed-phase stationary phases can be significantly improved through a classic ion pair mechanism and/or dynamic ion exchange mechanism (Figure 3.1B and C).[4]

In the classic ion pairing model (Figure 3.1B), charged analytes couple with oppositely charged ion pair reagents to form more hydrophobic neutral species, thus enabling hydrophobic interactions and improved retention on reversed-phase stationary phases.[8] In the dynamic ion-exchange model (Figure 3.1C), lipophilic ion pair reagents are absorbed onto the alky-bonded non-polar surface of the stationary phase and dynamically form charged ion layers at anion-exchange sites, which enable retention of charged analytes through ionic (or electrostatic) interactions.[9] While the dynamic ion-exchange mechanism is generally considered to be the dominating retention mechanism of IP-LC, the retention on IP-LC likely involves both mechanisms.[4]

The combination of both hydrophobic and electrostatic interaction retention mechanisms allows IP-LC to analyze many diverse metabolites and also to separate chemically similar metabolites. Typically, chemically related compounds have a similar retention time. Similar compounds that differ in the number of charged groups, such as nucleoside mono-, di-, and triphosphates, exhibit a positive correlation between number of charges and retention time (*i.e.*, AMP < ADP < ATP) because additional charges attract more ion pairing molecules thus conferring more hydrophobic interaction with the stationary phase.[6] On the other hand, identically charged compounds can be separated by differences in hydrophobic characteristics or differences in conformation (*e.g.*, isomers).[6]

3.2.2 Critical Parameters of IP-LC

Several key parameters need to be optimized in order to develop a sensitive, reliable, and reproducible IP-LC method. These include the selection of appropriate stationary phase and mobile phase (consisting of organic modifiers, ion pair reagents, counterions) as well as optimization of pH and gradient elution.

3.2.2.1 Stationary Phase

A variety of stationary phases are applied in IP-LC, among which the conventional alkyl silica C8 or C18 stationary phases are the most commonly used in metabolomics applications. When coupled with mass spectrometry detection, C18 stationary phase-based IP-LC offered excellent resolution and sensitivity in the quantitative analysis of oligonucleotides, nucleotides, nucleosides, and amino acids. For example, an IP-LC method using the C18 stationary phase (Inertsil ODS-4 and ODS-3) was developed for simultaneous determination of nucleotide sugars and nucleotides.[10] Other C18 columns (Atlantis T3, OST C18, BEH C18, and XBridge C18) were also successfully applied in the IP-LC analysis of oligonucleotides.[11–15] IP-LC using the C18 stationary phase (Agilent Zorbax SB-C18) provided adequate separation for natural amino acids as well as their metabolites and biogenic amines.[16–18]

In addition, unique stationary phases, such as the porous graphitic carbon stationary phase, have demonstrated utility in IP-LC based metabolomics for targeted or untargeted analysis of amino acid metabolism,[19–21] and chiral analysis of enantiomers of amino alcohols.[22,23] The porous graphitic carbon stationary phase has unique chemical surface properties.[24,25] Specifically, it is very insoluble and fully porous with a porosity of approximately 75%. Its highly ordered crystalline surface is made up of sheets of hexagonally arranged aromatic carbon atoms with delocalized electrons. In principle there are no functional groups on the surface, and it acts as a strongly retentive alkyl-bonded silica gel with different retention and selectivity properties toward polar and structurally related compounds. Comparing to classical reversed-phase columns, porous graphitic carbon columns tolerate a wider pH range (1–13) and require a shorter equilibrium time for ion pair reagent adsorption. However, oxidation may occur when fluorinated ion pair reagents are present, which limits the application of porous graphitic carbon stationary phases in IP-LC analysis.[20,21]

3.2.2.2 Mobile Phase

IP-LC employs a water-rich polar mobile phase that typically consists of aqueous buffer solution, organic modifier, ion pair reagent, and counter ions. Methanol and acetonitrile are the most commonly used organic modifiers because they are fully miscible with water and compatible with mass spectrometry. As compared to methanol, acetonitrile exhibits higher elution strength, lower UV absorbance, and lower pressure to the column when mixing with aqueous solution. Since a more aqueous mobile phase is favorable to the formation of ion pairs between charged ion pair reagents and ionized analytes, a shallow gradient slope (*i.e.*, slow increase of elution strength) is often required to achieve good chromatographic retention and resolution in IP-LC.[13] Given the weaker elution strength of methanol than acetonitrile, higher resolution for the separation of analytes with extremely similar properties (such as isomers) may be relatively easily achieved by fine-tuning the content of methanol (instead of acetonitrile) in the gradient. For example, it was shown that a better resolution for separation of isomeric sugar phosphates was achieved when applying 5% increase of methanol content in the gradient as compared to 2% increase of acetonitrile content.[26]

The ion pair reagent is a key component in IP-LC analysis. The typical structure of ion pair reagents consists of a bulky hydrophobic region and an ionic end (Figure 3.2). The hydrophobic region interacts with the stationary

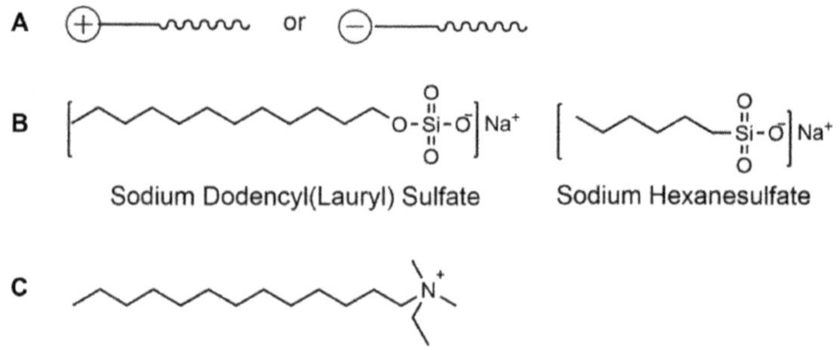

Figure 3.2 (A) General structure of ion-pairing reagents, which consists of two key components, a substantial hydrophobic region and an ionic end. (B) Structure of negatively charged (or anionic) ion pair reagents, such as sodium alkanesulfonate. (C) Structure of positively charged (or cationic) ion pair reagents, such as quaternary ammonium salts.

Table 3.1 Common ion pair reagents.

Classification	Ion pair reagents	Ref.
Positive-charged (cationic)		
Quaternary ammonium salts	Diisopropylamine, diisopropylethylamine, *N*′, *N*′-dimethylbutylamine, dimethylcyclohexylamine, dimethylhexylamine, hexylamine, *N*-methyldibutylamine, octylamine, propylamine, tetremethylethylenedinamine, *N*-methyldibutylamine, triethylamine, tributylamine, tripropylamine, triethylamine, tripropylamine, dibutylamine, *trans-N,N*-dimethylcyclohexane-1,2-diamine, cyclohexyldimethylamine	27–33
Negative-charged (anionic)		
Sodium alkyl sulfate	Alkanesulfonate	34–37
Others	Trifluoroacetic acid (TFA), phosphoric acid, pentafluoropropionic acid (PFPA), heptafluorobutyric acid (HFBA), perchlorate, trifluoroacetate,	38–40

phase while the ionic end with charged functional groups interacts with charged analytes. Ion pair reagents are generally classified as positively charged (or cationic) and negative charged (or anionic) (Table 3.1). Positively charged ion pair reagents (*e.g.*, quaternary ammonium salts) are commonly used to retain negatively charged compounds, while negatively charged reagents (*e.g.*, sodium alkyl sulfate) are used to retain positively charged compounds.

The chemical structure of an ion pair reagent impacts the retention of analytes. In general, the increase of the hydrophobic chain of ion pair reagents provides enhanced ion pair efficiency and renders better retention and separation.[38] When IP-LC is coupled with electrospray ionization mass spectrometry (ESI–MS), the concentration of ion pair reagent influences both chromatographic resolution and mass spectrometric signals.[38,41] Increasing ion pair reagent concentrations generally renders better chromatographic retention and resolution, albeit to different extents dependent on the hydrophobicity of analytes.[42] However, ion pair reagents often negatively impact ionization efficiency (*i.e.*, ion suppression). One reason is the formation of strong ion pairs between analytes and ion pair reagents, which are not easily broken down in the electrospray ionization source.[41] Therefore, the concentration of an ion pair reagent needs to be carefully optimized to balance chromatographic resolution and mass spectrometric sensitivity. It is recommended to use a minimum concentration of an ion pair agent that is capable for adequate chromatographic retention and resolution.

Counterions can be added into the mobile phase to modulate chromatographic retention and mass spectrometry sensitivity. Through interacting with ion pair reagents adsorbed on the stationary phase, counterions may reduce the net surface charges and potential of the stationary phase.[43] In addition, counterions may enhance volatility of the mobile phase, thus enabling rapid evaporation during desolvation and higher electrospray ionization efficiency.[41,44] The common counterions (such as acetate, bicarbonate, formate, and chloride), which are introduced from the buffer solution for pH adjustment, have no apparent effects on electrospray ionization efficiency.[43] In contrast, the addition of hexafluoroisopropanol (HFIP) into a trimethylamine (TEA)-based ion-pair mobile phase has demonstrated to significantly enhance electrospray ionization signals for the analysis of oligonucleotides and nucleotides.[13,45] The underlying mechanism is attributable to the dynamic adjustment of the pH of the electrospray droplet as a result of the preferential evaporation removal of the more volatile anionic counterion (*i.e.*, HFIP) from the droplet surface.[45] HFIP (boiling point, 57 °C) is more volatile, as compared to TEA (89 °C), acetic acid (118 °C), tributylamine (214 °C), tetrabutylamine (100 °C), or dimethylhexylamine (146 °C). As a buffer system in the mobile phase, the weak acid/weak base system of HFIP/TEA maintains a stable pH at ~8.3. However, as the column eluent is electrosprayed and desolvated in the ion source, the more volatile HFIP is depleted at the droplet surface, and consequently the pH rises toward 10 at the surface. This high pH promotes the dissociation of TEA-analyte ion pairs, thereby enhancing evaporation and ionization efficiency.[45] The application of HFIP as a counterion in IP-LC–ESI–MS analysis was a milestone, following which the combination of various ion pair reagents with other fluoroalcohols as counterions was developed.[15,46]

3.2.2.3 *pH*

The pH of the mobile phase is an important parameter that regulates hydrophobic and electrostatic interactions between the stationary phase and analytes by controlling the ionization and charge status of the analytes. Hydrophobic interaction is the dominate mechanism of retention in RP-LC, where neutral or un-ionized analytes have better retention than charged or ionized analytes. IP-LC enables the retention of charged or ionized analytes through electrostatic interactions with ion pair reagents, and thus the change of charge status has a different effect on the chromatographic retention of IP-LC and RP-LC.[47,48] As the pH is adjusted to increase the net charge of an analyte, the retention on IP-LC increases while decreasing on RP-LC.[37,49] This is because additional charges attract more ion pairing molecules, thus conferring more hydrophobicity of analyte-ion pair reagent complexes and better retention on reversed-phase stationary phases.

The adjustment of mobile phase pH is an effective strategy to improve chromatographic resolution of IP-LC. For example, in the IP-LC analysis of purine nucleotide monophosphates, while GMP and IMP were co-eluted at pH 4.0 to 6.25, they were chromatographically separated (GMP eluted earlier) when the mobile phase pH was reduced to $4.0-2.25$.[50] In IP-LC based metabolomics, a pH gradient, increasing from pH 6.3 (with 5 mM hexylamine in water) to pH 8.5 (with 90% methanol and 10% 10 mM ammonium acetate), was applied to achieve optimal retention and peak shapes for a number of polar and acidic metabolites including nucleotides, coenzyme A esters, sugar nucleotides, and sugar bisphosphates.[51]

3.2.2.4 Gradient Elution

Gradient elution, which means the steady change of elution strength by altering the mobile phase composition, is often applied to separate various analytes and interferences in a mixture. In RP-LC, the gradient normally starts with a low content of organic modifier in the mobile phase where the elution strength is low to allow poorly retained analytes to fully interact with the stationary phase, followed by the increase of organic modifier content that results in stronger eluting strength to facilitate faster elution of strongly retained analytes.[52] In IP-LC, a more aqueous mobile phase is favorable to the formation of ion pairs between charged ion pair reagents and ionized analytes, and hence a shallow gradient slope (*i.e.*, slow increase of elution strength) is often required to achieve good chromatographic retention and resolution. For example, using the TEA–HFIP ion-pair mobile phase, a gradient elution starting with aqueous mobile phase (*i.e.*, 0% acetonitrile) and with a shallow gradient slope (*i.e.*, 0.16% increase of acetonitrile per minute) was applied to achieve adequate chromatographic resolution of 8 nucleoside triphosphates (including ATP, CTP, GTP, UTP, dATP, dCTP, dGTP, and dTTP).[13] Similarly, a gradient elution starting with an aqueous mobile phase and followed by a shallow gradient slope was applied in the IP-LC analysis of 20 underivatized amino acids using pentadecafluorooctanoic acid as the ion pair reagent.[20]

In addition to the change of organic modifier concentrations in the mobile phase, gradient elution can also involve the change of pH, ion pair reagent composition, and flow rate. For instance, a dual mode gradient elution involving both organic modifier concentration gradient and the change of flow rate has shown to shorten the chromatographic run, especially the re-equilibrium time for the next injection.[53] A more complex gradient elution strategy that involves the organic modifier concentration gradient (0–90% methanol), pH gradient (pH 5.3–8.5), ion pair reagent concentration gradient (5–0 mM) has been successfully applied for simultaneous quantitative analysis of a variety of polar metabolites, including nucleotides, coenzyme A esters, sugar nucleotides, and sugar bisphosphates.[51]

3.3 Mass Spectrometry

When coupling with liquid chromatography, mass spectrometry (MS) provides the most versatile, sensitive detection, and broadest metabolite coverage for metabolomics applications. In LC–MS, following chromatographic separation, analytes are ionized in an ion source, separated by a mass analyzer, and detected. Most LC–MS methods for metabolomics rely on electrospray ionization (ESI), a soft ionization technique that minimizes analyte fragmentation and works effectively for charged metabolites. By combining chromatographic separation and MS signatures (which can be obtained from tandem MS or high-resolution MS), LC–MS based metabolomics offers the capability of quantitative analysis of targeted metabolites as well as global untargeted analysis with the potential for identification of unknown metabolites.

3.3.1 Targeted Analysis

Targeted metabolomics aims to quantitatively determine a predefined group of metabolites with known chemical structures. Reference standards are often available to define the chromatographic retention times and MS signatures for individual metabolites. By constructing calibration curves especially with stable isotope-labeled internal standards, targeted analysis can provide absolute metabolite quantification.[54] Targeted analysis is commonly performed on a triple quadruple mass spectrometer (MS/MS) using the multiple reaction monitoring (MRM) scan mode, where predefined specific parent-to-product ion transitions are monitored for individual metabolites. The main advantage of the MRM approach is high specificity and sensitivity. However, the MS/MS instrument has to be programmed to predefined parent-to-product ion scan events, and it has a finite scan capacity limiting the total number of transitions or metabolites that can be monitored and reliably determined. Typically, for current triple quadrupole mass spectrometers, a scan time of ~50 ms is adequate to provide reliable peak quantitation. Faster scanning (*i.e.*, less scan time) is needed to accommodate a large number of transitions (metabolites) monitored. However, less scan time allows for detection of fewer ions, resulting in lower signal-to-noise ratio and larger variation of the signal. This is particularly problematic for low abundant metabolites.

An alternative approach for targeted analysis is high-resolution full-scan LC–MS, where specificity is achieved by combining chromatographic separation and high-resolution mass separation. High-resolution mass spectrometers commonly used in metabolomics are time-of-flight (TOF) or Orbitrap-based mass analyzers. A TOF mass analyzer typically has a mass resolution of 10 000–40 000, and an Orbitrap mass analyzer can reach a resolution of 400 000. Resolution is defined as the ratio of an ion mass to the minimum mass difference that allows two ions to be distinguished. The high mass resolution enables effective mass-based separation of virtually all compounds except for isomers. With the knowledge of chromatographic retention time, the peaks of individual targeted metabolites can be identified and quantitated from high-resolution full-scan MS chromatograms. As compared to an MRM scan on a triple quadruple instrument, a high-resolution full-scan MS approach benefits from a relatively longer scan time, which allows reliable quantitation of a large number of metabolites. In addition, the high mass resolution and accuracy enable effective separation of targeted metabolites from background ions that have the same nominal mass but a different exact mass. These features make high-resolution full-scan MS the most straightforward approach for targeted analysis of a large number of metabolites and their labeled forms in parallel, which is particularly useful for determining metabolite isotopomer distribution patterns in stable isotope-assisted metabolomics (or flux analysis). The comparison of the quantitative performance of TOF full-scan MS and triple quadrupole MRM scan for targeted metabolomics applications suggests that high-resolution full-scan MS has adequate quantitative performance in terms of sensitivity, selectivity, dynamic range, and reproducibility, although it may not necessarily provide fully comparable sensitivity as an MRM scan.[55]

Uniquely, the high-resolution full-scan MS approach provides the opportunity to carry out targeted and untargeted metabolomics simultaneously. High-resolution full-scan MS data can be used for quantitative analysis of targeted metabolites and in the meantime identification of unanticipated peaks (or unknown metabolites) that show statistically significant changes across biological conditions. With the high-resolution MS/MS capability in a hybrid quadrupole-TOF or quadruple-Orbitrap instrument, further MS^n scans can be performed to facilitate fragmentation-based identification of structures for unknowns. In addition, the high-resolution MS/MS capability also enables distinguishing certain isomers and maximizing detection sensitivity for selected metabolites.

3.3.2 Untargeted Analysis

Untargeted metabolomics involves a large-scale analysis of all detectable metabolites in a system including chemically unknowns. Due to its comprehensive and unbiased nature, untargeted metabolomics provides an opportunity for novel discovery.[56–60] Unlike targeted metabolomics, untargeted analysis does not require metabolite reference standards or construction of calibration curves, and metabolite data are acquired as the ion abundances of individual chromatographic peaks.[59]

Two data acquisition approaches are commonly used in untargeted analysis. The first approach involves a MS1 full scan followed by data-dependent acquisition.[59] An MS1 full scan is carried out in all samples to determine the accurate mass of individual molecules and associated ion abundances, based on which the molecules with signal (ion) intensity statistically significantly different from the control are identified. Then, fragmentation patterns of these molecules are generated by MS/MS scan and subsequently used for metabolite identification. The second approach is data-independent acquisition, which carries both MS1 full scan and MS/MS scan of all precursor ions either simultaneously (MSE) or in finite mass ranges (SWATH).[59,61,62] In data-independent acquisition, fragmentation patterns data are acquired regardless of metabolite signal intensity, and fragment ions are later assigned to their precursors based on the corresponding retention time, mass, and drift time.[59] In the final step of both approaches, the identified features of both precursor ions and corresponding fragment ions are matched to particular metabolites using spectral databases such as Metlin[63,64] (https://metlin.scripps.edu/landing_page.php?pgcontent=mainPage) and metabolomics databases, such as human metabolome database[65] (https://hmdb.ca/), PhenoMeter[66] (https://www.metabolome-express.org/phenometer.php), The Livestock Metabolome Database[67] (http://lmdb.ca/), MetaboLights[68] (https://www.ebi.ac.uk/metabolights/index), and In Vivo/In Silico Metabolites Database[69] (http://metabolomics.pharm.uconn.edu/iimdb/). The identified metabolites distinguishable from the control can further be searched in pathway databases, such as Kyoto Encyclopedia of Genes and Genomes[70] (https://www.genome.jp/kegg/), and MetaCyc[71] (https://metacyc.org/), to gain insights into potential metabolic pathways altered by drugs or gene modifications.

3.4 Sample Preparation

Sample preparation is a critical step prior to metabolomic analysis. The purpose of sample preparation is to produce a stable extract that quantitatively reflects metabolites of interest presented in the biological system. As such, an appropriate sample preparation process should avoid perturbation of metabolite levels, quickly terminate enzyme metabolic activity, and minimize metabolite degradation or interconversion. It should also be fast and convenient. Depending on the chemical properties (*e.g.*, water-soluble *vs.* insoluble) of the metabolites of interest, different sample preparation processes are needed. For IP-LC–MS based metabolomics, sample preparation focuses on the retainment and extraction of hydrophilic metabolites. In generally, sample preparation involves quenching of metabolism and extraction of metabolites.

3.4.1 Quenching

Many biologically important metabolites (*e.g.*, ATP, glutamine, and glucose 6-phosphate) have a rapid turnover rate in live cells or tissues. The purpose of quenching is to quickly terminate metabolism with no perturbation of metabolite levels. Quenching is particularly important for highly metabolically active specimens such as cells and tissues and is of less concern for serum, plasma, or urine specimens. Care must be taken to avoid two potential problems during quenching: 1) perturbation of metabolite levels and 2) incomplete or insufficiently fast termination of enzyme activity.[72,73]

The first step in quenching for cell samples is to isolate cells from media. Pelleting cells by centrifugation is slow and perturbs the nutrient environment. Instead, fast filtration is recommended for suspension cell culture, and simple aspiration of medium is recommended for adherent cultures.[72] Washing cells with phosphate-buffered saline (PBS) should be avoided unless there are compelling reasons for it because washing may cause a metabolic perturbation and cold PBS (cold shock) may lead to leakage of intracellular metabolites.[74,75] Nevertheless, sometimes there are compelling reasons for washing. For instance, measurement of intracellular amino acids will be inaccurate without washing due to the high amino acid concentrations in media. In such cases, quick washing (<10 s) with warm PBS is recommended.[72]

Quenching of metabolism (*i.e.*, inactivation of enzymatic activity) is often achieved by some combination of snap freezing in liquid nitrogen, freeze clamping, organic solvent, acid, base, heat, or salt treatment.[72,73,76,77] For tissue and cell specimens, the typical first step is deep freezing in liquid nitrogen.[72,73] For cultured cells, cold or hot organic solvent can also be added directly for quenching and extraction.[72,73] Despite concerns about thermal degradation, hot organic solvent (commonly boiling ethanol) is more effective and faster for denaturing enzymes than cold organic solvent, and it works well for heat stable metabolites.[77–79] Acidic solvent, such as a mixture of methanol and water (80 : 20, *v*/*v*) with 0.1 M formic acid, can accelerate quenching and effectively minimize metabolite interconversion, although exposure to acid, even briefly, may cause degradation of some metabolites. After quenching metabolism, the extract is neutralized with ammonium bicarbonate to prevent further acid-catalyzed degradation.[72] In spite of potential acid lability of some metabolites, quenching in cold acidic organic solvent followed by neutralization with ammonium bicarbonate generally works well for many metabolites in tissue and cell specimens.[72]

3.4.2 Extraction

The purpose of extraction is to obtain quantitative yields of metabolites while removing proteins, lipids, and other interferences in biological specimens. Before extraction, tissue samples need to be pulverized into fine powders or homogenates using manual grinding or a vibration mill at a cold temperature, a refrigerated centrifugal homogenizer, or a cryomill machine operated at a liquid nitrogen temperature.[80–83] Common sample extraction methods prior to LC–MS analysis include protein precipitation, liquid–liquid extraction, solid-phase extraction, and ultrafiltration, each with advantages and disadvantages.[80–83] The most appropriate extraction procedure depends largely on the sample type and metabolites of interest. For IP-LC–MS based metabolomics applications, protein precipitation is the most commonly used, simple, fast, and effective approach for the extraction of a large number of water-soluble metabolites.

Extraction with protein precipitation is performed by mixing the biological specimen (*e.g.*, tissue/cell homogenate, plasma, or urine) and solvent on a shaker at a cold temperature for 10–20 min, followed by centrifugation and collection of the supernatant. Two rounds of extraction are recommended because substantial metabolite yields (typically 20–40% of the total) can be obtained in a second round of extraction.[84] There is no consensus on the best solvent system for metabolite extraction. The extraction methods, ranging from organic solvents and acids to heat denaturation, vary widely in terms of protein precipitation efficiency and the number of metabolites detected.[85] Among 14 protein precipitation methods, methanol protein precipitation was found to be the most effective, straightforward, and reproducible approach for extraction of serum metabolites; whereas, acetonitrile showed the lowest protein precipitation efficiency and the most variation in metabolite intensity.[86] Acids and heat treatments were more reproducible than acetonitrile, but resulted in fewer metabolites detected probably due to the acid- or heat-lability of some metabolites.[85]

Typically, for liquid specimens (*e.g.*, serum, plasma, urine, or cell medium), protein precipitation using cold (−80 °C) methanol is a simple but effective approach for metabolite extraction. The volume ratio of methanol to biological sample is usually 1 : 3 or 1 : 4 for serum or plasma and 1 : 1 for urine or cell medium.[87] For cell and tissue specimens, the mixture of methanol : water (80 : 20, *v*/*v*, −80 °C) is generally adequate for extraction of hydrophilic metabolites.[87] Modifications may be needed for extraction of particular metabolites from a particular matrix. For example, the mixture of methanol : acetonitrile : water (40 : 40 : 20, *v*/*v*) is more efficient than methanol : water alone for extraction of nucleotide triphosphates from *E. Coli* or yeast, but not from mammalian cells.[84] As mentioned earlier, cold acidic organic solvent is effective to prevent metabolite interconversion during quenching. Thus, the mixture of methanol : acetonitrile : water (40 : 40 : 20, *v*/*v*) with 0.1 M formic acid (and subsequent neutralization with ammonium bicarbonate) is generally an effective solvent system for both quenching and extraction of cell and tissue specimens, especially for the analysis of ATP and other high-energy phosphorylated metabolites.[72,84] Typically, 1 mL of solvent mixture is used to extract 25 mg of tissue specimen.[72]

3.4.3 Post-extraction

Solvents used for quenching and extraction may interfere with LC–MS analysis; in particular, they negatively influence retention of analytes on reversed-phase stationary phases. Removal of solvent by drying extract in a lyophilizer, N_2 evaporator, or vacuum evaporator (at 4 °C) can address this problem. Drying also concentrates metabolites and facilitates long-term storage.[80] However, during solvent removal, metabolite degradation or interconversion may occur. This can be a big concern for redox-active metabolites (such as NADPH and GSH) and some phosphorylated compounds (such as ATP and ADP).[72] In some cases, preservatives are needed, for example, addition of ascorbic acid in the extraction solvent to mitigate folate degradation and interconversion.[88] Thus, although most metabolites are well preserved quantitatively during drying, caution is needed for some particular metabolites.

3.5 Ion Pair Liquid Chromatography–Mass Spectrometry-based Metabolomics

IP-LC with mass spectrometry detection, either tandem MS or high-resolution MS, provides a powerful tool for metabolomic analysis of a wide range of hydrophilic and polar metabolites that are involved in the major anabolic and catabolic cellular metabolic pathways. Here, we provide an overview on the application of IP-LC–MS, with the focus on chromatographic and mass spectrometry methods, for metabolomic analysis of central carbon metabolism, energy transfer, as well as biosynthesis and catabolism of nucleotides and amino acids.

3.5.1 Central Carbon Metabolism

The central carbon metabolism is composed of the flow of carbon from nutrients to biomass synthesis and energy supply through glycolysis, the pentose phosphate pathway, and the tricarboxylic acid (TCA) cycle or Krebs cycle (see Figure 3.3).[89,90] Glycolysis is a universal metabolic pathway that converts glucose into pyruvate and generates energy by forming nicotinamide adenine dinucleotide (NADH) and adenosine triphosphate (ATP).[91] Glycolysis contains ten sequential enzyme-catalyzed reactions, is oxygen-independent, and primarily occurs in the cytoplasm of the cell.[91] The pentose phosphate pathway is an alternative glucose catabolism pathway that converts glucose-6-phosphate (G6P) into NADPH, pentoses, and ribose 5-phosphate (R5P) (which is a precursor for nucleotide synthesis).[92] There are two distinct phases in the pentose phosphate pathway, including the oxidative phase that generates NADPH and the non-oxidative phase that forms 5-carbon sugars.[92] The TCA cycle is a series of chemical reactions used by all aerobic organisms to release stored energy through the oxidation of acetyl-CoA derived from carbohydrates, fats, and proteins, into ATP and

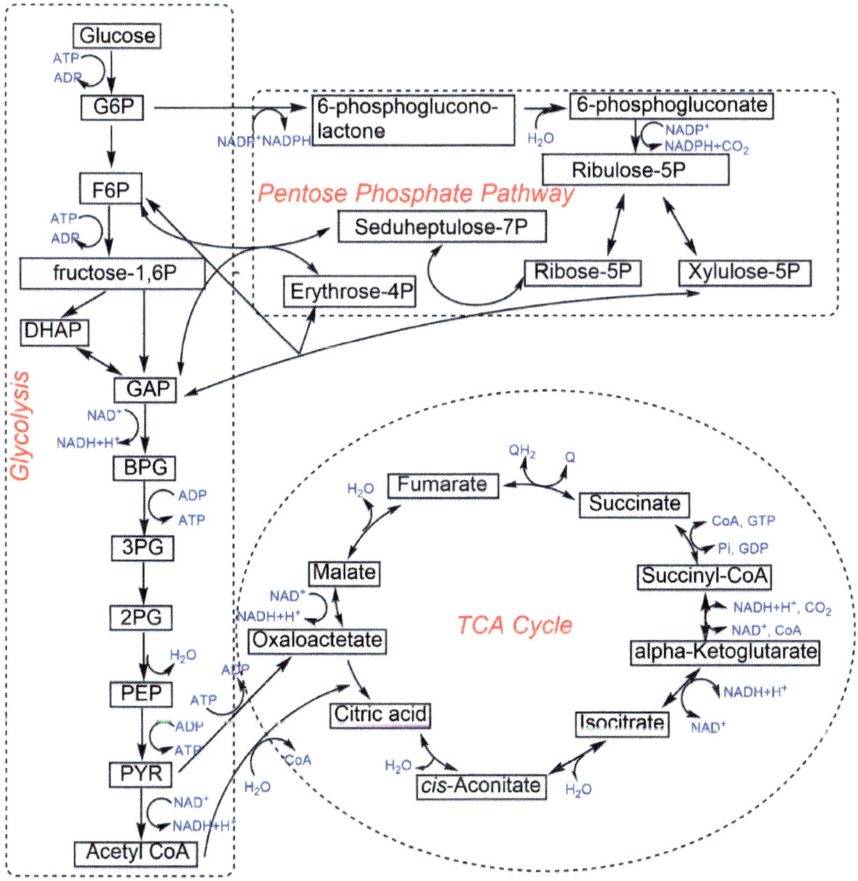

Figure 3.3 Schematic illustration of central carbon metabolism including glycolysis, pentose phosphate pathway, and tricarboxylic acid (TCA) cycle. Abbreviations: G6P, D-glucose-6-phosphate; F6P, D-fructose-6-phosphate; R5P, D-ribose-5-phosphate; E4P, D-erythrose-4-phosphate; DHAP: dihydroxyacetone phosphate; GAP, D-glyceraldehyde-3-phosphate; 3PG, glycerate-3P; PEP, phosphoenolpyruvate; PYR, pyruvate; BPG, glycerate-1, 3P

carbon dioxide.[89] In addition, the TCA cycle provides precursors of certain amino acids, as well as the reducing agent NADH that is fed into the oxidative phosphorylation (electron transport) pathway.[93] Overall, central carbon metabolism supports the largest metabolic fluxes in the cell and produces the most abundant metabolites, which ensures a balanced supply of biosynthetic precursors, energy, and reducing equivalents for cell growth.[89,93] Clearly, quantitative analysis of the metabolite abundances and metabolic flow in central carbon metabolism will provide important biochemical insights into disease progression and therapeutic response.

The majority of published IP-LC–MS methods for quantitative analysis of central carbon metabolites is based on the ion pair reagent, tributylamine, because it effectively improves the retention of highly polar anionic species on reversed-phase stationary phases, and it is compatible with electrospray ionization in the negative mode. Here, three representative IP-LC methods coupled with tandem MS or high-resolution MS for metabolomic analysis of central carbon metabolism are highlighted.

Luo *et al.* developed a highly selective and sensitive IP-LC–ESI–MS/MS method for simultaneous determination of intracellular metabolites.[26] Chromatographic separation of 29 negatively charged metabolites (including sugar phosphates, nucleotides, and carboxylic acids) was achieved on a reversed-phase C18 column, Phenomenex Synergi Hydro-RP (150 mm×2.1 mm, 4 μm particle size), using a mobile phase consisting of the volatile ion pair reagent tributylammonium acetate (10 mM) (adjusted pH to 4.95 with 15 mM acetic acid) and methanol as the organic modifier. Targeted metabolites were monitored using an MRM scan in the negative ionization mode. Of note, separation of sugar phosphates by LC–MS is challenging due to their extremely high polarity and similar structure including isomers. By employing a slow gradient elution (with a total running time of 90 min) and running 100% aqueous eluent for 15 min at the beginning of the gradient, Luo's method achieved good chromatographic resolution for not only redox cofactors (NADH, NAD, NADPH, and NADP) and energy cofactors (ATP, ADP, and AMP), but also structural isomers of sugar phosphates. This method was applied to determine intracellular metabolites in various cell samples, where cellular metabolism was quenched by cold (−50 °C) methanol : water (60 : 40, *v*/*v*), and metabolites were further extracted with 25% ethanol (containing 0.3 M KOH) and neutralized by glacial acetic acid.

Buescher *et al.* developed an ultrahigh performance IP-LC–ESI–MS/MS method that allowed comprehensive quantitative analysis of 138 metabolites involved in central carbon metabolism and related biosynthetic reactions, including carboxylic acids, amino acids, sugar phosphates, nucleotides, and functionalized aromatics.[6] Chromatographic separation was achieved on a Waters Acquity T3 end-capped reversed phase column (150 mm×2.1 mm, 1.8 μm particle size), using a gradient of mobile phase A (10 mM tributylamine, 15 mM acetic acid, 5% (*v*/*v*) methanol) and mobile phase B (2-propanol). Selective and sensitive detection was performed on a Thermo TSQ Quantum Ultra triple quadrupole mass spectrometer using heated electrospray ionization. Targeted metabolites were monitored using an MRM scan in the negative ionization mode. The use of a stationary phase with smaller particle size (1.8 μm) enabled improved separation and increased column capacity and thus enhanced sensitivity. In addition, 2-propanol was added as an organic modifier in the mobile phase, which effectively reduced peak tailing and improved separation. Overall, as compared to Luo's method, Buescher's method had a shorter gradient time without compromising reproducibility, enhanced sensitivity, and particularly enhanced separation of isomers of sugar phosphates and carboxylic acids. This method allows routine metabolic profiling in a diverse range of biological matrices of heterogeneous origin, where metabolites were extracted three times with 60% (*v*/*v*) ethanol buffered with 10 mM ammonium acetate (pH 7.2) at 78 °C for 1 min.

Nano flow RP-LC coupled with nano-ESI–MS is commonly used in LC–MS based proteomics. Nano flow allows nano-ESI, which generally improves the ionization process, reduces ion suppression effects, and increases tolerance toward salt contamination in the sample. Kiefer *et al.* combined nano-IP-LC with nano-ESI high-resolution MS for targeted and untargeted metabolomics analyses.[94] Chromatographic separation was achieved on a narrow-diameter C18 column (Thermo Hypersil gold 150 mm×0.1 mm, 5 µm), using a gradient of mobile phase A (1.7 mM tributyl ammonium in 1.5 mM acetic acid, pH adjusted to 9.4 with 6 M ammonium hydroxide) and mobile phase B (methanol), at a nano-flow rate of 500 nL per min. Mass acquisition was carried out in the negative ionization full scan mode at a unit resolution of 60 000 (at *m*/*z* 400). The limits of detection obtained were in the low femtomol down to hundreds of attomol range for all metabolites tested. The high sensitivity enabled a comprehensive metabolic profiling of central carbon metabolism and energy metabolism using a few hundred nanograms of cell extracts.[106] Of note, to efficiently form ion pairs between analytes and tributylamine (pKa 10.4), analytes need to be negatively charged and tributylamine positively charged. As many metabolites have a pK_a ~6 (*e.g.*, phosphorylated sugars) or even pK_a ~9 (*e.g.*, nucleotides), the pH was set to 9.4 in mobile phase A. In addition, to minimize ion suppression effects of ion pair reagents, a minimal concentration of tributylamine (1.7 mM) enabling adequate retention of metabolites was applied. The final IP-LC condition resulted in adequate chromatographic separation for the majority of metabolites including organic acids, phosphorylated compounds, sugar derivates, and coenzyme A (CoA) thioesters. Chemically related compounds are generally eluted in the same separation window, indicating similar retention behavior. However, separation of isomeric sugar phosphates (*e.g.*, glucose-6-phosphate and fructose-6-phosphate) was not achieved.

3.5.2 Amino Acids

As the building blocks of proteins, amino acids play an important role in any living organism. Amino acid profiling is an important biochemical approach in detecting inherited disorders of amino acid metabolism, which has diagnostic and/or therapeutic implications in diseases such as Alzheimer's disease, Parkinson's disease, and cancer.[95,96] A number of IP-LC–MS methods using various stationary phases, ion pair reagents, and mobile phase gradients have been reported for the targeted analysis of amino acids in biological matrices. These methods are summarized in Table 3.2, among which two methods are highlighted below.

The first validated method for quantitative analysis of proteinogenic amino acids in biological samples, without any kind of derivatization, was

Table 3.2 Representative ion pair liquid chromatography methods coupled with mass spectrometry detection for quantitative analysis of amino acids.[a]

Analytes	Stationary phase	Mobile phase	Gradient, flow rate, temperature	MS detection	Ref.
20 proteinogenic amino acids	Octyl or octadecyl bonded phases: Purospher RP-18e	Eluent A: 0.5 mM pentadecafluorooctanoic acid in water; Eluent B: acetonitrile	0–8 min, 0% B; 8–13 min, 0–17% B; 13–21 min, 17% B; 21–31 min, 17–28% B 1 mL per min; room temperature	Positive MRM	97
20 proteinogenic amino acids	Supelcosil ABZ Plus	Eluent A: 1 mM tridecafluoroheptanoic acid in water; Eluent B: acetonitrile	0–6 min, 0% B; 6–8 min, 0–20% B; 8–16 min, 20% B; 16–18 min, 20–25% B 1 mL per min; room temperature	Positive MRM	97
22 underivatized amino acids	Dikama Diamonsil C18	Eluent A: 0.8 mM perfluoroheptanoic acid and 0.05% trifluoroacetic acid in water; Eluent B: acetonitrile	0–6 min, 0% B; 6–14 min, 0–24% B; 14–21 min, 24% B, 21–29 min, 24–60% B 0.2 mL per min	Positive MRM	98
16 underivatized amino acids	Octadecyl-bonded silica gel column: Modulo-cart QS UptiSphere	Eluent A: 0.5 mM tridecafluoroheptanoic acid in water; Eluent B: acetonitrile	0–1 min, 0–15% B; 1–6 min, 15% B; 6–9 min, 15–25% B; 9–15 min, 25% B; 15–16 min, 25–0% B; 16–31 min, 0% B 0.2 mL per min; 26 °C	Positive and negative MRM	16,99
22 amino acids	Hypercarb® PGC column	Eluent A: 25 mM nonafluoropentanoic acid in water; Eluent B: acetonitrile	0–8 min, 0–10% B; 8–16 min, 10–22% B; 16–25 min, 22–50% B; 25–32 min, 50% B 0.25 mL per min; 10 °C	TOF full-scan MS at *m*/*z* 50–1000 in positive ion mode	19
33 amino acids	Agilent Zorbax SB-C18 column	Eluent A: 0.2% formic acid and 0.02% perfluorocarboxylic acids in water; Eluent B: methanol	0–6 min, 2–5% B; 6–9 min, 5–20% B; 9–12 min, 20–80% B; 12–13 min, 80–100% B 0.4 mL per min; 50 °C	Positive and negative MRM	18
20 amino acids and 5 metabolically related compounds	Waters Symmetry C18 column	Eluent A: 1 mM perfluoro-heptanoic acid in water; Eluent B: acetonitrile	0–0.1 min, 2–20% B; 0.1–2.3 min, 20–40% B; 2.3–4 min, 40% B; 4–4.1 min, 40–2% B; 4.1–6 min, 2% B 0.3 mL per min; 30 °C	Positive MRM	100

[a]Abbreviations: MS, mass spectrometry; MRM, multiple reaction monitoring.

achieved by IP-LC coupled with stable isotope dilution tandem MS.[98] Sufficient separation of 22 underivatized amino acids was achieved on a Dikama Diamonsil C18 column (150×2.1 mm, particle size 5 μm), using a gradient mobile phase consisting of perfluoroheptanoic acid (0.8 mM) –trifluoroacetic acid (0.05%) as the ion pair reagents and acetonitrile as the organic modifier. Plasma and blood sample preparation involved protein precipitation with methanol containing 0.5% (*v*/*v*) formic acid and internal standards. This method is highly selective due to the combination of sufficient chromatographic separation and specific MRM for characteristic transitions.

Interestingly, Chen *et al.* developed an IP-LC method coupled with high-resolution TOF MS for both targeted and untargeted analysis of amino acids biomarkers in *Candida albicans* biofilm formation.[19] Planktonic and biofilm cells were washed rapidly with ultrapure water, and the cell pellets were extracted twice with 80% methanol (−80 °C). IP-LC was performed on a porous graphic carbon column, using a gradient mobile phase consisting of nonafluoropentanoic acid (25 mM) as the ion pair reagent and acetonitrile as the organic modifier. Targeted quantitative analysis of 22 amino acids was performed with stable isotope-labeled internal standards, among which 9 amino acids were identified as the biomarkers for *C. albicans* biofilm formation. In addition, untargeted analysis was carried out with full positive scan mode at the mass range of *m*/*z* 50–1000, and 16 metabolites were identified as potential biomarkers, most of which are amino acids or related compounds. This study well demonstrates the capability of IP-LC coupled with high resolution MS for simultaneous targeted and untargeted analysis in biomarker studies.

3.5.3 Nucleotides

A nucleotide is composed of a nucleobase, a five-carbon sugar (either ribose or 2-deoxyribose), and one, two, or three phosphate groups.[101] In addition to serving as the precursors of RNA and DNA, nucleotides play a central role in cell metabolism, cellular signaling, and energy transfer.[101] Given their multiple functions, the levels of nucleotides in eukaryotic cells maintain a delicate balance under normal physiological conditions. Quantitative profiling of intracellular nucleotides is important for understanding complex biochemical changes in the nucleotide pool in response to genetic modifications, pathophysiological or environment stimuli. Because nucleotides are composed of phosphate groups bound to a five-carbon sugar that is linked to a purine or a pyrimidine base, these nucleotide molecules are negatively charged at pH > 2.

Several IP-LC–MS/MS methods have been reported for quantitative analysis of nucleotides in biological samples, where trimethylamine (TEA) –acetic acid or triethylammonium acetate (TEAA) and similar ion-pair reagents (*e.g.*, tributylammonium acetate, tetrabutylammonium acetate, and dimethylhexylamine) are commonly used.[102–106] However, all these methods suffer from limited sensitivity, with the lower limit of quantitation ranging from 50–5000 nM for individual nucleotides.[102–106] The lack of sensitivity was attributable to the fact that the ion pair reagents used in these methods are not particularly volatile, thereby resulting in inefficient electrospray ionization of the analytes. To overcome this limitation, an IP-LC–MS/MS method using TEA as the ion pair reagent and hexafluoroisopropanol (HFIP) as the counterion was developed and fully validated for the determination of intracellular nucleoside triphosphates.[13] The optimized mobile phase consisted of mobile phase A (*i.e.*, 100 mM HFIP and 8.6 mM TEA in water, final pH 8.3) and mobile phase B (*i.e.*, 10% acetonitrile in mobile phase A).[13] Superior to the less volatile ion-pair mobile phase, the TEA–HFIP ion-pair mobile phase produced efficient chromatographic separation as well as high-sensitivity electrospray ionization. The mechanism underlying enhanced sensitivity is based on the dynamic adjustment of the pH of the electrospray droplet as a consequence of the preferential evaporation removal of the more volatile anionic counterion (*i.e.*, HFIP) from the droplet surface.[45] As mentioned in Section 2.2.4, a more aqueous mobile phase is favorable to the formation of ion pairs between the positively charged ion pairing agent and the negatively charged phosphate group of nucleotides. Thus, a shallow gradient slope (*i.e.*, 0.16% increase of acetonitrile per minute) was applied to achieve good chromatographic resolution of nucleotides.[13] It should be mentioned that the formation of shallow gradients can place performance demands on the HPLC pumps and mixers, and therefore, it is recommended that the mobile phase B contains a premix blend of aqueous and organic solvents (*i.e.*, 10% of acetonitrile in aqueous mobile phase A) to minimize potentially inadequate solvent mixing that can compromise component resolution.[13] In addition to quantitation of nucleoside triphosphate, this method has a broad application potential for quantitative analysis of a variety of polar intracellular metabolites including nucleoside mono- and bi-phosphate, carboxylates, sugar phosphates, coenzymes, and fatty acids.[107]

3.6 Potential Pitfalls in IP-LC–MS-based Metabolomics

Metabolomics is intrinsically challenging due to the reactivity, physicochemical diversity, and broad concentration range of metabolites. In Section 3.4, we discuss the potential pitfalls (*i.e.*, perturbation of metabolite levels, incomplete termination of metabolism, metabolite degradation, and interconversion) and how to avoid them in the sample preparation process, which are nearly universal in metabolomic studies. Here, we focus on the potential pitfalls particularly associated with IP-LC–MS based metabolomics and provide practical solutions to mitigate these problems.

3.6.1 Separation of Isomers

Isomers frequently occur in the core cellular metabolism pathways, for example, isomers of sugar phosphates and carboxylic acids in central carbon metabolism. While sharing identical elemental composition and similar physicochemical properties, isomers often possess distinct biological functions. Efficient separation of isomers but without compromising broad coverage of metabolites is important, though challenging, in IP-LC–MS based metabolomics. Two strategies, chromatographic separation and/or selection of unique MS/MS transitions, could be applied to address this issue.

Chromatographic separation of isomers is challenging because of similar chemical structures and physicochemical properties. Nevertheless, chromatographic separation of isomers may be possible by the selection of appropriate stationary phase and mobile phase. In an IP-LC–MS/MS method reported by Buescher *et al.* for targeted analysis of more than 100 central metabolites, good chromatographic separation for isomers of different classes of metabolites, such as sugar phosphates (ribose 5-phosphate and ribulose 5-phosphate), carboxylic acids (citric acid and isocitric acid), and amino acid precursors (ketoisoleucine and ketoleucine), was achieved on a Waters Acquity T3 end-capped reversed phase column, using a gradient of mobile phase A (10 mM tributylamine, 15 mM acetic acid, 5% (*v*/*v*) methanol) and mobile phase B (2-propanol).[6] The retention of analytes on the end-capped C18 stationary phase depends on inherent and ion pairing medicated hydrophobic interactions. Because ion pairing mediated retention is sensitive to the conformation of acidic (phosphates, carboxyls) and hydroxylic groups, the conformational difference of isomers conveys additional selectivity with respect to the interaction with the stationary phase thus enabling the separation of isomers.[6]

Isomers are often indistinguishable by mass spectrometry, even with high-resolution mass spectrometry, because of their identical elemental composition and similar fragmentation pattern. Although isomers typically yield the same MS/MS fragments, significant differences in their relative intensities may occur, and therefore unique MS/MS transitions may be selected for the quantitation of isomers. For example, hexose phosphate isomers, including fructose 6-phosphate (F6P), glucose 1-phosphate (G1P), and mannose 1-phosphate (Ms1P), all have the MS/MS transition at *m*/*z* 259 > 79. However, selection of a unique MS/MS transition at *m*/*z* 259 > 169 enables quantitation of F6P only because other hexose phosphate isomers are not visible at this transition.[6] Similarly, selection of a unique MS/MS transition at *m*/*z* 259 > 241 allows the separation of G1P with other hexose phosphate isomers.[6]

3.6.2 Ion Suppression

Although inclusion of an ion pair reagent is beneficial for chromatographic retention and separations, the positively charged ion pair reagents such as tributyl ammonium ions in the mobile phase can cause severe ion suppression in the positive electrospray ionization mode. Based on the class of ion pair reagents (either amine or acid substituted structure), mass spectrometric detection is generally limited to either negative or positive ionization, respectively, to avoid ion suppression phenomena, high background and general interference. In addition, strong ion pairs formed between the analyte and ion pair reagent, which are not easily broken down in the electrospray ionization source, may significantly reduce ionization efficiency of the analyte.[41] On the other hand, increasing ion pair reagent concentrations often result in better chromatographic retention and resolution. Therefore, the concentration of an ion pair reagent needs to be carefully optimized to minimize ion suppression but without compromising chromatographic resolution. Furthermore, counterions (*e.g.*, volatile acids or fluoroalcohol) can be added into the mobile phase to enhance volatility of the mobile phase, which would enable rapid evaporation during desolvation, facilitate dissociation of excessive pairing between analytes and ion pair reagents, and ultimately enhance electrospray ionization efficiency.[41,44]

3.6.3 Detrimental Effect of Metal Cations

The majority of metabolites involved in central carbon metabolism have a high affinity for metal cations.[108] The formation of metal salts or chelates can reduce analytical performance of IP-LC coupled with ESI–MS by negatively influencing chromatographic retention (*e.g.*, *via* peak-tailing or broadening) or ionization efficiency (*e.g.*, *via* ion suppression). Analytes with extremely high metal affinities may actually escape detection altogether, as observed for the secondary fungal metabolite tenuazonic acid, for instance.[109] Metal cations can enter the analytical system by several routes, particularly through leakage (such as Al^{3+} and Fe^{3+}) from columns due to progressive corrosion or metal deposits introduced by biological samples.[110] An effective approach to mitigate detrimental effects of metal cations is the addition of volatile metal chelators, which form complexes with metal species and is ESI-compatible. Siegel *et al.* demonstrated that the addition of acetylacetone, a volatile metal chelator (boiling point, 138 °C), into the tributylamine-based mobile phase significantly improved the analytical performance for virtually all 52 targeted metabolites in IP-LC–ESI–MS/MS analysis.[110] On average, acetylacetone (2 mM) increased peak intensity by 2.5-fold in aqueous standards and by 5-fold in biological matrices.[110] The most significant improvement was seen for analytes with particularly high metal affinities, such as citrate, salicylate, coenzyme A, acetyl-coenzyme A, aspartate, and phenylalanine.[110] The positive effect of acetylacetone is attributed mainly to improvements in chromatographic performance (*i.e.*, improving peak shapes, column efficiency, and separation) because the complexation of acetylacetone with metal cations efficiently alleviates analyte–metal interactions that are otherwise detrimental to analytical performance. Acetylacetone at the concentration of 2 mM could be routinely added to the well-established tributylamine ion pair mobile phase. If other ion pair mobile phase systems are used, the concentrations of the metal-chelating eluent modifier should be optimized.

3.6.4 System Contamination

A well-known disadvantage associated with IP-LC–MS analysis is the long-term contamination of chromatographic and mass spectrometry instrumentation due to the high affinity of ion pair reagents to the components of the LC–MS system. Effective decontamination of the LC instrument can usually be achieved with aggressive washes with either strong acids or bases. In the case of persistent contamination, the replacement of contaminated LC components is an effective solution. The clean-up of ion pair reagents from the LC column, even with extensive column flushing, is difficult due to strong hydrophobic interactions between the ion pair reagent and stationary phase. Thus, a dedicated column is recommended to use for a particular IP-LC method and even for the analysis of samples with the same matrix.[106]

Decontamination of the mass spectrometer is even more challenging. Complete removal of ion pair reagents (*e.g.*, tributylamine) from the mass spectrometer may not be possible even following repeated washes with strong organic solvents and detergent (in a sonic bath) of every single ion path component (from the cone up to Q3) and plus replacing the whole ion source block, curtain plate, and orifice.[106] The complete replacement of the whole ion path and detector is the most effective, but expensive, strategy for decontamination of the mass spectrometer. Thus, the application of IP-LC–MS in practice needs careful consideration if there are alternatives.

References

1. S. Rochfort, *J. Nat. Prod.*, 2005, **68**, 1813–1820.
2. B. F. O'Connor and P. M. Cummins, *Methods Mol. Biol.*, 2017, **1485**, 355–363.
3. A. J. Alpert, *J. Chromatogr.*, 1990, **499**, 177–196.
4. T. Cecchi, *Crit. Rev. Anal. Chem.*, 2008, **38**, 161–213.
5. W. Naldong, *J. Chromatogr. B: Anal. Technol. Biomed. Life Sci.*, 2003, **796**, 209–224.
6. J. M. Buescher, S. Moco, U. Sauer and N. Zamboni, *Anal. Chem.*, 2010, **82**, 4403–4412.
7. H. A. Claessens and M. A. van Straten, *J. Chromatogr. A*, 2004, **1060**, 23–41.
8. J. H. Knox and J. Jurand, *J. Chromatogr.*, 1976, **125**, 89–101.
9. J. C. Kraak, K. M. Jonker and J. K. Huber, *J. Chromatogr.*, 1977, **142**, 671–688.
10. K. Nakajima, S. Kitazume, T. Angata, R. Fujinawa, K. Ohtsubo, E. Miyoshi and N. Taniguchi, *Glycobiology*, 2010, **20**, 865–871.
11. L. Gong and J. S. McCullagh, *Rapid Commun. Mass Spectrom.*, 2014, **28**, 339–350.
12. L. Gong, *Rapid Commun. Mass Spectrom.*, 2015, **29**, 2402–2410.

13. J. Wu, Y. Zhang, R. Wiegand, J. Wang, G. Bepler and J. Li, *J. Chromatogr. B: Anal. Technol. Biomed. Life Sci.*, 2015, **1006**, 167–178.
14. N. Li, N. M. El Zahar, J. G. Saad, E. R. E. van der Hage and M. G. Bartlett, *J. Chromatogr. A*, 2018, **1580**, 110–119.
15. R. Liu, Y. Ruan, Z. Liu and L. Gong, *Rapid Commun. Mass Spectrom.*, 2019, **33**, 697–709.
16. M. Piraud, C. Vianey-Saban, K. Petritis, C. Elfakir, J. P. Steghens and D. Bouchu, *Rapid Commun. Mass Spectrom.*, 2005, **19**, 1587–1602.
17. U. Harder, B. Koletzko and W. Peissner, *J. Chromatogr. B: Anal. Technol. Biomed. Life Sci.*, 2011, **879**, 495–504.
18. T. Xia, S. Fu, Q. Wang, Y. Wen, S. A. Chan, S. Zhu, S. Gao, X. Tao, F. Zhang and W. Chen, *Biomed. Chromatogr.*, 2018, **32**, e4198.
19. X. Chen, H. Wu, Y. Cao, X. Yao, L. Zhao, T. Wang, Y. Yang, D. Lv, Y. Chai, Y. Cao and Z. Zhu, *Mol. BioSyst.*, 2014, **10**, 74–85.
20. P. Chaimbault, K. Petritis, C. Elfakir and M. Dreux, *J. Chromatogr. A*, 2000, **870**, 245–254.
21. S. Rinne, A. Holm, E. Lundanes and T. Greibrokk, *J. Chromatogr. A*, 2006, **1119**, 285–293.
22. A. Karlsson and O. Karlsson, *Chirality*, 1997, **9**, 650–655.
23. A. Karlsson and O. Karlsson, *J. Chromatogr. A*, 2001, **905**, 329–335.
24. C. West, C. Elfakir and M. Lafosse, *J. Chromatogr. A*, 2010, **1217**, 3201–3216.
25. L. Pereira, *J. Liq. Chromatogr. Relat. Technol.*, 2008, **31**, 1687–1731.
26. B. Luo, K. Groenke, R. Takors, C. Wandrey and M. Oldiges, *J. Chromatogr. A*, 2007, **1147**, 153–164.
27. A. C. McGinnis, E. C. Grubb and M. G. Bartlett, *Rapid Commun. Mass Spectrom.*, 2013, **27**, 2655–2664.
28. C. G. Huber and H. Oberacher, *Mass Spectrom. Rev.*, 2001, **20**, 310–343.
29. H. Oberacher, W. Parson, R. Muhlmann and C. G. Huber, *Anal. Chem.*, 2001, **73**, 5109–5115.
30. A. Premstaller, H. Oberacher and C. G. Huber, *Anal. Chem.*, 2000, **72**, 4386–4393.
31. P. J. Oefner and C. G. Huber, *J. Chromatogr. B: Anal. Technol. Biomed. Life Sci.*, 2002, **782**, 27–55.
32. V. K. Sharma, J. Glick and P. Vouros, *J. Chromatogr. A*, 2012, **1245**, 65–74.
33. R. Erb, K. Leithner, A. Bernkop-Schnurch and H. Oberacher, *AAPS J.*, 2012, **14**, 728–737.
34. B. A. Bidlingmeyer, S. N. Deming, W. P. Price, B. Sachok Jr and M. Petrusek, *J. Chromatogr.*, 1979, **186**, 419–434.
35. L. M. S. Zappoli and F. Osti, *J. Chromatogr. A*, 1996, **721**, 269–277.
36. T. Cecchi, F. Pucciarelli and P. Passamonti, *Anal. Chem.*, 2001, **73**, 2632–2639.
37. R. LoBrutto, A. Jones, Y. V. Kazakevich and H. M. McNair, *J. Chromatogr. A*, 2001, **913**, 173–187.

38. M. Shibue, C. T. Mant and R. S. Hodges, *J. Chromatogr. A*, 2005, **1080**, 68–75.
39. R. LoBrutto, A. Jones and Y. V. Kazakevich, *J. Chromatogr. A*, 2001, **913**, 189–196.
40. C. T. Mant and R. S. Hodges, *J. Chromatogr. A*, 2006, **1125**, 211–219.
41. A. Apffel, S. Fischer, G. Goldberg, P. C. Goodley and F. E. Kuhlmann, *J. Chromatogr. A*, 1995, **712**, 177–190.
42. M. Shibue, C. T. Mant and R. S. Hodges, *J. Chromatogr. A*, 2005, **1080**, 58–67.
43. C. G. Huber and A. Krajete, *Anal. Chem.*, 1999, **71**, 3730–3739.
44. A. V. Willems, D. L. Deforce, W. E. Lambert, C. H. Van Peteghem and J. F. Van Bocxlaer, *J. Chromatogr. A*, 2004, **1052**, 93–101.
45. A. Apffel, J. A. Chakel, S. Fischer, K. Lichtenwalter and W. S. Hancock, *Anal. Chem.*, 1997, **69**, 1320–1325.
46. B. Basiri, H. van Hattum, W. D. van Dongen, M. M. Murph and M. G. Bartlett, *J. Am. Soc. Mass Spectrom.*, 2017, **28**, 190–199.
47. H. Hashem and T. Jira, *J. Chromatogr. A*, 2006, **1133**, 69–75.
48. F. Tang, L. Tao, X. Luo, L. Ding, M. Guo, L. Nie and S. Yao, *J. Chromatogr. A*, 2006, **1125**, 182–188.
49. S. Wybraniec, *J. Chromatogr. A*, 2006, **1127**, 70–75.
50. D. O. Mack, V. L. Reed and L. D. Smith, *J. Liq. Chromatogr.*, 1985, **8**, 591–602.
51. L. Coulier, R. Bas, S. Jespersen, E. Verheij, M. J. van der Werf and T. Hankemeier, *Anal. Chem.*, 2006, **78**, 6573–6582.
52. K. Robards, P. R. Haddad and P. E. Jackson, High-performance Liquid Chromatography—Instrumentation and Techniques, *Principles and Practice of Modern Chromatographic Methods*, 2004, pp. 227–303.
53. Y. Yokoyama, O. Ozaki and H. Sato, *J. Chromatogr. A*, 1996, **739**, 333–342.
54. B. D. Bennett, J. Yuan, E. H. Kimball and J. D. Rabinowitz, *Nat. Protoc.*, 2008, **3**, 1299–1311.
55. W. Lu, B. D. Bennett and J. D. Rabinowitz, *J. Chromatogr. B: Anal. Technol. Biomed. Life Sci.*, 2008, **871**, 236–242.
56. A. Ribbenstedt, H. Ziarrusta and J. P. Benskin, *PLoS One*, 2018, **13**, e0207082.
57. S. K. Manier, A. Keller, J. Schaper and M. R. Meyer, *Sci. Rep.*, 2019, **9**, 2741.
58. N. Vinayavekhin and A. Saghatelian, *Curr. Protoc. Mol. Biol.*, 2010, 30, 30.1.1–30.1.24.
59. A. C. Schrimpe-Rutledge, S. G. Codreanu, S. D. Sherrod and J. A. McLean, *J. Am. Soc. Mass Spectrom.*, 2016, **27**, 1897–1905.
60. L. Cui, H. Lu and Y. H. Lee, *Mass Spectrom. Rev.*, 2018, **37**, 772–792.
61. R. S. Plumb, K. A. Johnson, P. Rainville, B. W. Smith, I. D. Wilson, J. M. Castro-Perez and J. K. Nicholson, *Rapid Commun. Mass Spectrom.*, 2006, **20**, 1989–1994.

62. L. C. Gillet, P. Navarro, S. Tate, H. Rost, N. Selevsek, L. Reiter, R. Bonner and R. Aebersold, *Mol. Cell. Proteomics*, 2012, **11**, O111. 016717.
63. C. Guijas, J. R. Montenegro-Burke, X. Domingo-Almenara, A. Palermo, B. Warth, G. Hermann, G. Koellensperger, T. Huan, W. Uritboonthai, A. E. Aisporna, D. W. Wolan, M. E. Spilker, H. P. Benton and G. Siuzdak, *Anal. Chem.*, 2018, **90**, 3156–3164.
64. C. A. Smith, G. O'Maille, E. J. Want, C. Qin, S. A. Trauger, T. R. Brandon, D. E. Custodio, R. Abagyan and G. Siuzdak, *Ther. Drug Monit.*, 2005, **27**, 747–751.
65. D. S. Wishart, Y. D. Feunang, A. Marcu, A. C. Guo, K. Liang, R. Vazquez-Fresno, T. Sajed, D. Johnson, C. Li, N. Karu, Z. Sayeeda, E. Lo, N. Assempour, M. Berjanskii, S. Singhal, D. Arndt, Y. Liang, H. Badran, J. Grant, A. Serra-Cayuela, Y. Liu, R. Mandal, V. Neveu, A. Pon, C. Knox, M. Wilson, C. Manach and A. Scalbert, *Nucleic Acids Res.*, 2018, **46**, D608–D617.
66. A. J. Carroll, P. Zhang, L. Whitehead, S. Kaines, G. Tcherkez and M. R. Badger, *Front. Bioeng. Biotechnol.*, 2015, **3**, 106.
67. S. A. Goldansaz, A. C. Guo, T. Sajed, M. A. Steele, G. S. Plastow and D. S. Wishart, *PLoS One*, 2017, **12**, e0177675.
68. K. Haug, K. Cochrane, V. C. Nainala, M. Williams, J. Chang, K. V. Jayaseelan and C. O'Donovan, *Nucleic Acids Res.*, 2020, **48**, D440–D444.
69. L. C. Menikarachchi, D. W. Hill, M. A. Hamdalla, I. I. Mandoiu and D. F. Grant, *J. Chem. Inf. Model.*, 2013, **53**, 2483–2492.
70. K. Moutselos, I. Kanaris, A. Chatziioannou, I. Maglogiannis and F. N. Kolisis, *BMC Bioinf.*, 2009, **10**, 324.
71. G. Wu, R. C. Anafi, M. E. Hughes, K. Kornacker and J. B. Hogenesch, *Bioinformatics*, 2016, **32**, 3351–3353.
72. W. Lu, X. Su, M. S. Klein, I. A. Lewis, O. Fiehn and J. D. Rabinowitz, *Annu. Rev. Biochem.*, 2017, **86**, 277–304.
73. K. Dettmer, P. A. Aronov and B. D. Hammock, *Mass Spectrom. Rev.*, 2007, **26**, 51–78.
74. M. R. Mashego, K. Rumbold, M. De Mey, E. Vandamme, W. Soetaert and J. J. Heijnen, *Biotechnol. Lett.*, 2007, **29**, 1–16.
75. C. Wittmann, J. O. Kromer, P. Kiefer, T. Binz and E. Heinzle, *Anal. Biochem.*, 2004, **327**, 135–139.
76. U. Theobald, W. Mailinger, M. Reuss and M. Rizzi, *Anal. Biochem.*, 1993, **214**, 31–37.
77. R. P. Maharjan and T. Ferenci, *Anal. Biochem.*, 2003, **313**, 145–154.
78. C. L. Winder, W. B. Dunn, S. Schuler, D. Broadhurst, R. Jarvis, G. M. Stephens and R. Goodacre, *Anal. Chem.*, 2008, **80**, 2939–2948.
79. D. Gonzalez, J. Francois and M. Renaud, *Yeast*, 1997, **13**, 1347–1355.
80. D. Vuckovic, *Anal. Bioanal. Chem.*, 2012, **403**, 1523–1548.
81. R. El Rammouz, F. Letisse, S. Durand, J. C. Portais, Z. W. Moussa and X. Fernandez, *Anal. Biochem.*, 2010, **398**, 169–177.

82. E. J. Want, P. Masson, F. Michopoulos, I. D. Wilson, G. Theodoridis, R. S. Plumb, J. Shockcor, N. Loftus, E. Holmes and J. K. Nicholson, *Nat. Protoc.*, 2013, **8**, 17–32.
83. W. Weckwerth, K. Wenzel and O. Fiehn, *Proteomics*, 2004, **4**, 78–83.
84. J. D. Rabinowitz and E. Kimball, *Anal. Chem.*, 2007, **79**, 6167–6173.
85. E. J. Want, G. O'Maille, C. A. Smith, T. R. Brandon, W. Uritboonthai, C. Qin, S. A. Trauger and G. Siuzdak, *Anal. Chem.*, 2006, **78**, 743–752.
86. W. Chan, Y. Zhao and J. Zhang, *Rapid Commun. Mass Spectrom.*, 2019, **33**, 561–568.
87. M. Yuan, S. B. Breitkopf, X. Yang and J. M. Asara, *Nat. Protoc.*, 2012, **7**, 872–881.
88. W. Lu, Y. K. Kwon and J. D. Rabinowitz, *J. Am. Soc. Mass Spectrom.*, 2007, **18**, 898–909.
89. W. Zou and M. Al-Rubeai, *Biotechnol. Appl. Biochem.*, 2016, **63**, 642–651.
90. B. M. Cumming and A. J. Steyn, *Proc. Natl. Acad. Sci. U. S. A.*, 2015, **112**, 13135–13136.
91. A. H. Romano and T. Conway, *Res. Microbiol.*, 1996, **147**, 448–455.
92. L. Jin and Y. Zhou, *Oncol. Lett.*, 2019, **17**, 4213–4221.
93. I. Marin-Valencia, S. K. Cho, D. Rakheja, K. J. Hatanpaa, P. Kapur, T. Mashimo, A. Jindal, V. Vemireddy, L. B. Good, J. Raisanen, X. Sun, B. Mickey, C. Choi, M. Takahashi, O. Togao, J. M. Pascual, R. J. Deberardinis, E. A. Maher, C. R. Malloy and R. M. Bachoo, *NMR Biomed.*, 2012, **25**, 1177–1186.
94. P. Kiefer, N. Delmotte and J. A. Vorholt, *Anal. Chem.*, 2011, **83**, 850–855.
95. E. Siminska and M. Koba, *Amino Acids*, 2016, **48**, 1339–1345.
96. E. Socha, M. Koba and P. Koslinski, *Amino Acids*, 2019, **51**, 367–371.
97. P. Chaimbault, K. Petritis, C. Elfakir and M. Dreux, *J. Chromatogr. A*, 1999, **855**, 191–202.
98. J. Qu, Y. Wang, G. Luo, Z. Wu and C. Yang, *Anal. Chem.*, 2002, **74**, 2034–2040.
99. M. Piraud, C. Vianey-Saban, K. Petritis, C. Elfakir, J. P. Steghens, A. Morla and D. Bouchu, *Rapid Commun. Mass Spectrom.*, 2003, **17**, 1297–1311.
100. L. Gu, A. D. Jones and R. L. Last, *Anal. Chem.*, 2007, **79**, 8067–8075.
101. L. Stryer, *Biochemistry,* 3rd edn, 1988, W. H. Freeman, New York.
102. S. Cohen, M. Megherbi, L. P. Jordheim, I. Lefebvre, C. Perigaud, C. Dumontet and J. Guitton, *J. Chromatogr. B: Anal. Technol. Biomed. Life Sci.*, 2009, **877**, 3831–3840.
103. R. L. Claire, 3rd, *Rapid Commun. Mass Spectrom.*, 2000, **14**, 1625–1634.
104. P. Chen, Z. Liu, S. Liu, Z. Xie, J. Aimiuwu, J. Pang, R. Klisovic, W. Blum, M. R. Grever, G. Marcucci and K. K. Chan, *Pharm. Res.*, 2009, **26**, 1504–1515.
105. T. Qian, Z. Cai and M. S. Yang, *Anal. Biochem.*, 2004, **325**, 77–84.
106. F. Michopoulos, N. Whalley, G. Theodoridis, I. D. Wilson, T. P. Dunkley and S. E. Critchlow, *J. Chromatogr. A*, 2014, **1349**, 60–68.

107. X. Bao, J. Wu, S. Kim, P. LoRusso and J. Li, *J. Clin. Pharmacol.*, 2019, **59**, 20–34.
108. N. Stern, D. T. Major, H. E. Gottlieb, D. Weizman, A. H. Sayer, E. Blum and B. Fischer, *J. Biol. Inorg. Chem.*, 2012, **17**, 861–879.
109. D. Siegel, T. Rasenko, M. Koch and I. Nehls, *J. Chromatogr. A*, 2009, **1216**, 4582–4588.
110. D. Siegel, H. Permentier and R. Bischoff, *J. Chromatogr. A*, 2013, **1294**, 87–97.

CHAPTER 4

Hydrophilic Interaction Chromatography–Mass Spectrometry (HILIC–MS) Approaches for Probing the Polar Metabolome

H. GALLART-AYALA, T. TEAV AND J. IVANISEVIC*

Metabolomics Platform, Faculty of Biology and Medicine, University of Lausanne, Switzerland
*Email: julijana.ivanisevic@unil.ch

4.1 HILIC–MS: A Powerful Analytical Approach for the Qualitative and Quantitative Analysis of the Polar Cellular Metabolome

Qualitative and quantitative study of the molecular network constituting cellular metabolism, including the small molecule – metabolite component, is essential for advancing the current knowledge on metabolism, its energy and signaling pathways and their regulation. The majority of cellular metabolites implicated in energy producing (glycolysis, pentose phosphate pathway, TCA cycle) and biosynthetic pathways (amino acids, nucleotides, vitamins and their precursors) of central carbon metabolism are highly hydrophilic, inherently transient and of relatively low abundance.[1,2] While the challenge of the

New Developments in Mass Spectrometry No. 10
Advanced Mass Spectrometry-based Analytical Separation Techniques for Probing the Polar Metabolome
Edited by Rawi Ramautar

Published by the Royal Society of Chemistry, www.rsc.org

described polar metabolome analysis remains, liquid chromatography (LC), and specifically hydrophilic interaction chromatography (HILIC) coupled to electrospray ionization (ESI) mass spectrometry (MS), has become one of the leading analytical technologies for its analysis. Together with capillary electrophoresis (CE), HILIC provides an efficient retention and separation for a broad spectrum of ionic compounds (*e.g.* amino acids and other organic acids, nucleotides, energy currency metabolites), and ESI-MS offers significantly higher sensitivity (compared to other ionization modes) for the water-soluble cellular metabolites that are naturally charged in solution. HILIC coupled to MS was introduced for targeted screening and global metabolite profiling over the past ten years as a versatile alternative to derivatization-assisted GC–MS and not-MS suitable ion pairing reversed phase (RP) LC methods.[3–6] With the orthogonal retention mechanism to reversed-phase liquid chromatography (RPLC), the highly polar analytes that otherwise elute in the void volume (using RP stationary phases), can be efficiently separated using HILIC while avoiding MS contamination with ion pairing agents and fastidious derivatization techniques (that also introduce bias during sample preparation).[7] This high chromatographic resolution of HILIC methods for diverse polar metabolites considerably improves their identification and quantification.[8,9] The metabolite annotation, with high level of confidence,[10] can be significantly enhanced by good chromatographic resolution reducing ion suppression by abundant metabolites and facilitating the acquisition of higher number and higher quality MS/MS fragmentation spectra.[11] Finally, the HILIC separation also maximizes the measurement specificity by providing the retention time (RT) identifiers and the ability to differentiate, based on retention time, isobaric compounds.[2,8]

HILIC was introduced 30 years ago by Alpert as a promising solution for the separation of amino acids, peptides and nucleic acids.[12] It has become increasingly popular following the milestone review by Hemström and Irgum in 2006[13] and is extensively used today in different metabolomics approaches, including untargeted discovery profiling, broad-spectrum targeted screening, isotopic profiling and targeted quantification with a focus on one specific pathway or class of metabolites.[14]

4.2 From Stationary Phases to Analytical Conditions and Retention Mechanisms

The chemistry of column stationary phases, mobile phase modifiers (*i.e.* buffers), elution gradient and sample diluent determine the success of HILIC separation. In addition, appropriate column conditioning and equilibration time are essential for reproducible analysis.

4.2.1 Stationary Phases According to Chemical Structure

Stationary phase or HILIC column chemistry was ranked as the most important factor governing the selectivity in HILIC.[15] Globally, stationary phases can be classified, based on their chemical structure, as ***neutral*** (*e.g.* amide, cyano and diol), ***positively charged*** (*e.g.* amine, imidazole), ***negatively charged*** (*e.g.* bare silica, polyaspartic acid) and ***zwitterionic*** (*e.g.* sulfobetaine, phosphocholine). A great variety of bonded HILIC stationary phases were obtained by binding ligands on the surface of silica or polymeric materials to improve the efficiency and robustness of polar analyte retention, through multiple mechanisms (see Section 4.2.3 below). Silica-based bonded phases are characterized by a limited pH-stability range, from pH 2 to 8, while the polymer-supported phases have an extended pH range up to pH 11. The functional group chemistry, column examples and separation efficiency of bonded HILIC stationary phases were extensively reviewed by Guo and Gaiki[16] and Qiao *et al.*[17] The most commonly used stationary phases in HILIC, especially in metabolomics applications (where the aim is broad coverage) are the polymeric zwitterionic (ZIC) and amide stationary phases, such as ZIC-pHILIC and BEH (glycan ethylene bridged hybrid) amide showing the highest hydrophilic selectivity.[4,15] The zwitterionic stationary phase is used for the simultaneous analysis of anionic and cationic species through weaker ionic interactions (compared to conventional ion exchangers).[7,18] Zwitterionic columns are considered as the HILIC all-purpose phases, for the separation of neutral, acidic and basic analytes (*e.g.* di- and tri-carboxylic acids, phosphorylated sugars, nucleosides) due to their hydrophilicity, their modest ion-exchange properties and wide pH range stability (from 2 to 11).[7,19] They offer highly reproducible separations and particularly broad application ranges, including the measurement of biologically highly relevant but highly hydrophilic metabolites, such as phosphorylated energy currency metabolites (*e.g.* ADP/ATP) for which there is no adequate alternative.

4.2.2 Gradient Elution, Buffers and Ionic Strength of Mobile Phases

The elution gradient applied in HILIC spans from the highly organic down to highly aqueous solvent mixture. Mobiles phases typically have high acetonitrile (from 97% down to 60%) and low aqueous buffer content (from 3% up to 40%) that display low viscosity and therefore low back pressures.[15] It was demonstrated that a minimum of 2–3% water is needed to allow for the formation of the water layer on the surface of the polar stationary phase,[20,21] in addition to ensuring buffer solubility. Acetonitrile is the solvent of choice for HILIC separations while protic solvents (such as methanol or isopropanol) with a similar affinity (to water) for the silica surface prevent the formation of a water layer and adsorption retention mechanism (see below).[22] Together with the organic solvent proportion, the mobile phase pH and buffer concentration constitute the second most important factor influencing HILIC selectivity.[23] The most widely used buffers to carry out HILIC separations are ammonium acetate and ammonium formate. These salts are used to stabilize the charge on the analytes and the stationary phase ligands, they are soluble even in high proportions of acetonitrile and volatile allowing the use with ESI–MS interfaces.[15] In addition, the buffer concentration is particularly important for the maintenance of sufficient ionic strength of the mobile phase, necessary for the elution of retained analytes. The elution strength of the mobile phase is determined by the polarity of the solvent mixture and the ability to ensure the proton–donor and acceptor interactions.[23] The pH of the mobile phases can be regulated by the addition of ammonium hydroxide, and formic or acetic acid. The pH has a profound effect on the HILIC selectivity by regulating the retention of acidic, neutral and basic analytes, depending on their charge and ionic interactions with the stationary phase. In "omic" scale profiling, the basic pH ≥ 9 was reported to produce the best results (in terms of retention, selectivity and reliable elution of strongly retained triphosphate species) for the analysis of the broadest spectrum of primarily negatively charged metabolites involved in energy pathways (glycolysis intermediates, carboxylic acids in the TCA cycle, energy currency metabolites).[24–26] For comprehensive guidelines on how to screen different parameters for HILIC method development the reader is advised to consult a report by Periat *et al.*[27]

4.2.3 What is Known About Retention Mechanisms in HILIC

In brief, the main mechanisms of analyte retention in HILIC comprise i) the adsorption on polar groups on the column's stationary phase ligands, ii) analyte partitioning between the water-poor bulk mobile phase and water-rich layer on the stationary phase and ii) hydrogen bonding and dipole–dipole (cation/anion) exchange interactions.[4,15] The understanding of these complex retention mechanisms remains limited and makes HILIC method optimization more compelling. It is recognized in general, that derivatization of silica and polymeric stationary phases enhances retention of polar metabolites by hydrogen bonding (donor/acceptor) or ionic interactions.[28] Due to the complexity of retention mechanisms the retention characteristics are less predictable, compared to RPLC, for which reliable RT modeling is often applied.[15,29] For more details and a comprehensive discussion on different retention mechanisms in HILIC please refer to a comprehensive review by McCalley (2017).[15]

4.3 Selectivity, Sensitivity and Reproducibility of HILIC–MS approaches

When it comes to analysis of polar and charged analytes, the HILIC–MS strategy has multiple advantages over other separation modes and specifically RPLC with respect to selectivity, but also the measurement sensitivity and repeatability.[6,23,30–32] This is particularly compelling for metabolomic assays where the broadest coverage possible of chemically highly diverse polar metabolites is of utmost interest.

Globally, HILIC selectivity is complementary to RPLC owing to superior retention of polar and charged metabolites (see Figure 4.1 for the analysis of a wide panel of amino acids). Due to this orthogonal selectivity, HILIC and RPLC are often combined in untargeted metabolomic assays to maximize the polar, moderately polar and lipid metabolome coverage (see Section 4.4

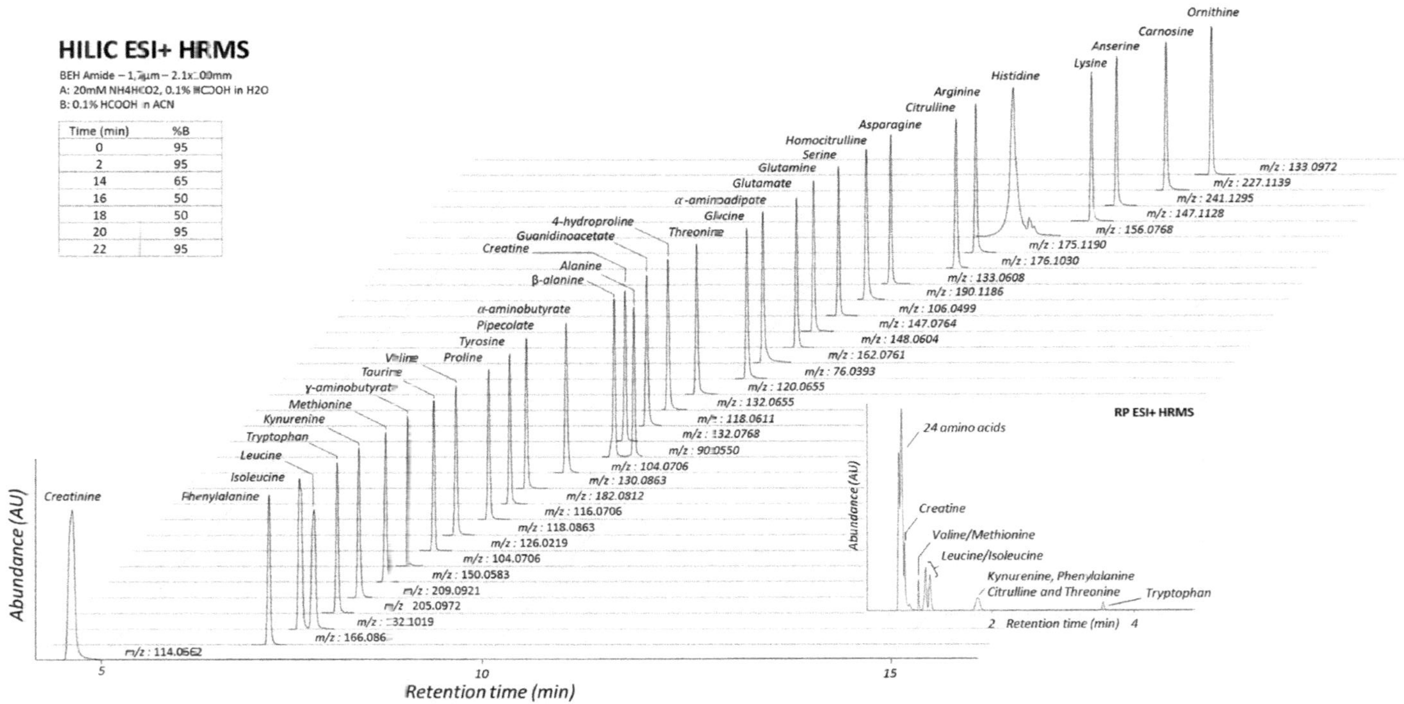

Figure 4.1 Selectivity of HILIC–HRMS for amino acid profiling using a BEH-Amide column. Chromatographic resolution represents great added value to the HRMS for the separation of isobaric compounds (leucine and isoleucine, alanine and beta-alanine). HILIC–HRMS method for merged targeted quantification and untargeted profiling was published by Teav *et al.* in Analytical Chemistry.[33] RPLC–HRMS analysis was performed on a Zorbax RRHD C18 column (2.1 mm×100 mm, 1.8 µm), mobile phases : water and ACN, both with 0.1% formic acid at 400 µl min^{-1}.

below). In addition, the introduction of UHPLC columns with sub-2-μm particles (packing material) has significantly increased the HILIC separation efficiency and sensitivity as a result of narrower chromatographic peaks. When compared to RPLC, as a gold standard among liquid chromatography techniques, HILIC provides a considerably higher MS sensitivity for polar metabolites, in general. This gain in sensitivity (which can vary from 2 to more than 100-fold depending on the analyte) was explained by higher acetonitrile content favoring analyte desolvation. Moreover, this high proportion of organic solvent combined with the ionic character of analytes (depending on their pK_a values and mobile phase pH) was also demonstrated to favor the ionization.[30] Finally, numerous studies (depicted in Section 4.4 below) have proven the robustness of HILIC (also when applied to large-scale human population studies) with respect to retention time and peak definition in general, also determined by the MS detection (subject to an inherent drift over time).[7,31,33] Minimal or no retention time shift is observed with an appropriate usage of state-of-the-art columns and standardization of mobile phase preparation protocols.

The appropriate usage, to achieve excellent reproducibility in HILIC, comprises column conditioning with biological samples (prior to analysis of real sample batch) and relevant equilibration time (typically 10–20 column volumes) at the end of each analysis. It is also important to pay attention to the sample diluent composition (and injection volume) which might have an important impact on peak shape. Considering that the content of organic solvent in sample extracts (following the protein precipitation or SPE-preparation protocols) matches the highly organic HILIC mobile phase composition (at the initial gradient), HILIC can actually improve the analysis throughput and automation (bypassing the evaporation and reconstitution unless necessary for sample pre-concentration).

Overall, the HILIC stationary phases have been considerably improved during the past 10–15 years since the first application in the metabolomics field. The analytical conditions were also extensively investigated and the HILIC–MS approaches reached the separation efficiency and reproducibility equivalent to the RPLC-based methods with complementary selectivity and significantly higher sensitivity for the analysis of polar metabolome. This being said, the challenge of specific polar metabolite analysis remains, including the chromatographic separation of biochemically relevant isomers, accurate measurement of inherently transient metabolites and broad coverage of metabolite classes and/or pathways which is constantly being improved.

4.4 Applications: From Single-pathway Targeted Quantification to Multiple Pathway Targeted and Untargeted Profiling

HILIC–MS is extensively applied today in a wide variety of analytical sciences, from pharmaceutical and clinical environments for the targeted quantification of selected drugs[30,34,35] or endogenous metabolites (*e.g.* amino acids and other organic acids)[36–38] to broad-scale profiling in metabolomics,[39] lipidomics[40] and gylcomics.[41] For polar metabolome analysis the applications range from targeted quantification with a focus on one specific pathway or class of metabolites to broad-spectrum targeted screening and untargeted discovery profiling, including stable isotope-assisted assays.[14]

HILIC-based methods for targeted quantification of specific classes of metabolites (*e.g.* amino acids, amines, organic acids, nucleotides sugars) or specific pathways (*e.g.* tryptophan–kynurenine pathway, NAD+ *de novo* synthesis and salvage pathways, folate metabolism, one carbon metabolism) are becoming a more common practice (Table 4.1) although with a significant time lag, compared to the gold standard RPLC-based methods. This delay is explained by the lower reproducibility of HILIC retention mechanisms and less stable stationary phases which were significantly improved by the introduction of polymeric support materials and UHPLC columns. The validation of HILIC-based targeted quantification methods (using stable isotope-dilution or internal standard spike) demonstrated their high precision, accuracy and reproducibility, according to FDA guidelines, which warrants their application range from small-scale studies on model organisms to large-scale human population studies (Tables 4.1–4.3).

Beyond targeted quantification focused on one specific pathway or class of metabolites, the ***broad-coverage*** or ***multiple pathway targeted analysis*** was recently introduced as an alternative to untargeted profiling (Table 4.2). Although untargeted analysis has been heavily applied over the past two decades as a discovery approach for metabolite profiling without an *a priori* hypothesis, the targeted LC tandem MS (or MS/MS) analysis in multiple reactions monitoring (MRM) acquisition mode is gaining attention due to its unprecedented scan speed, sensitivity and the fact that the polar cellular metabolome (involved in central carbon metabolism) is well characterized and can be readily targeted. For this reason, the targeted approaches can also save a lot of time usually devoted to fastidious annotation (in untargeted assays) of well-known metabolites – intermediates in energy pathways. Due to its advantageous selectivity and sensitivity for the analysis of polar metabolome, HILIC is widely employed for the development of these multiple pathway targeted methods whose coverage often largely exceeds the metabolome coverage in untargeted profiling, with respect to the size of truly annotated metabolome (Table 4.2). This is a result of superior sensitivity when using HILIC coupled with targeted MRM-based MS detection, compared to HRMS scanning in a full mass range, *m/z* 50–1200. Among different methods applied to the analysis of biofluids (*i.e.* plasma and urine) and cell or tissue extracts, we can appreciate the particularly comprehensive coverage of the polar metabolome (up to 400 metabolites excluding lipids) that can be obtained by combining two analytical runs, one in the positive ESI mode and one in the negative ESI mode (Table 4.2).

Table 4.1 Targeted HILIC ESI–MS/MS methods with a focus on the quantification of intermediates in one specific biochemical pathway or one specific class of metabolites (*this list might not be exhaustive*).

Metabolite class or pathway	Metabolite coverage	Sample type	Column	Analytical conditions(mobile phases, pH, ESI mode)	Reference
NAD+ metabolism	10	Cell and tissue lysates, biofluids	XBridge BEH Amide (150 mm×2.1 mm, 2.5 µm)	20 mM ammonium carbonate in water : ACN (95 : 5), pH = 9.8, and 100% ACN – positive mode	Liu *et al*. Cell Metabolism, 2018[42]
Amino acids and derivatives, acylcarnitines	46	Cell and tissue lysates, biofluids	Acquity BEH Amide (100 mm×2.1 mm, 1.7 µm)	20 mM ammonium formate 0.1% formic acid in water and 0.1% formic acid in ACN – positive mode	Teav *et al*. Analytical Chemistry, 2019[33]
Folate species	7	Cell lysates	SeQuant ZIC-pHILIC (150 mm×2.1 mm, 5 µm)	20 mM ammonium carbonate in water, pH = 9.2, and 100% ACN – positive mode	Schittmayer *et al*. Analytical Chemistry, 2018[43]
Nucleoside triphosphates	8	Cell lysates	SeQuant ZIC-pHILIC (150 mm×4.6 mm, 5 µm)	10 mM ammonium bicarbonate, 0.05% ammonium hydroxide in water, and 100% ACN – positive mode	Matsuda *et al*. Genes and Environment, 2018[44]
Histamine and its main metabolites	6	Biofuild (Urine)	Acquity BEH HILIC (100 mm×2.1 mm, 1.7 µm)	10 mM ammonium formate in water, pH = 3.0, and 100% ACN – positive mode	Nelis *et al*. Talanta, 2020[45]
Nucleotide sugars	12	Plant leaf	SeQuant ZIC-HILIC (150 mm×1 mm, 3.5 µm)	100% water and 10 mM ammonium acetate in water/ACN (1/9), pH = 7.0 – negative mode	Ito *et al*. Analytical Biochemistry, 2014[46]
Methylated amines and amino acids	30	Biofuilds	Acquity BEH Amide (100 mm×2.1 mm, 1.7 µm)	50 mM ammonium formate in water/ACN (1/1), pH = 3.2, and 0.05% formic acid in ACN – positive mode	Roggensack *et al*. Rapid communications in Mass Spectrometry, 2019[47]
Organic acids	8	Biofluids	SeQuant ZIC-HILIC (250 mm×2.1 mm, 5 µm)	100 mM ACN/ammonium acetate (80/20), pH = 4.7 (isocratic conditions) – negative mode	Sriboonvorakul *et al*. Journal of Chromatography B, 2013[48]
Organic acids	7	Biofluids	SeQuant ZIC-HILIC (250 mm×4.6 mm, 5 µm)	10 mM ammonium acetate in water and ACN – negative mode	Huang *et al*. Journal of Chromatography B, 2012[49]

Table 4.2 Broad-coverage targeted HILIC ESI–MS/MS or ESI–HRMS methods (targeting >100 polar metabolites) applied for the analysis of the cellular metabolome implicated in central carbon metabolism.

Sample type	Column	Metabolite coverage	Analytical conditions (mobile phases, pH)	ESI mode	Reference
Plasma	Polymer-based NH2 Asahipak NH2P-40 2E Shodex (250 mm × 2.0 mm, 4 µm)	610 metabolites including 231 lipids[a] (95 internal standards)	20 mM ammonium carbonate in water : ACN (95 : 5), pH = 9.8, and 100% ACN	Single MRM analysis with polarity switching	Li *et al.* Metabolomics, 2017[31]
Serum, urine, amniotic fluid	Acquity BEH Amide (100 mm × 2.1 mm, 3.5 µm)	100 polar metabolites	10 mM ammonium formate in water : ACN (70 : 30) and 10 mM ammonium formate in ACN : water (95 : 5)	Single MRM analysis with polarity switching	Virgiliou *et al.* Methods in Molecular Biology, 2018[32,50]
Cell and tissue lysates, biofluids	Acquity BEH Amide (150 mm × 2.1 mm, 1.7 µm)	130 metabolites	0.1% formic acid and 0.075% ammonium hydroxide in ACN : water (95 : 5), and 0.2% formic acid and 0.1% ammonium hydroxide in water : ACN (98 : 2)	Two separate MRM analyses in the positive and in the negative mode	Gika *et al.* J. Chromtogr. A, 2012[51]
Mammalian cell lysates	XBridge BEH Amide (150 mm × 2.1 mm, 2.5 µm)	339 metabolites (including polar lipids[a])	20 mM ammonium acetate and 15 mM ammonium hydroxide in water : ACN (97 : 3), pH = 9.0, and 100% ACN	Single analysis with polarity switching in HRMS and MS/MS mode	Liu *et al.* Analytical Chemistry 2014[52]
Cell and tissue lysates, biofluids	Acquity BEH Amide (150 mm × 2.1 mm, 1.7 µm)	400 metabolites (including polar lipids[a])	Positive mode: 10 mM ammonium formate and 0.1% formic acid in water (A) and in ACN (B) Negative mode: 10 mM ammonium acetate/0.04% NH_4OH (pH 9.7) in water (A) and in ACN (95%, B)	Two separate MRM analyses in the positive and in the negative mode	Cao *et al.* Analytica Chimica Acta, 2020[53]
Cell and tissue lysates, biofluids	Acquity BEH Amide (100 mm × 2.1 mm, 1.7 µm)	200 metabolites	25 mM ammonium acetate and 25 mM ammonium hydroxide in water (A), pH = 9.6, and 100% ACN (B)	Single MRM analysis with polarity switching	Cai and Zhu, Methods in Molecular Biology, 2019[54]
Serum	SeQuant ZIC-cHILIC (150 mm × 2.1 mm, 3 µm)	595 metabolites (including lipids[a])	5 mM Ammonium acetate and 0.2% acetic acid in water : ACN (90 : 10, A) and in ACN : water (90 : 10, B)	Four separate MRM analyses: three in the positive and one in the negative mode	Gu *et al.* Analytical Chemistry, 2015[55]

Table 4.2 (*Continued*)

Sample type	Column	Metabolite coverage	Analytical conditions (mobile phases, pH)	ESI mode	Reference
Cell and tissue lysates, biofluids	XBridge BEH Amide (100 mm×4.6 mm, 3.5 μm) or Luna NH2 (50 mm×4.6 mm, 5 μm)	258 polar metabolites	20 mM ammonium acetate and 20 mM ammonium hydroxide in water : ACN (95 : 5, A), pH = 9.45, and 100% ACN (B)	Single MRM analysis with polarity switching	Yuan *et al.* Nature Protocols, 2013[56]
Urine	Acquity BEH Amide (50 mm×1 mm, 1.7 μm)	Number of identified metabolites not specified	0.1% formic acid in water (A) and in ACN (B)	Single analysis in the positive ionization mode IMS–HRMS and MS/MS	King *et al.* Metabolomics, 2019[57]
Feces	HILICON AB iHILIC-Fusion SS, (100 mm× 2.1 mm, 1.8 μm)	150 metabolites	5 mM ammonium acetate in ACN (95%), 5–25 mM ammonium acetate in water (95%), pH 4.6, 6.8 and 9.0 tested (adjusted with acetic acid of 25% ammonium hydroxide)	Two HRMS analyses in the positive and in the negative mode	Sillner *et al.* J Chromtogr. B, 2019[58]
Cell and tissue lysates	Luna Aminopropyl NH2 (250 mm× 2.1 mm, 5 μm)	141 polar metabolites	20 mM ammonium acetate and 20 mM ammonium hydroxide in water : ACN (95 : 5, A), pH = 9.45, and 100% ACN (B)	Two MRM analyses in the positive and in the negative mode	Bajad *et al.* J.Chromtogr. A, 2006[25]
Cell and tissue lysates, biofluids	XBridge BEH Amide XP (150 mm× 2.1 mm, 2.5 μm)	600 metabolites including polar lipids[a]	20 mM ammonium acetate and 20 mM ammonium hydroxide in water : ACN (95 : 5, A), pH = 9.45, and 100% ACN (B)	Two HRMS analyses in the positive and in the negative mode	Jang *et al.* Cell Metabolism, 2019[59]
Cell and tissue lysates, biofluids	Acquity BEH Amide (100 mm×2.1 mm, 1.7 μm) and ZIC-pHILIC (100 mm× 2.1 mm, 5 μm)	374 polar metabolites	Positive mode: 20 mM Ammonium formate and 0.1% formic acid/0.1% formic acid in ACN Negative mode: 20 mM ammonium acetate/20 mM NH_4OH (pH 9.7) and ACN	Two MRM analyses in the positive and in the negative mode	Medina *et al.* Metabolites, 2020[26]

[a]Profiled lipids include mainly polar acylcarnitines, fatty acids and some phospholipids (this list might not be exhaustive).

Table 4.3 Human population phenotyping studies that applied HILIC–ESI–HRMS or HILIC ESI–MS/MS methods for untargeted and targeted analysis of polar circulatory metabolites across different biofluids (*this list might not be exhaustive*).

Biofluid	Sample number	Type of study	Column	Analytical conditions (mobile phases, pH)	Reference
Urine	$N = 15$ (total sample number = 51)	Longitudinal	Luna HILIC 200 A (150 mm × 2.0 mm, 3 µm)	50 mM ammonium acetate and 50 mM ammonium acetate : ACN (5 : 95)	Gevi *et al.* Metabolites 2020[68]
Urine	$N = 60$	Cross-sectional	Luna HILIC 200 A (150 mm × 2.0 mm, 3 µm)	50 mM ammonium acetate and 50 mM ammonium acetate : ACN (5 : 95)	Gevi *et al.* Molecular Autism 2016[69]
Plasma	$N = 96$	Cross-sectional	ZIC-HILIC (150 mm × 4.6 mm, 5 µm)	ACN : water both with 0.1% formic acid	Kamleh *et al.* Journal of Proteome Research 2015[70]
Plasma	$N = 139$	Cross-sectional	BEH HILIC (100 mm × 2.1 mm, 1.7 µm)	5 mM ammonium formate : ACN with 0.025% formic acid	Graham *et al.* Plos One 2014[71]
Plasma and CSF	$N = 74$	Cross-sectional	Acquity BEH Amide (100 mm × 2.1 mm, 1.7 µm) and ZIC-pHILIC (100 mm × 2.1 mm, 5 µm)	Positive mode: 20 mM ammonium formate and 0.1% formic acid/0.1% formic acid in ACN Negative mode: 20 mM ammonium acetate/20 mM NH_4OH (pH 9.7) and ACN	Van der Velpen *et al.* Alzheimer's Research & Therapy 2019,[72] Gallart-Ayala *et al.* Analytica Chimica Acta, 2018[7]
Plasma	$N = 118$	Cross-sectional	ZIC-HILIC (150 mm × 4.6 mm, 5 µm)	ACN : water both with 0.1% formic acid	Naz *et al.* Eur Respir J, 2017[73]
Serum	$N = 78$	Cross-sectional	ZIC-HILIC (150 mm × 4.6 mm, 5 µm)	ACN : water both with 0.1% formic acid	Reinke *et al.* Eur Respir J 2017[74]
Serum	$N = 47$	Cross-sectional	Acquity BEH HILIC (100 mm × 2.1 mm, 1.7 µm)	20 mM ammonium formate with 0.1% formic acid and ACN with 0.1% formic acid	Alsaleh *et al.* Journal of clinical and experimental hepatology, 2020[75]

Table 4.3 (*Continued*)

Biofluid	Sample number	Type of study	Column	Analytical conditions (mobile phases, pH)	Reference
Plasma	$N = 22$	Cross-sectional	Kinetex HILIC (100 mm × 2.1 mm, 2.6 μm)	ACN : 5 mM ammonium formate (pH = 3) (50 : 50) and ACN : 5 mM ammonium formate (pH = 5) (90 : 10)	Abela *et al.* J Inherit Metab Dis 2016[76]
Plasma	$N = 12$ (9 points and total sample number = 108)	Longitudinal	Atlantis HILIC (150 mm × 2.1 mm, 3 μm) and Luna NH2 (150 mm × 2.0 mm, 3 μm)	Positive mode: 10 mM ammonium formate and 0.1% formic acid ACN 0.1% formic acid Negative mode: 20 mM ammonium acetate/ammonium hydroxide and 10 mM ammonium hydroxide in ACN : MeOH	Steinhauser *et al.* JCI insight 2018[77]
Plasma	$N = 694$	Cross-sectional and longitudinal	Atlantis HILIC (150 mm × 2.1 mm, 3 μm) and Luna NH2 (150 mm × 2.0 mm, 3 μm)	10 mM ammonium formate, 0.1% formic acid and ACN 20 mM ammonium acetate/ ammonium hydroxide and ACN	Yu *et al.*, Clinical chemistry 2018[78]

Single injection analysis using polarity switching can also be employed with a somewhat lower coverage.[31,32,56] Among different HILIC stationary phases, the polymer-based zwitterionic (ZIC-cHILIC, ZIC-pHILIC), BEH amide and amino columns showed the best performance in achieving the broadest metabolite coverage and the highest analytical reproducibility.

It is important to note that HILIC–MS approaches allow for the profiling of both the polar metabolites and lipids from the same extract, including the complex and highly hydrophobic lipid species (Figure 4.2).[26] An almost infinite lipid diversity, with multitude of variations on the same theme (within each lipid class) makes the coverage of HILIC methods attain more than a thousand species in a single run with polarity switching or combining two independent analyses to allow for the enhanced sensitivity.

Applications of HILIC in the field of metabolomics were launched with untargeted assays and the aim to extend the metabolome coverage to polar and ionic compounds.[60,61] While RPLC was used for profiling of mid-polar and non-polar metabolites, the major challenge in metabolomics represented the separation of highly hydrophilic central carbon metabolites,[62] specifically to understand the metabolic shifts in cellular energy metabolism, under different physiological or disease conditions. To enhance the poor retention of hydrophilic metabolites by RPLC, the alternative strategies using ion pairing agents,[63] multimodal C18 columns (*e.g.* HSS T3, Scherzo SM-C18), and in particular HILIC were developed and became increasingly robust for polar compound retention.[4,7] Among specified strategies, HILIC showed the best selectivity for polar metabolites while being MS-compatible (contrary to ion pairing agents causing important contamination of the LC–MS system). For an untargeted metabolomics experiment, one would ideally maximize the metabolome coverage by sequential analyses using HILIC and RPLC, in both positive and negative ESI modes.[64–66] HILIC analysis in positive ionization mode, in acidic conditions, usually allow for the assessment of amino acid metabolism, beta-oxidation (*i.e.* acylcarnitines) and NAD+ metabolism[33] while the analysis in basic conditions in the negative ion ization mode provide insight into glycolysis, the TCA cycle, OXOPHOS and purine and pyrimidine metabolism.[7]

HILIC–MS assisted untargeted and targeted assays are applied from model systems to human population studies, thus proving their robustness and complementarity to RPLC. HILIC–MS approaches were already applied in several large-scale population studies for metabolic phenotyping of plasma and urine in the context of biomarker discovery but also for better understanding of alterations associated with the ageing process and the onset and progression of specific metabolic diseases (Table 4.3). These studies have unambiguously demonstrated the reproducibility and high-throughput of HILIC analysis (Figure 4.3). One of the largest proof-of-principle studies was reported by Lewis *et al.* in Analytical Chemistry (2016) using two independent epidemiological sample sets, comprising 650 and 1000 participant urine samples.[67]

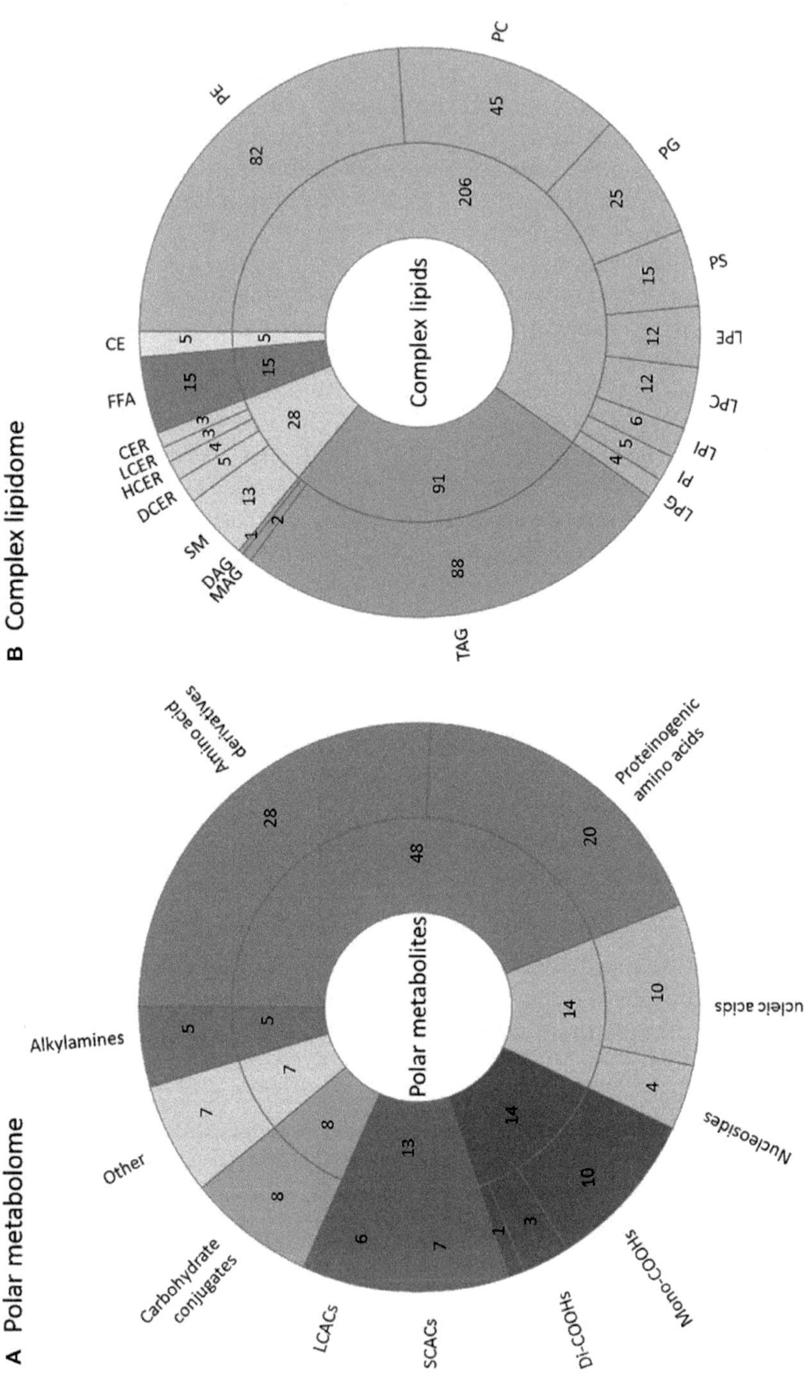
A Polar metabolome
Polar metabolites
Amino acid derivatives
Proteinogenic amino acids
Nucleic acids
Nucleosides
Mono-COOHs
Di-COOHs
SCACs
LCACs
Carbohydrate conjugates
Other
Alkylamines
48
28
20
14
10
4
14
10
3
1
13
7
6
8
8
7
7
5
5
B Complex lipidome
Complex lipids
PE
PC
PG
PS
LPE
LPC
LPI
PI
LPG
TAG
MAG
DAG
SM
DCER
HCER
LCER
CER
FFA
CE
206
82
45
25
15
12
12
6
5
4
91
88
2
1
28
13
5
4
3
3
15
15
5
5

4.5 Isotopic Profiling

Excellent HILIC performance for polar metabolome analysis is also extensively exploited for stable isotope-assisted tracing experiments. These isotopic profiling assays can be performed by coupling HILIC with high or low resolution MS depending on the biological question but also the analysis setup.[78] The aim of these experiments is to track the fate of a labeled nutrient (^{13}C, ^{15}N or deuterium) in different conditions (*e.g.* WT *vs.* mutant, or CTRL *vs.* treated) in the steady-state, following the label incorporation into the metabolites – intermediates in the pathways used for nutrient consumption (Figure 4.4). The consumption is tracked *via* labeled carbon or nitrogen transfer through chemical transformation reactions.[14] HILIC selectivity and retention of polar metabolites make this technique suitable for isotopic tracing experiments to identify the label enriched polar metabolites and therefore conclude about the pathways used for the 'metabolization' of the isotopically labeled substrate.[1,19,42,59,80] One recent remarkable example of isotope tracing analysis with HILIC (by Rabinowitz's team) depicts the metabolite consumption and release across 11 different organs (*e.g.* kidney, liver, spleen, muscle, *etc.*) by measuring arterial-venous differences in metabolite concentrations to provide a quantitative atlas of inter-organ metabolite exchange.[81]

Figure 4.2 Diversity and size of measurable human plasma A) polar metabolome and B) complex lipidome combining two targeted HILIC–ESI–MS/MS analyses in the positive and in the negative ionization mode. The presented coverage includes the measurable (and not only detectable) polar ($n = 109$) and lipid ($n = 345$) metabolites for which the coefficient of variation (CV) was lower than 30% across 20 independently extracted aliquots (using isopropanol) of pooled plasma sample by 4 different operators ($n = 5$ replicates per operator). Lipid classification abbreviations: SM—sphingomyelin, CER—ceramides, DCER—dihydroceramides, LCER—lactosylceramides, HCER—hexosylceramides, MAG—monoacylglycerol, DAG—diacylglycerol, TAG—triacylglycerol, CE—cholesterol esters, FFA—free fatty acids, PI—phosphatidylinositol, PC—phosphatidylcholine, PG—phosphatidylglycerol, PE—phosphatidylethanolamine, PS—phosphatidylserine, LPC—lysophosphatidylcholine, LPG—lysophosphatidylglycerol, LPE—lysophosphatidylethanolamine, LPI—lysophosphatidylinositol, LPS—lysophosphatidylserine. Class abbreviations for polar metabolites: short (SCACs) and long chain (LCACs) acylcarnitines, mono- (Mono-COOHs)-, di- (Di-COOHs)-, and tri- (Tri-COOHs) carboxylic acids and others (comprising glycocholate, hydroquinone, hydroxyphenyllactate, pyridoxine, salsolinol, trigonelline, and tryptamine). Reproduced from ref. 26, https://doi.org/10.3390/metabo10120495, under the terms of the CC BY 4.0 license, https://creativecommons.org/licenses/by/4.0/.

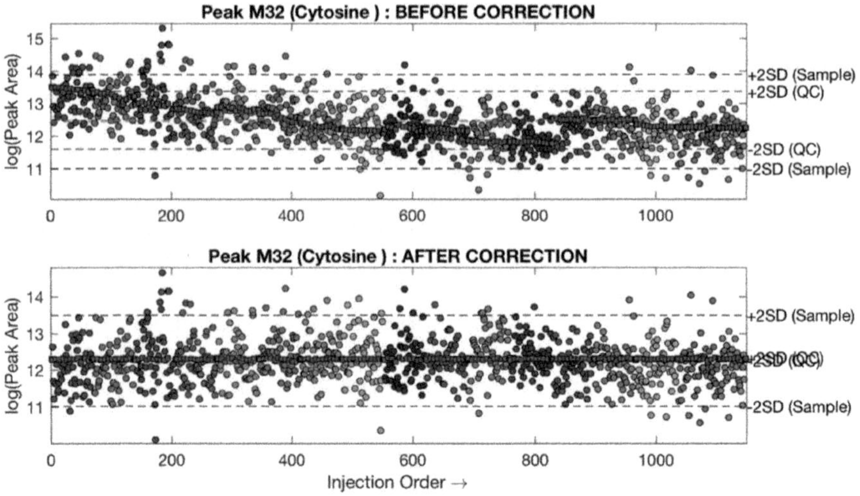

Figure 4.3 Signal intensity drift for cytosine analyzed using HILIC–ESI HRMS across 1200 injections of 900 urine samples from 600 patients (U-BIOPRED study, for 300 patients' urine was sampled at two time points). The given example illustrates the reproducibility of HILIC profiling in a large-scale human population phenotyping study (samples analyzed across 17 batches). The observed systematic bias in peak area over time ("signal drift") is primarily due to the inherent variability of MS detection over time and potential variation in chromatographic retention mechanism.[79,83] Relative standard deviation (RSD) across 234 quality control (QC) pooled samples was 28.6% before, and 1.6% after, signal drift correction using an adaptive cubic smoothing spline algorithm (QC-RSC).[84] A HILIC–HRMS method was published by Naz *et al.* in Analytical Chemistry.[18] Data were kindly provided by Prof. Craig Wheelock and we thank Prof. David Broadhurst and Dr. Stacey Reinke for figure preparation (extracted from a publication in preparation).

4.6 Conclusive Remarks

While the chromatographic resolution and the efficiency and sensitivity of LC–MS analysis are being constantly improved, the generation of the sample extract (*i.e.* cell or tissue lysate) that accurately reflects the true intracellular metabolic content continues to be, although often ignored, a key challenge in the probing of the polar metabolome.[2,82] This challenge is, due to the inherently transient nature of cellular metabolites, impacted by the residual enzymatic activity and spontaneous chemical transformations. Fast and appropriate quenching of metabolism is therefore essential. Finally, the latest emerging challenge of metabolomics applications reposes on data integration and interpretation in a biologically and physiologically relevant context.

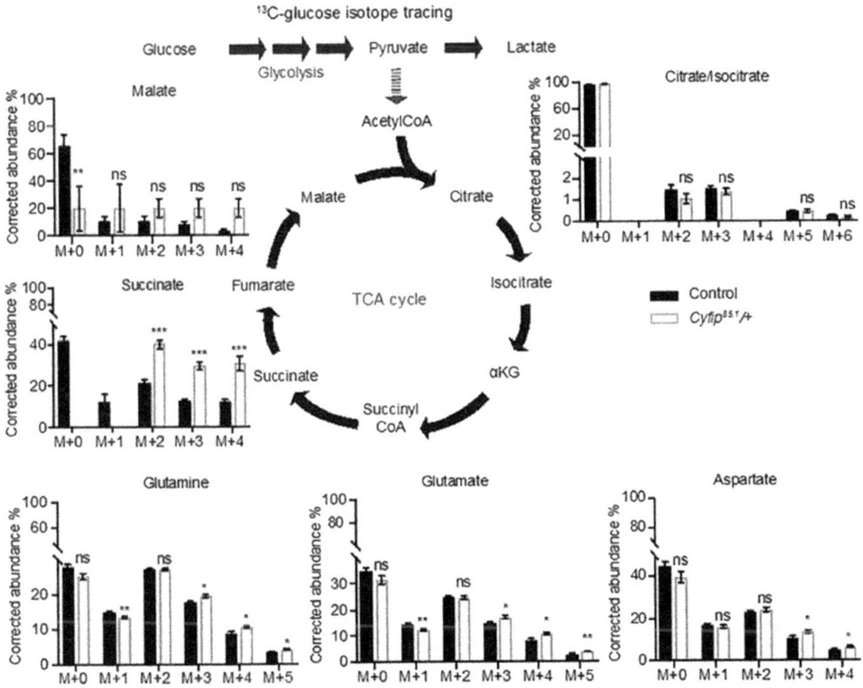

Figure 4.4 Results of stable isotope-assisted HILIC–HRMS profiling of the cellular metabolome. The figure represents the contribution of [U-$^{13}C_6$] glucose to TCA cycle metabolites in *Drosophila* brain tissue after 6 h of feeding with [^{13}C] glucose. The *x* axis denotes the ^{13}C isotopologue profile (*i.e.* the mass isotopologue distribution [MID]) of labeled metabolites after incorporation of the [^{13}C] glucose skeleton. The catabolism of ^{13}C-labeled glucose leads to M + 3 pyruvate. Decarboxylation of pyruvate makes M + 2 acetyl-coenzyme A (CoA) in mitochondria, which is then incorporated into M + 2 citrate. Values shown are mean ± SEM. $n = 5$ (10 brains were pooled per replicate). ***$p < 0.001$, **$p < 0.01$, *$p < 0.05$, Sidak's multiple comparisons test. The HILIC–HRMS method was previously published by Gallart-Ayala *et al.* in Analytica Chimica Acta.[7]

Acknowledgements

Many thanks to all our current and past colleagues of the Metabolomics Platform team at the University of Lausanne for their precious contributions to method developments. This work was supported by funds from the Faculty of Biology and Medicine (FBM) at the University of Lausanne (UNIL) and Swiss National Science Foundation (SNF) grant 316030_183377.

References

1. C. Jang, L. Chen and J. D. Rabinowitz, *Cell*, 2018, **173**, 822–837.
2. J. D. Rabinowitz, *Expert Rev. Proteomics*, 2007, **4**, 187–198.
3. X. Liu, Z. Ser and J. W. Locasale, *Anal. Chem.*, 2014, **86**, 2175–2184.
4. S. Wernisch and S. Pennathur, *Anal. Bioanal. Chem.*, 2016, **408**, 6079–6091.
5. S. U. Bajad, W. Lu, E. H. Kimball, J. Yuan, C. Peterson and J. D. Rabinowitz, *J. Chromatogr. A*, 2006, **1125**, 76–88.
6. D. Q. Tang, L. Zou, X. X. Yin and C. N. Ong, *Mass Spectrom. Rev.*, 2016, **35**, 574–600.
7. H. Gallart-Ayala, I. Konz, F. Mehl, T. Teav, A. Oikonomidi, G. Peyratout, V. van der Velpen, J. Popp and J. Ivanisevic, *Anal. Chim. Acta*, 2018, **1037**, 327–337.
8. J. Ivanisevic and E. J. Want, *Metabolites*, 2019, **9**, 308.
9. G. J. Patti, *J. Sep. Sci.*, 2011, **34**, 3460–3469.
10. D. J. Creek, W. B. Dunn, O. Fiehn, J. L. Griffin, R. D. Hall, Z. Lei, R. Mistrik, S. Neumann, E. L. Schymanski, L. W. Sumner, R. Trengove and J.-L. Wolfender, *Metabolomics*, 2014, **10**, 350–353.
11. H. P. Benton, J. Ivanisevic, N. G. Mahieu, M. E. Kurczy, C. H. Johnson, L. Franco, D. Rinehart, E. Valentine, H. Gowda, B. K. Ubhi, R. Tautenhahn, A. Gieschen, M. W. Fields, G. J. Patti and G. Siuzdak, *Anal. Chem.*, 2015, **87**, 884–891.
12. A. J. Alpert, *J. Chromatogr.*, 1990, **499**, 177–196.
13. P. Hemström and K. Irgum, *J. Sep. Sci.*, 2006, **29**, 1784–1821.
14. H. Gallart-Ayala, T. Teav and J. Ivanisevic, *BioEssays*, 2020, **42**, e2000052.
15. D. V. McCalley, *J. Chromatogr. A*, 2017, **1523**, 49–71.
16. Y. Guo and S. Gaiki, *J. Chromatogr. A*, 2011, **1218**, 5920–5938.
17. L. Qiao, X. Shi and G. Xu, *TrAC, Trends Anal. Chem.*, 2016, **81**, 23–33.
18. S. Naz, H. Gallart-Ayala, S. N. Reinke, C. Mathon, R. Blankley, R. Chaleckis and C. E. Wheelock, *Anal. Chem.*, 2017, **89**, 7933–7942.
19. I. Roci, H. Gallart-Ayala, A. Schmidt, J. Watrous, M. Jain, C. E. Wheelock and R. Nilsson, *Anal. Chem.*, 2016, **88**, 2707–2713.
20. P. Jandera, *Anal. Chim. Acta*, 2011, **692**, 1–25.
21. M. R. Gama, R. G. da Costa Silva, C. H. Collins and C. B. G. Bottoli, *TrAC, Trends Anal. Chem.*, 2012, **37**, 48–60.
22. S. M. Melnikov, A. Höltzel, A. Seidel-Morgenstern and U. Tallarek, *J. Phys. Chem. C*, 2015, **119**, 512–523.
23. L. Nováková, L. Havlíková and H. Vlčková, *TrAC, Trends Anal. Chem.*, 2014, **63**, 55–64.
24. A. Tan and J. C. Fanaras, *Biomed. Chromatogr.*, 2019, **33**, e4409.
25. S. U. Bajad, W. Lu, E. H. Kimball, J. Yuan, C. Peterson and J. D. Rabinowitz, *J. Chromatogr. A*, 2006, **1125**, 76–88.
26. J. Medina, V. van der Velpen, T. Teav, Y. Guitton, H. Gallart-Ayala and J. Ivanisevic, *Metabolites*, 2020, **10**, 495.

27. A. Periat, B. Debrus, S. Rudaz and D. Guillarme, *J. Chromatogr. A*, 2013, **1282**, 72–83.
28. Y. Guo and S. Gaiki, *J. Chromatogr. A*, 2011, **1218**, 5920–5938.
29. I. Sampsonidis, M. Witting, W. Koch, C. Virgiliou, H. G. Gika, P. Schmitt-Kopplin and G. A. Theodoridis, *J. Chromatogr. A*, 2015, **1406**, 145–155.
30. A. Periat, J. Boccard, J.-L. Veuthey, S. Rudaz and D. Guillarme, *J. Chromatogr. A*, 2013, **1312**, 49–57.
31. K. Li, J. C. Naviaux, A. T. Bright, L. Wang and R. K. Naviaux, *Metabolomics*, 2017, **13**, 122.
32. C. Virgiliou, H. G. Gika and G. A. Theodoridis, *Methods Mol. Biol.*, 2018, **1738**, 65–81.
33. T. Teav, H. Gallart-Ayala, V. van der Velpen, F. Mehl, H. Henry and J. Ivanisevic, *Anal. Chem.*, 2019, **91**, 11757–11769.
34. U. L. Kathriarachchi, S. S. Vidhate, N. Al-Tannak, A. H. Thomson, M. J. J. da Silva, Neto and D. G. Watson, *J. Chromatogr. B*, 2018, **1089**, 78–83.
35. N. Antonopoulos, G. Machairas, G. Migias, A. Vonaparti, V. Brakoulia, C. Pistos, D. Gennimata and I. Panderi, *Molecules*, 2018, **23**, 1548.
36. H.-A. K. Lakso, P. Appelblad and J. R. Schneede, *Clin. Chem.*, 2008, **54**, 2028–2035.
37. J. Martens-Lobenhoffer and S. M. Bode-Böger, *Clin. Chem.*, 2006, **52**, 488–493.
38. W. H. de Jong, K. S. Graham, J. C. van der Molen, T. P. Links, M. R. Morris, H. A. Ross, E. G. de Vries and I. P. Kema, *Clin. Chem.*, 2007, **53**, 1684–1693.
39. K. Spagou, H. Tsoukali, N. Raikos, H. Gika, I. D. Wilson and G. Theodoridis, *J. Sep. Sci.*, 2010, **33**, 716–727.
40. A. Li, K. M. Hines and L. Xu, *Methods Mol. Biol.*, 2020, **2084**, 119–132.
41. G. Zauner, A. M. Deelder and M. Wuhrer, *Electrophoresis*, 2011, **32**, 3456–3466.
42. L. Liu, X. Su, W. J. Quinn, 3rd, S. Hui, K. Krukenberg, D. W. Frederick, P. Redpath, L. Zhan, K. Chellappa, E. White, M. Migaud, T. J. Mitchison, J. A. Baur and J. D. Rabinowitz, *Cell Metab.*, 2018, **27**, 1067–1080, e1065.
43. M. Schittmayer, R. Birner-Gruenberger and N. Zamboni, *Anal. Chem.*, 2018, **90**, 7349–7356.
44. S. Matsuda and T. Kasahara, *Genes Environ.*, 2018, **40**, 13.
45. M. Nelis, L. Decraecker, G. Boeckxstaens, P. Augustijns and D. Cabooter, *Talanta*, 2020, **220**, 121328.
46. J. Ito, T. Herter, E. E. Baidoo, J. Lao, M. E. Vega-Sánchez, A. M. Smith-Moritz, P. D. Adams, J. D. Keasling, B. Usadel, C. J. Petzold and J. L. Heazlewood, *Anal. Biochem.*, 2014, **448**, 14–22.
47. T. Roggensack, B. Merz, N. Dick, A. Bub and R. Krüger, *Rapid Commun. Mass Spectrom.*, 2020, **34**, e8646.

48. N. Sriboonvorakul, N. Leepipatpiboon, A. M. Dondorp, T. Pouplin, N. J. White, J. Tarning and N. Lindegardh, *J. Chromatogr. B: Anal. Technol. Biomed. Life Sci.*, 2013, **941**, 116–122.
49. Y. Huang, Y. Tian, Z. Zhang and C. Peng, *J. Chromatogr. B: Anal. Technol. Biomed. Life Sci.*, 2012, **905**, 37–42.
50. C. Virgiliou, I. Sampsonidis, H. G. Gika, N. Raikos and G. A. Theodoridis, *Electrophoresis*, 2015, **36**, 2215–2225.
51. H. G. Gika, G. A. Theodoridis, U. Vrhovsek and F. Mattivi, *J. Chromatogr. A*, 2012, **1259**, 121–127.
52. X. Liu, Z. Ser and J. W. Locasale, *Anal. Chem.*, 2014, **86**, 2175–2184.
53. G. Cao, Z. Song, Y. Hong, Z. Yang, Y. Song, Z. Chen, Z. Chen and Z. Cai, *Anal. Chim. Acta*, 2020, **1125**, 144–151.
54. Y. Cai and Z. J. Zhu, *Methods Mol. Biol.*, 2019, **1859**, 263–274.
55. H. Gu, P. Zhang, J. Zhu and D. Raftery, *Anal. Chem.*, 2015, **87**, 12355–12362.
56. M. Yuan, S. B. Breitkopf, X. Yang and J. M. Asara, *Nat. Protoc.*, 2012, **7**, 872–881.
57. A. M. King, L. G. Mullin, I. D. Wilson, M. Coen, P. D. Rainville, R. S. Plumb, L. A. Gethings, G. Maker and R. Trengove, *Metabolomics*, 2019, **15**, 17.
58. N. Sillner, A. Walker, E. M. Harrieder, P. Schmitt-Kopplin and M. Witting, *J. Chromatogr. B: Anal. Technol. Biomed. Life Sci.*, 2019, **1109**, 142–148.
59. C. Jang, S. Hui, X. Zeng, A. J. Cowan, L. Wang, L. Chen, R. J. Morscher, J. Reyes, C. Frezza, H. Y. Hwang, A. Imai, Y. Saito, K. Okamoto, C. Vaspoli, L. Kasprenski, G. A. Zsido, 2nd, J. H. Gorman, 3rd, R. C. Gorman and J. D. Rabinowitz, *Cell Metab.*, 2019, **30**, 594–606, e593.
60. A. Kamleh, M. P. Barrett, D. Wildridge, R. J. S. Burchmore, R. A. Scheltema and D. G. Watson, *Rapid Commun. Mass Spectrom.*, 2008, **22**, 1912–1918.
61. D. G. Watson, *Parasitology*, 2010, **137**, 1409–1423.
62. O. Yanes, R. Tautenhahn, G. J. Patti and G. Siuzdak, *Anal. Chem.*, 2011, **83**, 2152–2161.
63. W. Lu, M. F. Clasquin, E. Melamud, D. Amador-Noguez, A. A. Caudy and J. D. Rabinowitz, *Anal. Chem.*, 2010, **82**, 3212–3221.
64. F. Fei, D. M. E. Bowdish and B. E. McCarry, *Anal. Bioanal. Chem.*, 2014, **406**, 3723–3733.
65. J. Ivanisevic, Z.-J. Zhu, L. Plate, R. Tautenhahn, S. Chen, P. J. O'Brien, C. H. Johnson, M. A. Marletta, G. J. Patti and G. Siuzdak, *Anal. Chem.*, 2013, **85**, 6876–6884.
66. J. Haggarty, M. Oppermann, M. J. Dalby, R. J. Burchmore, K. Cook, S. Weidt and K. E. Burgess, *Metabolomics*, 2015, **11**, 1465–1470.
67. M. R. Lewis, J. T. Pearce, K. Spagou, M. Green, A. C. Dona, A. H. Yuen, M. David, D. J. Berry, K. Chappell, V. Horneffer-van der Sluis, R. Shaw, S.

Lovestone, P. Elliott, J. Shockcor, J. C. Lindon, O. Cloarec, Z. Takats, E. Holmes and J. K. Nicholson, *Anal. Chem.*, 2016, 88, 9004–9013.
68. F. Gevi, A. Meloni, R. Mereu, V. Lelli, A. Chiodo, A. Ragusa and A. M. Timperio, *Metabolites*, 2020, **10**, 290.
69. F. Gevi, L. Zolla, S. Gabriele and A. M. Persico, *Mol. Autism*, 2016, **7**, 47.
70. M. A. Kamleh, S. G. Snowden, D. Grapov, G. J. Blackburn, D. G. Watson, N. Xu, M. Ståhle and C. E. Wheelock, *J. Proteome Res.*, 2015, **14**, 557–566.
71. S. F. Graham, O. P. Chevallier, C. T. Elliott, C. Hölscher, J. Johnston, B. McGuinness, P. G. Kehoe, A. P. Passmore and B. D. Green, *PLoS One*, 2015, **10**, e0119452.
72. V. van der Velpen, T. Teav, H. Gallart-Ayala, F. Mehl, I. Konz, C. Clark, A. Oikonomidi, G. Peyratout, H. Henry, M. Delorenzi, J. Ivanisevic and J. Popp, *Alzheimer's Res. Ther.*, 2019, **11**, 93.
73. S. Naz, J. Kolmert, M. Yang, S. N. Reinke, M. A. Kamleh, S. Snowden, T. Heyder, B. Levänen, D. J. Erle, C. M. Sköld, M. Å. Wheelock and C. E. Wheelock, *Eur. Respir. J.*, 2017, **49**, 1602322.
74. S. N. Reinke, H. Gallart-Ayala, C. Gómez, A. Checa, A. Fauland, S. Naz, M. A. Kamleh, R. Djukanović, T. S. Hinks and C. E. Wheelock, *Eur. Respir. J.*, 2017, **49**, 1601740.
75. M. Alsaleh, Z. Leftley, T. A. Barbera, L. K. Koomson, A. Zabron, M. M. E. Crossey, H. L. Reeves, M. Cramp, S. Ryder, S. Greer, M. Prince, P. Sithithaworn, M. Shariff, N. Khuntikeo, W. Loilome, P. Yongvanit, Y. L. Shen, I. J. Cox, R. Williams, C. A. Wadsworth, E. Holmes, K. Nash and S. D. Taylor-Robinson, *J. Clin. Exp. Hepatol.*, 2020, **10**, 17–29.
76. M. M. Mielke, *Psychiatr. Times*, 2018, **35**, 14–17.
77. M. L. Steinhauser, B. A. Olenchock, J. O'Keefe, M. Lun, K. A. Pierce, H. Lee, L. Pantano, A. Klibanski, G. I. Shulman, C. B. Clish and P. K. Fazeli, *JCI Insight*, 2018, **3**.
78. E. Yu, C. Papandreou, M. Ruiz-Canela, M. Guasch-Ferre, C. B. Clish, C. Dennis, L. Liang, D. Corella, M. Fitó, C. Razquin, J. Lapetra, R. Estruch, E. Ros, M. Cofán, F. Arós, E. Toledo, L. Serra-Majem, J. V. Sorlí, F. B. Hu, M. A. Martinez-Gonzalez and J. Salas-Salvado, *Clin. Chem.*, 2018, **64**, 1211–1220.
79. W. B. Dunn, D. Broadhurst, P. Begley, E. Zelena, S. Francis-McIntyre, N. Anderson, M. Brown, J. D. Knowles, A. Halsall, J. N. Haselden, A. W. Nicholls, I. D. Wilson, D. B. Kell and R. Goodacre, *Nat. Protoc.*, 2011, **6**, 1060–1083.
80. A. K. Kanellopoulos, V. Mariano, M. Spinazzi, Y. J. Woo, C. McLean, U. Pech, K. W. Li, J. D. Armstrong, A. Giangrande, P. Callaerts, A. B. Smit, B. S. Abrahams, A. Fiala, T. Achsel and C. Bagni, *Cell*, 2020, **180**, 1178–1197, e1120.
81. C. Jang, S. Hui, X. Zeng, A. J. Cowan, L. Wang, L. Chen, R. J. Morscher, J. Reyes, C. Frezza, H. Y. Hwang, A. Imai, Y. Saito, K. Okamoto, C. Vaspoli, L. Kasprenski, G. A. Zsido, J. H. Gorman, R. C. Gorman and J. D. Rabinowitz, *Cell Metab.*, 2019, **30**, 594–606, e593.

82. W. Lu, X. Su, M. S. Klein, I. A. Lewis, O. Fiehn and J. D. Rabinowitz, *Annu. Rev. Biochem.*, 2017, **86**, 277–304.
83. D. Broadhurst, R. Goodacre and S. N. Reinke, *et al.*, Guidelines and considerations for the use of system suitability and quality control samples in mass spectrometry assays applied in untargeted clinical metabolomic studies, *Metabolomics*, 2018, **14**, 72.
84. J. A. Kirwan, *et al.*, Characterising and correcting batch variation in an automated direct infusion mass spectrometry (DIMS) metabolomics workflow, *Anal. Bioanal. Chem.*, 2013, **405**(15), 5147–5157.

CHAPTER 5

Advances in Solid Phase Microextraction (SPME) for Metabolomics

E. BOYACI,[a,†] E. GIONFRIDDO,[b,†] A. ROSZKOWSKA[c,†] AND B. BOJKO*[d]

[a] Department of Chemistry, Middle East Technical University, Ankara, Turkey; [b] Department of Chemistry and Biochemistry, The University of Toledo, Toledo, Ohio, USA; [c] Department of Pharmaceutical Chemistry, Medical University of Gdansk, Department of Pharmaceutical Chemistry, Medical University of Gdańsk, Gdańsk, Poland; [d] Department of Pharmacodynamics and Molecular Pharmacology, Faculty of Pharmacy, Collegium Medicum in Bydgoszcz, Nicolaus Copernicus University in Toruń, Bydgoszcz, Poland
*Email: bbojko@cm.umk.pl

5.1 Introduction to Solid Phase Microextraction

First introduced in the late 1980s by Pawliszyn and co-workers, solid phase microextraction (SPME) is a non-exhaustive sample preparation approach that is capable of integrating several steps of the analytical workflow such as sampling, extraction, and preconcentration into one solvent-free step. The extraction mechanism of SPME is based on the partitioning of analytes between the sample matrix and extractive phase.[1] The name "microextraction" originates from the

†These authors contributed equally to this work.

New Developments in Mass Spectrometry No. 10
Advanced Mass Spectrometry-based Analytical Separation Techniques for Probing the Polar Metabolome
Edited by Rawi Ramautar

Published by the Royal Society of Chemistry, www.rsc.org

very small amount of the extracting phase. Generally speaking, only a diminutive portion of the analyte is extracted from the matrix *via* SPME, rendering SPME a non-exhaustive technique. In addition to its non-exhaustive and integrative properties, SPME can be implemented using several different geometries with the same fundamental principles, the most common of them being sorbent coated fiber, thin-film, mesh, blade, and in-tube SPME.[1,2] Initially, SPME was widespread as a method for analysis of volatile and semi-volatile compounds in the headspace mode, since this mode circumvented the need for direct contact of the sorbent immobilized on the solid support with the sample, thus solving the lack of sorbent compatibility with complex matrices characteristic of early SPME sorbents. In early headspace mode applications, analytes were sorbed into extracting phases then thermally desorbed from the fiber and analyzed with GC coupled to various detectors.[1] However, the introduction of new coating chemistries and manufacturing processes have since enabled direct immersion SPME in complex matrices, facilitating its use for a much wider variety of applications. Moreover, the compatibility of new sorbents with organic solvents used in liquid chromatography has further broadened the range of analysis *via* SPME to non-volatile compounds, as analytes extracted from the sample can be desorbed to an organic–water mixture and subsequently injected into an LC system. LC-compatible SPME coatings generally consist of sorbents used in solid phase extraction (SPE) or liquid chromatography itself immobilized by biocompatible binder onto the support.[2]

The majority of SPME applications to date have focused on targeted analysis, particularly of drugs and environmental or food contaminants. In recent years, however, SPME has garnered increased attention as an alternative or complementary method for determination of selected endogenous metabolites and untargeted screening. Such a rise in its popularity in this and other areas can be said to be primarily due to the unique and complementary features of the technique: simple and time-saving sampling and sample preparation, minimum or no invasive sampling, suitability for *in vivo* sampling, which enables the capture of unstable and short-lived species, temporally and spatially resolved sampling, as well as limited matrix effects. In this regard, several review articles have summarized developments to date in SPME-based metabolomics.[3–7] The main goal in metabolomics is to extract the highest possible number of compounds present in a system at a given time, thus allowing for a realistic chemical ‘snapshot’ of said system; to achieve this feat, different extractive and analytical methods, from the most common ones such as protein precipitation to multi-step solvent extraction and solid phase extraction, need to be considered. The more comprehensive a protocol is, the wider the range of metabolites of different properties that can be extracted; since there is no single method currently capable of covering the wide spectrum of analytes present in living systems, a combination of various complementary strategies provides the best chance of attaining a comprehensive metabolome profile.

Since extraction *via* SPME affects only the free concentration of analytes in a system, phenomena such as coating saturation and analyte replacement can occur in headspace (HS) analysis, where enrichment into the headspace of hydrophobic compounds occurs because of high Henry Constants, and no binding molecules are present, and competition for active sites on the coating is driven primarily by the affinity of the analyte for the sorbent. In the direct immersion (DI) mode, the established multi-phase equilibrium results in balanced coverage of analytes, especially when extraction is performed from complex matrices containing binding molecules. However, devices used for direct extraction from a complex matrix need to be compatible with the given matrix, which can be achieved by using an appropriate polymer to immobilize sorbent particles on the solid support. The porosity of the polymer should enable penetration of small molecules into the sorbent and restrict access of macromolecules and other interfering matrix components. If necessary, an additional layer of the polymer can be applied on the coating surface of the phase to provide extra protection and increase the smoothness of the coating, which also aids in decreasing physical attachment of sample components. Such a solution has been proven to work well in complex matrices such as food and whole blood.[1–6] More detailed comparisons of HS and DI modes are discussed in Section 5.2. In aqueous environments such as biological samples, a significant portion of hydrophobic molecules remains bound to matrix components. Therefore, the concentration of the free fraction of such analytes available for SPME is much lower as compared to that of hydrophilic analytes, which remain mainly unbound. The other factor that determines the amount of analyte extracted by a SPME device is the affinity between a given analyte and the selected sorbent. In general, SPME coatings have higher affinity toward hydrophobic compounds and lower affinity toward polar ones; as such, the aforementioned concentration of the free fraction of analytes can be compensated by SPME coating–analyte affinity, thus providing well-balanced overall coverage. The relationship between all factors determining the extraction process in SPME is represented in the fundamental equation below (eqn (5.1)):

$$n_{\mathrm{e}}^{\mathrm{eq}} = \frac{K_{\mathrm{es}} V_{\mathrm{e}} V_{\mathrm{s}}}{K_{\mathrm{es}} V_{\mathrm{e}} + V_{\mathrm{s}}} C_{s}^{0} \tag{5.1}$$

where K_{es} $(=C_{\mathrm{e}}^{\mathrm{eq}}/C_{\mathrm{s}}^{\mathrm{eq}})$ is the partition coefficient, or distribution coefficient, that describes the partition of an analyte between the SPME extraction phase and the sample. K_{es} is defined as the ratio of the equilibrium concentration of analyte in the extractant ($C_{\mathrm{e}}^{\mathrm{eq}}$) to that in the sample matrix ($C_{\mathrm{s}}^{\mathrm{eq}}$), V_{e} is the volume of the extraction phase, and V_{s} is the sample volume.[1,2]

The small, often negligible amount of analyte extracted *via* SPME carries certain advantages such as elimination or minimalization of matrix effects, balanced coverage, and no or minimum depletion, which consequently affords minimum disturbance to the biological equilibrium in the studied system. While the small recoveries afforded by SPME might be considered problematic when it comes to sensitivity, it should be emphasized that sensitivity is not only related to the amount of analyte extracted, but also to the enrichment factor and extraction efficiency (recovery) of a given application. In SPME, the extraction phase volume is very small, resulting in a high concentration of the analyte in the sorbent reached with enrichment corresponding to appropriate distribution constant. Subsequent desorption of the analytes directly to the instrument *e.g.* GC or with the use of a small volume of desorption solvent prior to instrumental analysis enables high enrichment factors and thus good method sensitivity. This advantage of SPME runs counterpoint to exhaustive methods, where recovery is high but the enrichment factor and concentration of the analytes in the final extract are low.[2]

As mentioned before, when optimizing an SPME protocol for analysis of non-volatile species, existing SPE protocols can provide useful information, such as on testing of the same extractive phases and desorption solvents; however, the fundamental differences between the two approaches need to be understood. SPE is an exhaustive method, where extraction efficiency is largely dependent on the capacity of the sorbent and the affinity between the analyte and the extraction phase, among other factors. Owing to its selective features, SPE is a favorable method to improve cleanup of extracts prior to instrumental analyses, and in metabolomics, it is mostly used for targeted investigations rather than a global approach. SPE selectivity is related not only to the chemistry of the sorbent, but also the phenomenon of breakthrough, where a portion of analytes is not retained on the cartridge and is lost during the sample load/extraction process. This might consequently lead to false biological interpretation of the results. Comparative studies of SPE and SPME in its thin film format, both utilizing the same chemistry of extraction phase, were conducted on eighteen compounds with log *P* values ranging between 0.14 and 4.98. The data confirmed that for SPE, breakthrough volume is a limiting factor in multi-residue analysis, while the balanced coverage aspect of SPME permits simultaneous analysis of compounds with a wide range of properties. Details of the analysis with theoretical considerations can be found elsewhere.[8] For both SPE and SPME, selectivity related to coating chemistry can be addressed by using a few cartridges/fibers with complementary sorbents, mixing particles of different properties when preparing the extraction device (*e.g.* mixed-mode fibers consisting of an alkyl group such as C8 or C18 and a strong cation exchanger such as benzenesulfonic acid), or by using a multi-layer coating, where each layer contains different particles (*e.g.* DVB/Car/PDMS). On the other hand, the flexibility in coating selection afforded by such techniques seems to be particularly convenient in tuning the method when targeted analysis is the goal. Examples of selective extraction of polar compounds include the analysis of peptides performed with restricted access material (RAM) coating,[9] proton-donor chemically bonded silica for MALDI detection,[10] C18/PAN thin-film modified with lithium perchlorate,[11] adenosine nucleotide with aptamer coating,[12] and sarcosine with molecularly imprinted polymers.[13]

Standard sample preparation approaches often offer automation and high throughput workflows. This is also the case for SPME for analysis of collected samples, living cells, or bacteria culture in vials or multi-well plates.[1,3–6,14–16] In studies of collected samples, standard quenching of metabolism is required to protect metabolites from degradation. When *in vivo* SPME sampling is performed, the quenching step is combined with the extraction step. The restricted access of macromolecules to the sorbent and selected penetration of small molecules to the coating result in a protection of the latter from enzymatic degradation. In other words, *in vivo* extraction helps to effectively stop interactions between enzymes and their substrates, while storage of biological samples, even under ultralow temperatures, does not inhibit enzymatic activity completely.[17] The first work demonstrating SPME's ability to capture unstable metabolome pertained to the analysis of blood and included a comparison of *in vivo* SPME with common approaches, including two protein precipitation (PP) protocols, ultrafiltration (UF), and *ex vivo* SPME.[18] PP enables determinations of total concentrations of metabolites, while UF targets free molecules. A dedicated calibration approach of SPME can offer both total and free concentrations.[19] Although this work presenting a comparison of coverage of features detected by SPME, PP, and UF is somewhat restricted since only particular methods were used, the results clearly indicated that PP provided the highest number of features and broadest coverage over a range of polarities, while UF and SPME yielded a similar number of features. However, UF showed losses of less hydrophilic species (retention time > 10 min), whereas SPME permitted more balanced extraction. Subsequent studies showed similar findings in plants[20,21] and animal tissues[22,23] with respect to the capture of unstable compounds. All *in vivo* studies revealed instability of metabolites across the polarity range, from very polar compounds such as β-NAD or adenosine[18] to lipid species such as lysophosphatidylcholines[24] or oxylipins.[25] Extraction of other labile compounds, *i.e.* dissolved organic sulfur species, was also reported during biogeochemical profiling of deep-sea hydrothermal vents.[26] Moreover, *in vivo* SPME is able to provide fully quantitative results, even in cases where the sample volume is not known/cannot be estimated, as long as the sample volume is much larger than the product of the fiber volume and the distribution coefficient (eqn (5.1)). The quality of data attainable *via* SPME has been shown to be comparable with standard methods.[18] Currently, only a limited number of methods are applicable for *in vivo* analysis. The most commonly used *in vivo* method, or 'gold standard', is microdialysis (MD), particularly for measurements of brain neurotransmitters and other polar metabolites. The MD method is based on the perfusion of a solution resembling extracellular fluid through a semi-permeable tubing consisting of a few hundreds of micrometers in diameter. The tip or part of the tubing is equipped with a semi-permeable membrane that is brought into contact with the sampling site. The analytes then proceed to distribute themselves between the perfusion solution and the sample matrix through the semi-permeable membrane. One of the main limitations of MD is that the perfusion solution and the dialysate have a high saline content, thus presenting a high ion concentration that may eventually result in analyte signal suppression during subsequent mass spectrometric analyses. To address the sensitivity issue resulting mainly from the presence of ions coeluting with analytes of interest, MD is frequently combined with another additional sample preparation step(s). For instance,

Kim *et al.* proposed on-line SPE combined with LC–MS/MS for further preconcentration and sensitive detection of monoamine neurotransmitters following MD sampling.[27] The need for pump(s) and tubings to remain connected to the subject during sampling is another disadvantage of the method that limits its ease of use. The analyte recoveries achievable *via* MD are strongly dependent on the perfusate flow rate, semi-permeable membrane composition, analytes, and sampling temperature. Sufficient sensitivity requires slow flow rates; therefore, the amount of dialysate collected is often limited by the flow rate. MD membranes are often made of cellulose, polyacrylonitrile, and polycarbonate-ether, and both the choice of membrane material and its porosity have an effect on analyte recoveries. While membrane cut-off values theoretically suggest that only macromolecules are restricted to permeate, usually under experimental conditions the membrane is selective mostly toward very small molecules. The main reason behind this phenomenon is the hindered diffusivity of relatively large molecules in the membrane; therefore, under commonly selected conditions, MD is often limited to the extraction of small molecules. A more detailed comparison of MD and SPME is provided in the later section describing extractive phases (Section 5.5), but general findings confirm that SPME provides broader analyte coverage, including a wider range of less polar and non-polar compounds, but is less effective in extraction of very polar and very small molecules.[28,29]

One of the challenges in SPME *in vivo* studies concerns the size of the sampling device. In this regard, miniaturization of the SPME probe has favored the technique for *in vivo* applications. Recently, this advantage was used in a few applications: single cell targeted analysis,[30] single cells untargeted analysis (caviar),[31] bioaccumulation kinetics and distribution of targeted compounds in *Daphnia magna*,[32] and temporally and spatially resolved measurement of neurotransmitters in monkey brains.[33,34] However, the smaller volume of the sorbent equates to lower method sensitivity. While this can be partially compensated by extended sampling times, in many cases, particularly for animal and human studies, serious restrictions are imposed on sampling times due to the nature of such studies. In the aforementioned monkey brain study, sampling was performed for 20 min in three locations simultaneously and repeated twice at specific time intervals; SPME probes were inserted *via* microdrivers through the cannulas;[33,34] human brain sampling performed prior to tumor biopsy in awake patients permitted a 4 min extraction with 3 mm coatings, enabling untargeted metabolomic and lipidomic profiling of gray and white matter.[35] For this particular case, the extraction device was modified to perform simultaneous sampling of both matters, and the extraction efficiency was increased by doubling the number of fibers in each sampled location, followed by desorption of duplicate fibers in one vial. The described application shows that SPME is an extremely flexible tool, given that its size, shape, and coating chemistry, as well as its protocol can be tailored for a particular application. The simplicity of the devices and easiness of their use open the door to applications of *in vivo* SPME in locations where other *in vivo* techniques such as MD cannot be applied, *e.g.* underwater biosystems (sponges, corals, fish, *etc.*), or where access is restricted and the sampling device has to be operated by specially designed vehicles *e.g.* remotely operated underwater vehicles (ROV)[26] or drones.[36]

This flexibility can also be exploited for applications limited by very small samples. One such option for fast analysis is spot analysis. Similar to the well-known Dried Blood Spot (DBS) method,[37] the "Extracted Blood Spot" (EBS) method was proposed for the first time for analysis of drugs from a small blood volume deposited on a coated mesh and quantitively measured by LC–MS/MS and DART–MS/MS.[38] Later, small sample volume analysis by spot sampling or direct immersion was tested on different SPME devices and different matrices, demonstrating that low limits of detection and quantitation, wide dynamic range, as well as very good precision and accuracy can be achieved with proper experimental design and method optimization.[30,31,39] To take advantage of the enrichment of compounds on the coating and to avoid diminishment of concentrations of analytes extracted from small sample volumes, a minimum desorption solvent volume should be used or direct desorption in the analytical instrument should be exploited. To address this, many strategies of direct coupling of SPME with analytical instruments, mainly mass spectrometers, have been developed.[40] Some of these strategies are focused on interfaces that can facilitate previously developed SPME devices *e.g.* nano-ESI,[32,39,41] open port probe (OPP),[42] or microfluidic open interface (MOI),[43] whereas others modify existing devices to propose new ways of introducing extracted compounds to the instrument *e.g.* coated blade spray (CBS),[44] SPME–DART,[38,45] or SPME–DESI.[46,47] While approaches to date have mainly focused on targeted analysis, recently, lipidomics analyses were conducted *in vivo* in zebrafish, *in situ* in *Daphnia magna*, and at the microscale level by sampling single eukaryotic cells.[48] The authors stated that lipid profiles are similar to those obtained with shotgun MS lipidomics, while additionally providing some advantages such as simpler sample pre-treatment, less sample consumption, and shorter analytical times. These promising results indicate the potential of SPME toward global analysis, but an in-depth evaluation of all factors that influence multi-compound analyses such as matrix effect, time of signal stability, and duration, to name just a few, remains to be performed. A new avenue in spatially resolved analysis is also foreseen in direct coupling of SPME with MS, but *via* a desorption electrospray ionization (DESI) interface. DESI is a well-known imaging strategy applicable to surface analysis.[49] Recently, coupling of SPME extraction with DESI–MS desorption and detection enabled fast probing of three-dimensional objects without the need of slicing.[47] The experiment was based on targeted analysis, but DESI and SPME can both be used for untargeted studies; therefore, future applications can be extended to global screening.

5.2 Application of SPME in Foodomics

Metabolomics of food samples has been widely studied using various extraction approaches, including SPME. The majority of such studies, including both targeted and untargeted analysis of food samples by SPME, have been performed using HS-SPME and GC due to the abundance of volatile compounds often characteristic of these matrices.[2,4,50,51] HS-SPME has the advantage of protecting the fiber coating from direct exposure to complex food matrices that should not be further processed to guarantee the integrity of their chemical composition. Risticevic *et al.* performed a comprehensive study to assess the suitability of HS-SPME, hyphenated to both GC–ToF/MS and GC×GC–ToF/MS, to explore the volatile fraction of the metabolome of ice wines and apples.[52,53] The methods demonstrated the capability of SPME to capture a broad range of volatiles emitted from the analyzed samples, and the use of comprehensive gas-chromatography enhanced the chromatographic space available for their separation. Although HS-SPME can often be recognized as a convenient option for food metabolomics *via* SPME, users must be aware of some shortcomings of this technique and properly tune their *modus operandi* to guarantee sampling procedures capable of providing an accurate chemical snapshot of the samples investigated.[2] One of the main effects that can affect the results obtained by HS-SPME for targeted and non-targeted metabolomic applications is the occurrence of competitive adsorption phenomena on the extraction phase, especially when extraction takes place *via* adsorption.[54,55] These effects are mainly related to mass transfer of metabolites into the SPME extraction phase: in the case of HS extraction, the mass transfer rate is proportional to the metabolites Henry's law constants; this implies that most volatile analytes will easily enrich the headspace and be extracted by the SPME device. However, these metabolites, which often present at high concentration and have a high affinity for the extraction phase, can saturate the active adsorption sites and start displacing metabolites with lower affinity for the extraction phase, usually the more polar ones. When this happens, the partitioning of the analytes onto the SPME device does not reflect the chemical distribution in the sample. Saturation and displacement phenomena are often noted due to an altered trend of adsorption isotherms and narrower linear dynamic range for the analytes displaced.[54,55] Usually, disruption of the food tissue by homogenization can promote the release of metabolites that can potentially saturate the SPME extraction phase.[55] To avoid the occurrence of saturation and displacement phenomena, careful optimization of extraction time, extraction temperature, sample agitation, and extraction phase must be performed for a particular matrix and desired application (analyte range, *etc.*).[54,55] Moreover, analysis of polar metabolites by HS-SPME–GC–MS can be challenging due to poor partitioning of these analytes onto the HS. However, a few strategies can be implemented for improved recovery of polar metabolites, such as tuning of the selectivity of the extraction phase and optimization of extraction parameters. Tuning the extraction phase selectivity for enhanced recovery of polar metabolites can be achieved with the use of task-specific extraction phases such as ionic liquids (ILs).[56] Gionfriddo *et al.* tested the recovery of various fruit metabolites *via* HS-SPME by two different polymeric ionic liquid (PILs) SPME fibers, namely poly(1-4-vinylbenzyl-3-hexadecylimidazolium) bis[(trifluoromethyl)sulfonyl] imide (poly([VBHDIM]-[NTf2])) (PIL 1) and *N*,*N*-didecyl-*N*-methyl-D-glucaminium poly(2-methyl-acrylic acid 2-[1-(3-{2-[2-(3-trifluoromethanesulfonylamino-propoxy)-ethoxy]-ethoxy}-propylamino)-vinylamino]-ethyl ester) (poly([DDMGlu][MTFSI])) (PIL 2), and

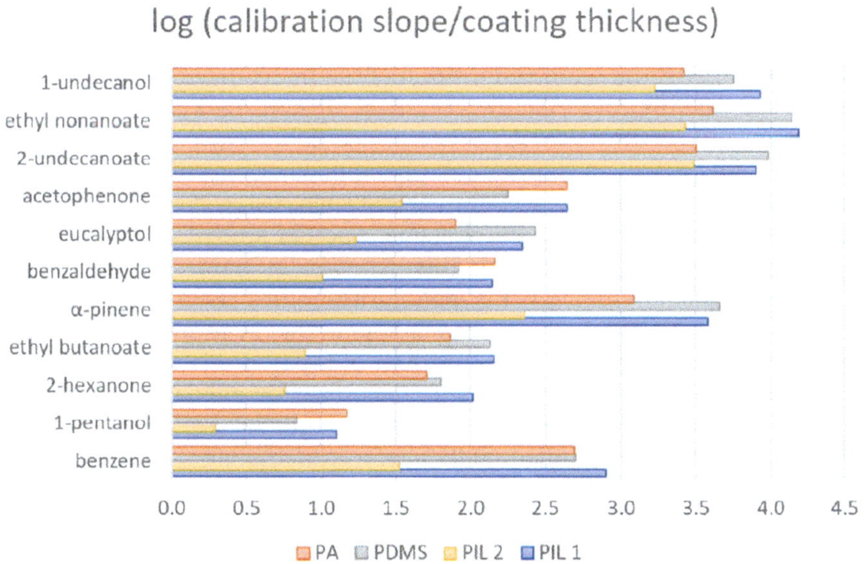

Figure 5.1 Extraction efficiencies of PIL1 and PIL2 compared to PDMS and PA SPME fibers, assessed as the ratio between respective sensitivities toward the targeted analytes and coating thicknesses.
Reproduced from ref. 57 with permission from Elsevier, Copyright 2018.

two commercially available SPME fibers with different hydrophobicity, polymethylsiloxane (PDMS) and polyacrylate (PA).[57] The calculated partition coefficients for the fruit metabolites revealed that the PIL SPME fibers presented comparable or enhanced selectivity toward fruit metabolites (see Figure 5.1). Moreover, the metabolite extraction coverage was investigated *via* HS-SPME–GC×GC–ToF/MS with grape homogenate as the model matrix: the PIL-based coatings showed excellent performances for untargeted extraction of fruit metabolites characterized by various physicochemical properties.[57] In a different study, Tang and Duan used an SPME fiber coated with 1-vinyl-3-(4-vinyl-benzyl)imidazolium chloride for the extraction of endogenous acids found in grape wines.[58] Comparison of the performance of the new coating with commercial polar coatings such as PA, Car/PDMS, and DVB/Car/PDMS showed that the novel SPME coating has superior extraction capabilities for polyatomic acids while delivering a similar performance for less polar monoacids.

The partition of polar food metabolites into the HS of the sample, and consequently the extraction efficiency of SPME, can be enhanced by performing vacuum-SPME, an alternative and convenient approach proposed by Brunton *et al.*[59] and developed practically by Psillakis *et at.*[60] Using this approach, Trujillo-Rodríguez *et al.* investigated the recovery of fatty acids and phenols, typical metabolites constituting the aroma of milk and milk derivatives[56]. The study revealed that the low-pressure sampling conditions used for Vac-HS-SPME enabled enhanced recovery of metabolites at lower temperatures and extraction times compared to conventional HS-SPME. This finding is particularly important since the use of relatively higher temperatures during extraction from certain food matrices could induce the production of artifacts or decomposition of the matrix itself. This study also showed that the improved recovery of metabolites using Vac-HS-SPME, due to faster analyte-loading onto the SPME device, could cause saturation of the extraction phase and displacement phenomena; thus, the use of short extraction times is highly recommended when using this approach for extraction from multicomponent systems. Another study involving Vac-HS-SPME for monitoring of the volatile fraction of hard cheese revealed enhanced extraction efficacy for polar metabolites such as 1-pentanol, butanoic acid, 2,3-butanedione and 2-heptanone.[61] As an alternative to HS-SPME extraction, the direct immersion of SPME devices (DI-SPME) into food samples advantageously allows for the extraction of analytes that do not easily partition into the headspace, providing broader extraction coverage. This phenomenon was studied by Risticevic *et al.* by untargeted metabolite profiling of apple samples: as shown in Figure 5.2, the extraction coverage obtained by DI-SPME is broader when compared to HS-SPME, enabling the extraction of more polar metabolites as well higher molecular weight ones.[53]

Moreover, it has been demonstrated that due to the balanced coverage effects that take place during DI-SPME, saturation and displacement phenomena are less likely to take place when using this extraction mode.[55] However, while the use of DI-SPME for food analysis with traditional PDMS-based extraction phases has been attempted, poor antifouling performance has often been observed. The roughness of the outer surface of traditional SPME coatings promotes accumulation of matrix constituents that irreversibly attach to the device during thermal desorption, thus affecting extraction performance and producing artifacts that will populate the chromatograms, potentially hiding the presence of metabolites. To overcome these challenges, a new matrix compatible extraction phase was developed by Souza-Silva *et al.* by applying a smooth and thin layer of pure PDMS onto a conventional DVB/PDMS fiber.[62] The efficacy of this new SPME device, namely PDMS/DVB/PDMS, was attentively tested in food samples, revealing excellent antifouling performances and enhanced ability to avoid saturation and displacement phenomena.[55,63,64] To enhance extraction coverage for metabolomics applications, a PDMS/DVB/Car/PDMS SPME fiber was also developed and used for apple metabolomics studies, revealing the excellent ability of this new fiber to avoid the occurrence of artifacts, in particular Maillard reaction products, during thermal desorption (see Figure 5.3).[21]

Such SPME approaches for metabolomic studies evolved into applications of the technology for *in-vivo* sampling of food matrices.[20,65] In this application, matrix compatible SPME devices were used to sample apple fruits during their ripening on the tree.[20] This unique and minimally-invasive approach successfully captured a broad metabolic fingerprint, catching unique metabolites and detecting changes in metabolic profiles associated with fruit maturation (see Figure 5.4).

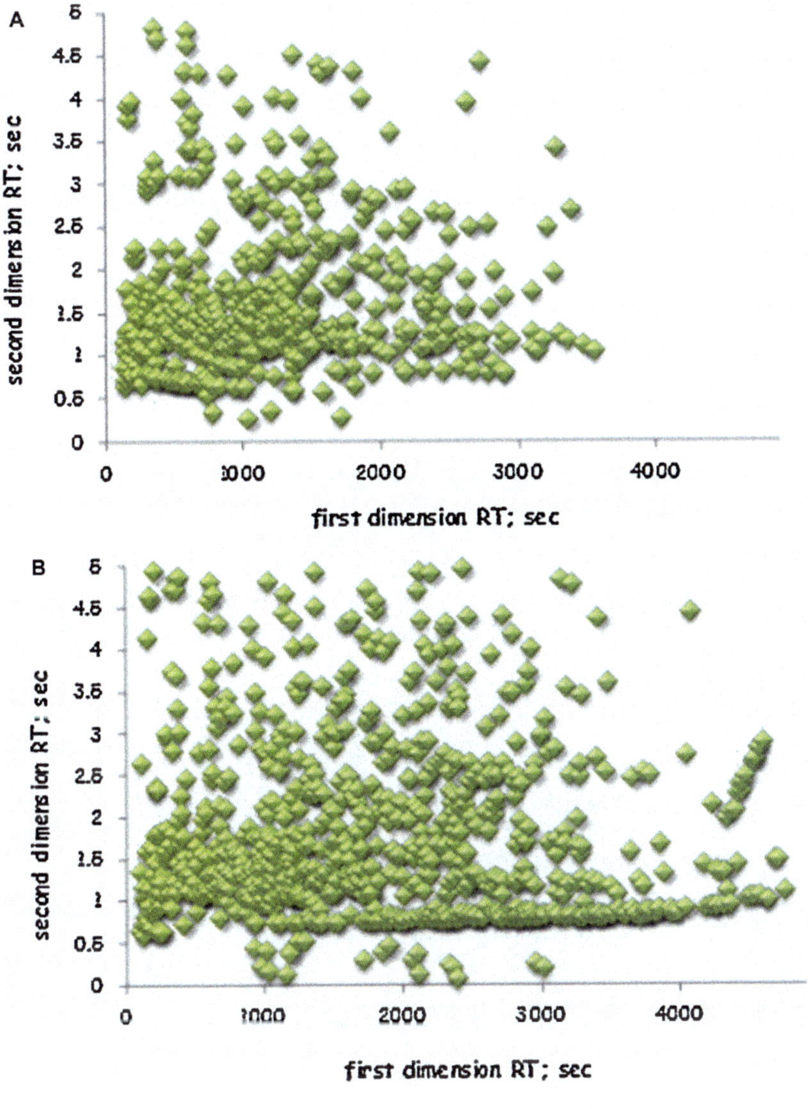

Figure 5.2 A comparison of HS-SPME (A) and DI-SPME (B) extraction modes for metabolite profiling in apples.
Reproduced from ref. 53 with permission from Elsevier, Copyright 2012.

The use of *in vivo* DI-SPME for food metabolomics presents an innovative strategy that can pave the way to further understanding of the chemical composition and metabolomic changes of complex matrices such as food specimens, guaranteeing an accurate chemical snapshot of the system under investigation. Several advantages of the use of *in vivo* DI-SPME in comparison to HS-SPME are listed in Table 5.1.

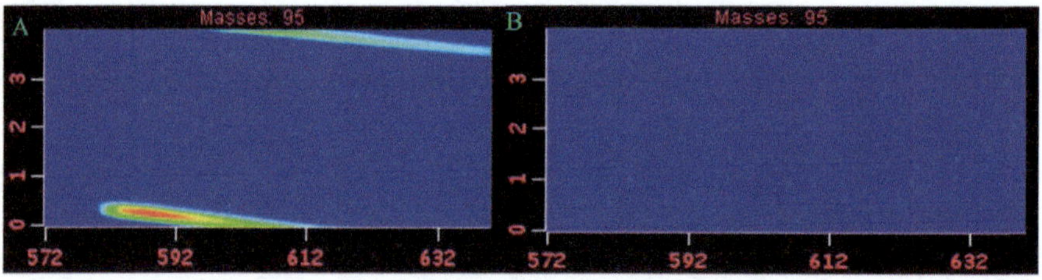

Figure 5.3 GC×GC extracted ion chromatograms corresponding to elution windows of furfural in *in vivo* metabolic profiles obtained by (A) DVB/Car/PDMS and (B) PDMS-overcoated DVB/Car/PDMS fiber coatings.
Reproduced from ref. 21, https://pubs.acs.org/doi/10.1021/acs.analchem.5b03684, with permission from American Chemical Society, Copyright 2015.

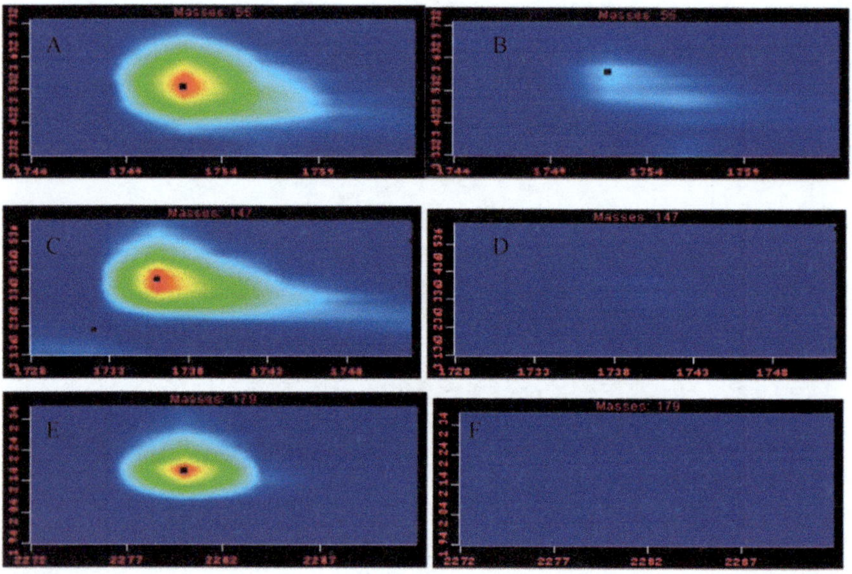

Figure 5.4 Contour plots of GC×GC extracted ion chromatograms corresponding to elution windows of metabolites unique to the *in vivo* approach in *in vivo* (left plots) and *ex vivo* (right plots) extracts. (A and B) Unidentified analyte 1 (hit #1 : 1-Hepten-4-ol), (C and D) 1,4-Diacetylbenzene (*p*-Acetylacetophenone), and (E and F) unidentified analyte2 (hit #1 : 2-(4-tert-Pentylphenoxy)-ethanol).
Reproduced from ref. 20, https://doi.org/10.1038/s41598-020-63817-8, under the terms of the CC BY 4.0 license, http://creativecommons.org/licenses/by/4.0/.

SPME-based food applications have also been extended to LC analysis, which has enabled a shift of focus to very polar metabolites.[4] As an example, Mirnaghi *et al.* screened the relative amounts of phenolic compounds present in wine, berry, and grape samples using the 96-blade solid phase

Table 5.1 Comparison between *ex vivo* and *in vivo* SPME. Reproduced from ref. 21, https://pubs.acs.org/doi/10.1021/acs.analchem.5b03684, with permission from American Chemical Society, Copyright 2015.

Consideration	*Ex vivo* SPME	*In vivo* DI-SPME
Sample preparation	Requires metabolism quenching and more extensive sample preparation	Metabolism quenching requirements eliminated
Extraction method	Requires extraction phase chemistries compatible with complex matrix and allowing clean-up of the coating (DI-SPME)	Requires extraction phase chemistries compatible with complex matrix and allowing clean-up of the coating
	Matrix effects and production of volatile end-products of the Maillard reaction in GC inlet (DI-SPME)	Matrix effects and enhanced production of volatile end-products of the Maillard reaction in GC inlet
Reproducibility in extraction	Unstable metabolites require fresh sample preparation	Improved reproducibility for unstable metabolites
	Lower reproducibility due to adsorptive losses of hydrophobic metabolites	Improved reproducibility for hydrophobic metabolites
		Extraction efficiency related to sampling position
Metabolome coverage	Biased against non-volatile and polar metabolites (HS-SPME)	Rich metabolite coverage
	Capture of polar and hydrophobic classes of metabolites (DI-SPME)	Capture of polar and hydrophobic classes of metabolites
Instrument	Less enhanced introduction of non-volatile matrix results in cleaner chromatograms, improved metabolite detectability and faster data processing	More enhanced introduction of non-volatile matrix results in complex chromatographic profiles
		Multi-dimensional instruments required to improve metabolite deconvolution from matrix and data processing
	Introduction of water and contamination due to matrix requires frequent replacement of second dimension column (DI-SPME)	Introduction of water and contamination due to matrix requires frequent replacement of second dimension column
	Higher thickness of stationary phase required in the second dimension (DI-SPME)	Higher thickness of stationary phase required in the second dimension

microextraction system.[66] The use of PAN-based SPME extraction phases also prevented fouling by food matrix components, resulting in a significant reduction of ion suppression/enhancement *via* SPME in comparison to traditional solvent extraction techniques. Moreover, Chen *et al.* proposed a novel SPME extraction phase composed by carbon nanotubes functionalized with phenylboronic acid to monitor carbohydrates *in vivo* in different biological matrices, including Malabar spinach and aloe vera.[67]

To create a versatile antifouling SPME extraction phase, Gionfriddo *et al.* proposed an SPME fiber based on amorphous fluoropolymers (AF-PTFE) compatible with thermal and solvent desorption, thus amenable to both GC and LC analysis.[68] To facilitate the use of this extraction phase for untargeted analysis, HLB particles were used as a sportive material. The extraction efficiency of this novel SPME coating was tested for various volatile food metabolites and compared to the conventional SPME fiber used for untargeted screening by GC, DVB/Car/PDMS. Calculations of the partition coefficients (K_{es}) for these analytes revealed improved selectivity for the more polar food metabolites.[68]

5.3 Application of SPME in Metabolomic Studies of Exposome

Increased awareness and concern for ecosystem protection have extended the implementation of pro-ecological technologies such as SPME in this broad field, with applications ranging from single environmental studies to wide-scale analyses of the fate of contaminants, including their metabolism and accumulation, along with consequences of such exposure to living systems.[69] Changes at the molecular level occurring in organisms under the influence of various environmental stimuli, such as pollutants and xenobiotics, are the basis of the 'exposome', a theory first introduced by Wild in 2005 that assumes that multiple factors may evoke non-genomic alterations at every stage of the lifecycle of living systems.[70] As a holistic perception, the exposome comprises the total exposure of the living system to different exogenous molecules from water, air, and soil, among others, along with the effects related to such influences at the level of proteome and metabolome. Recently, metabolomics technologies in combination with exposome analysis have become increasingly popular in environmental, toxicological, and clinical studies, as such a comprehensive approach can provide insight into the complex responses of organisms to the exogenous environment.[71] According to recent research, many chronic diseases are attributable to environmental factors through the interaction of exogenous compounds with endogenous molecules (*e.g.* low molecular weight metabolites), resulting in disturbances in biochemical pathways.[72,73] However, precise analysis of such compounds in the surrounding environment and in living systems may present a challenge given that these molecules (exogenous and endogenous) are often present at very low levels and/or can be highly reactive; hence, only implementation of advanced analytical procedures and instrumentation may guarantee their adequate extraction and detection.

As an environmentally-friendly and non-invasive technology, SPME was initially introduced for extraction of different pollutants, including persistent organic pollutants (POPs), pesticides, and pharmaceuticals and personal care products (PPCPs) from complex elements of aquatic environment, providing useful information about the ecosystem dynamics.[74] For instance, *in situ* SPME with the use of mixed-mode (octadecyl and benzenesulfonic acid groups) fibers was applied for simultaneous underwater sampling and extraction of small molecules from corals in order to analyze the response of aquatic organisms to climate changes and pollution.[75] Multiple endogenous compounds, including amino acids, eicosanoids, and different lipid species involved in various biological functions were identified in the water surrounding corals. Among the polar lipids detected with the use of this SPME–LC–MS method were glycerophospholipids (*e.g.* phosphatidyl choline, phosphatidyl ethanolamine, and phosphatidyl serine) and sphingophospholipids (*e.g.* sphingomyelin), which are important constituents of cell membranes. Moreover, short-lived intermediates of biological processes, such as thiocysteine, 5-transtetradecanoid, and 5,6-DHET were also tentatively identified with the use of SPME devices. These results have provided valuable insight into the functioning of living organisms in their natural habitats as many of the identified compounds are involved in normal growth and development and are also part of fundamental metabolic pathways essential for the survival of corals.

In order to attain a better understanding of the impact of exogenous substances on the functioning of the organisms, *in vivo* SPME has been introduced as a new technique for simultaneous extraction of a wide range of small molecules (polar and non-polar compounds) directly from living systems.[18] As mentioned previously, *in vivo* SPME combines sampling, metabolite extraction, and quenching in one step, enabling the capture of short-lived and unstable molecules of exogenous and endogenous origins. The feasibility of SPME devices to stabilize highly reactive metabolites or transient intermediates of biochemical processes directly from a living system has been confirmed in different metabolomic studies. Several lipid species, including low-abundant and short-lived metabolites, such as *N*-acyl amides and fatty aldehydes, present in their unbound form were captured with the use of *in vivo* SPME probes from fish muscles. This technology was also applied in exposome-wide associated studies (EWAS) in combination with metabolomics studies in order to track the behavior of certain toxins and to monitor biochemical responses after exposure to those factors, mainly in the aquatic environment.[76] In this study, a diversity of lipids was identified across organisms and environments, each playing a critical role in membrane organization, structure, and signaling response to environmental conditions. This *in vivo* technology also facilitated extraction of several unstable molecules, such as 1-oleoyl-*sn*-glycero-3-phosphocholine and 1-palmitoleoyl-glycero-3-phosphocholine from fish tissue samples, which could not be detected in *ex vivo* extracts due to auto-oxidation of those compounds during sample collection, transportation, and storage.

The capacity of SPME to extract a wide spectrum of metabolites was also exploited in an analysis of the composition of small molecules present in the sponge (*Sarcotragusfoetidus*) holobiont.[77] In this study, three types of SPME devices with different geometries were used for *in situ* and *in vivo* sampling of different sections of sponges to extract compounds associated with quorum sensing (signaling activity among bacteria) as well as to track the metabolism of organic pollutants, such as monocyclic aromatic hydrocarbons (MAHs), polycyclic aromatic hydrocarbons (PAHs), and other bioactive compounds taken up by sponges from the surrounding water. Interestingly, acyl-homoserine lactone, a signaling molecule related to cell-to-cell communication, quorum sensing among Gram-negative bacteria, and also bioluminescence or biofilm formation was tentatively identified in the sponge's exhalent pores. Moreover, different pollutants, including pesticides, drugs, MAHs, PAHs, and also 3-isochromanone – a metabolite associated with PAH degradation – were detected with the use of this SPME–LC–MS method. The presence of these compounds in sponge and/or its holobiont can be possibly attributed to the constant pumping of large amounts of seawater and to the exposure to pollutants that may potentially accumulate in the sponges and affect their metabolism.

In another exposome *in vivo* study utilizing SPME devices, the potential impact of exogenous factors on metabolomic changes was investigated by analyzing the exposure of fish muscles to benzo[a]pyrene (BaP).[23] BaP belongs to a group of PAHs and is known to disrupt cellular metabolism while also exerting strong carcinogenic, mutagenic, and cytotoxic effects. In this study, a case–control model consisting of three groups of fish – one control group and two experimental groups exposed to sub-lethal BaP doses, *i.e.* low dose group (1 ng L^{-1} of BaP) and a high dose group (10 ng L^{-1} of BaP) – was utilized to determine metabolomic changes according to exposure level. *In vivo* SPME extraction was carried out under controlled (laboratory) conditions using mixed-mode fibers with a thickness of 45 μm. After 14 days of the experiment, the dose- and time-dependent changes to BaP exposure were detected in fish metabolome. In comparison to the control group of fish, the presence of BaP in the surrounding environment had a significant impact on energy metabolism and osmoregulation, causing alterations in the profile of amino acids and lipids participating in those processes. Moreover, carnitine and its ester – acetylcarnitine, which play an important role in β-oxidation of fatty acids, were present only in the control group of fish, indicating that BaP may disrupt lipid metabolism in fish exposed to this toxin. The use of *in vivo* SPME fibers also enabled the extraction of a number of intermediate metabolites from many biochemical pathways, such as glucose-6-phosphate and its secondary messenger – cyclic cytidine monophosphate. However, both compounds were present only in the control group. Conversely, the presence of a labile and trace endogenous compound, namely 2-arachidonylglycerol, was detected solely in fish muscles exposed to the higher concentration of BaP. The presence of this endocannabinoid after the first day of the experiment enabled statistical separation of this group of fish from other studied groups. The observed changes also indicated the presence of adaptive mechanisms at the molecular level in fish muscles to lower BaP concentrations in the surrounding waters: after 14 days of the experiment, the metabolic profile of this group of fish resembled the profile of the control group of fish.

However, laboratory studies of the effects of single toxicants, while providing important information about their fate and influence on the living system, may not comprehensively unravel concerns regarding the impact of the total exposome on the functioning of organisms. In natural environmental conditions, organisms are exposed to many exogenous factors, which in various ways (*e.g.* synergistic, additive, antagonistic) can affect metabolism. Therefore, *in vivo* SPME was also applied for monitoring of toxicants along with an assessment of biochemical response among sixty white suckers (*Catastomus commersonii*) exposed to potential contaminants in selected fish habitats (the oil sands development region, outside the deposit, and in a pulp and paper mill discharge region).[22] SPME technology allows for simultaneous extraction and determination of both toxic compounds and endogenous metabolites from fish muscles and further facilitates a comprehensive analysis of the correlation of fish habitats along with the type of toxins present in their muscles and biochemical changes. The effects of exposome in wild fish revealed the presence of petroleum-related compounds in the fish muscle tissue. Several organic compounds potentially related to industrial and workplace toxins, such as aliphatic hydrocarbons and PAHs, pesticides and PPCPs, including unstable and highly-reactive compounds, were extracted *via* SPME probes. Observed alterations at the metabolite level concerned mainly transformations of eicosanoids and other metabolites involved in inflammatory reactions of the body, such as metabolites of arachidonic acid (*e.g.* hydroxyeicosatetraenoic acids, HETE) and linoleic acid (*e.g.* hydroxyoctadecadiene acids, HODE). Conversely, fish living outside the oil industry region were found to have an overall lower content of toxic compounds and a lack of substances associated with bitumen. Fewer changes in the metabolomic profile in this group of fish were also observed in comparison to the other two groups, which may suggest a correlation between the degree of environmental pollution and the metabolomic profile of a living organism. By interacting with endogenous metabolites, exogenous molecules may alter signaling processes and cell membrane functions, and thus affect the metabolism of living organisms; however, more studies are needed to precisely indicate whether the observed effects of various pollutants have an additive and/or cumulative impact on the metabolomic profile of fish muscles.

Exposome studies along with untargeted metabolomic profiling of short-lived and unstable tissue components offer multifactorial exploration of the cause–effect relationships between exogenous and endogenous molecules in living organisms. *In vivo* SPME is a useful tool for direct monitoring of the effects of exposome on the metabolism of living systems.[78,79] This *in vivo* technology offers an interesting alternative for biomonitoring studies aiming to track alterations in organisms exposed to environmental contaminants. Important compounds derived from natural or anthropogenic sources of contamination along with endogenous metabolites could be extracted and identified with the use of biocompatible SPME devices. However, more exploration and longitudinal studies are needed to track the exposome and its relation to biochemical profiles following exposure to contaminants. In the future, such an approach may allow for identification of novel biomarkers of exposure, or toxicants and their metabolites, and may provide crucial information necessary for a better understanding of the persistence and effects of pollutants in the environment.

5.4 Metabolomics with SPME in Biomedical and Clinical Applications

The number of biomedical or clinical applications with the use of SPME is constantly increasing. A significant proportion of applications published to date have been mostly dedicated to targeted analysis, mainly drug monitoring in high throughput mode or fast determination by direct coupling to MS.[2–7] However, for the past few years more studies have focused on the discovery of the metabolome and lipidome of living systems at the targeted or global levels. Since SPME protocols permit analysis of the volatilome and non-volatile metabolome (and lipidome) *in vitro*, *in vivo*, *in situ*, and *ex vivo*, opportunities to extend in-depth exploration of a variety of biological systems *via* SPME are vast. Particularly, the volatilome carries significant information regarding living systems; it reflects the biochemical status of a given organism, tissue, biofluid, cell, or bacteria, as organisms emit characteristic compounds that present significant changes in levels following exposure to different stimuli, including pathological factors, xenobiotics, or natural agents.[2–7] Bacteria and cells cultured *in vitro* are quite frequently subjected to analysis of their volatile metabolome by SPME since the technique can capture compounds released by living cells without the need to cease their metabolism. Therefore, the same cells can be sampled multiple times, thus providing temporal profiles, while preventing inter-culture variability, which is otherwise introduced when new cell seeding/bacteria culturing has to be used at each time point. The approach was successfully used in model cells[16] and bacteria,[14,80] as well as in isolates from human specimens.[2–7] Some studies have exploited the non-invasive nature of breath sampling to compare the biosignature of isolates or model cell cultures to the volatilome of exhaled breath or breath condensate of patients with a certain disease (*e.g.* lung cancer or bacterial infection of the respiratory track) with the aim to ascertain whether the metabolites found *in vitro* are also present *in vivo* and if they can serve as potential biomarkers of the disease.[3] A similar concept has been used in skin analysis aiming at the discovery of biomarkers of different skin diseases *e.g.* melanoma.[81] In such applications, the thin film geometry of SPME is preferable due to the larger surface area, higher sensitivity, and simplicity of use that this configuration affords. Headspace SPME can also be used for analysis of VOCs of all kinds of collected samples, including saliva, urine, blood, sweat, feces, and tissues.[2–7] Again, the flexibility afforded by SPME devices and protocols enables a wide range of possibilities; however, analysts must ensure all steps of the analytical procedure are carefully optimized and validated for the desired application in order to ensure reproducible and reliable results.

In contrast to the well-established SPME field of volatile metabolome analysis, the investigation of semi- and non-volatile compounds *via* SPME represents a new field in bio-SPME. As mentioned before, direct extraction from complex biological matrices has only become feasible since the introduction of biocompatible coatings. While the utilization of high-throughput systems, either commercial and in-house built platforms, has expanded the application of SPME to investigations of *in vitro* models (*i.e.* cells, bacteria and collected biofluid samples), the small size of the traditional SPME fiber has opened the door for *in vivo* studies, enabling minimally-invasive and non-destructive sampling procedures as compared to traditional, invasive procedures such as biopsy or blood draw. Indeed, the first *in vivo* SPME metabolomics studies have not only revealed a clear and significant difference between the *in vivo* and *ex vivo* metabolome of blood in mouse models, but also emphasized the unique opportunity of the method to downsize the number of animals used in biological experiments involving time-course investigations.[18] Furthermore, enabling observations of the same individual before and after stimuli effectively eliminates inter-animal variability from the experiment, hence producing superior results. Subsequent animal studies have since expanded to the analysis of tissues, particularly brain studies. As mentioned previously, the main technique used for *in vivo* brain studies is microdialysis; thus, this particular method was used for cross validation of SPME results when changes in the level of four neurotransmitters were measured in rat brains over time after administration of fluoxetine.[28] Both methods showed similar trends, with SPME exhibiting better precision (lower RSDs) and sensitivity compared to MD. In addition to targeted profiling, untargeted screening of SPME extracts and MD dialysates has demonstrated that MD is superior in capturing very polar and small compounds while SPME covers the range of less polar to non-polar species. These findings were later confirmed by Boyaci *et al.* in untargeted metabolomics and lipidomics analysis of rat brains with SPME and MD before and after administration of the same drug.[29] These studies showed dysregulation of hippocampal metabolite levels associated mainly with fatty acid transportation and purine metabolisms, confirming the complementary nature of these two methods. A similar experimental design of SPME sampling was employed by Reyes-Garces *et al.*, but this time authors monitored the influence of deep brain stimulation (DBS) on metabolome and lipidome in rat hippocampi.[82] Indirectly, the results indicated that one of the mechanisms of DBS could be related to the release of nitric oxide, a gaseous neurotransmitter. Other observations included changes of corticosterone levels, as well as phospho- and glycosphingolipids levels. In other experiments, miniaturized SPME probes were used for measurement of a range of neurotransmitters in macaque monkey brains during goal-directed behavior.[33,34] Quantitative analysis results revealed that the most pronounced differences in concentrations of dopamine *vs.* acetylcholine occurred in the striatum (ten-fold higher) during the performance of tasks assigned to the animals. No significant changes were observed for other analyzed neurotransmitters in other studied brain areas.

Metabolomic analyses of other organs have also been carried out *via* SPME. Profiling of four organs, namely brain, liver, kidney, and muscle of three different mice strains was performed immediately after euthanasia.[83] The strains selected for this study represent the most commonly-used mice strains in biomedical research, sometimes interchangeably. The results showed that the metabolome of these organs is not always identical between the strains; for instance, distinct inter-strain differences were revealed for brain and liver tissue, whereas the metabolome of the kidneys and muscles of BALB/c mice differed greatly from that of C57BL/6 and CD1 strains. The results of this investigation suggest that baseline metabolome might influence further observations and thus result in differences in experimental outcome when various strains are compared.

With the progress in SPME technology and new reports of its application in the biomedical area, new directions in bio-SPME have more strongly focused on practical aspects of medicine and medical diagnostics, since the simplicity and ease of use of SPME enables its use as a tool for on-site sampling. Preliminary proof-of-concept studies demonstrated that SPME can be a useful tool for on-site sampling of organs dedicated for transplantation to evaluate changes occurring during their preservation under different conditions.[84,85] The minimum invasiveness of sampling afforded by SPME fibers enables the unique opportunity to perform repeatable analyses of the organs, starting with *in vivo* sampling in the donor body, continuing throughout the preservation protocol (either static or dynamic using perfusion), and again, *in vivo* after/during transplantation to the recipient (see Figure 5.5). Standard assessments of organ quality adopted today are mostly based on a visual inspection of the graft, the results of the donor's body fluids, and, occasionally, tissue biopsy. While a biopsy can yield more reliable results regarding organ quality, its invasiveness limits the number of times it can be performed on a single organ; therefore, changes occurring during organ preservation are not monitored this way. In this regard, a protocol containing details of *in situ* and *in vivo* sampling of kidney grafts throughout the peri-transplant period was recently published.[85] The authors provided some examples of the metabolomic and lipidomic data collected at the aforementioned timepoints. The results demonstrated that the presented strategy enables the capture of metabolites that alter significantly during organ perfusion, thus enabling discernment of changes that have taken place, from the initial assessment of the organ in the donor's body, until transplantation in the recipient is completed. Another simple SPME experiment performed on dissected rabbit kidneys stored for 21 h under cold static conditions showed that significantly different metabolite patterns were found during the first 4 h of storage as compared to the later time points *i.e.* 6 and 21 h.[86] The observed changes were in line with literature reports describing early changes (2 and 4 h) associated with ischemic

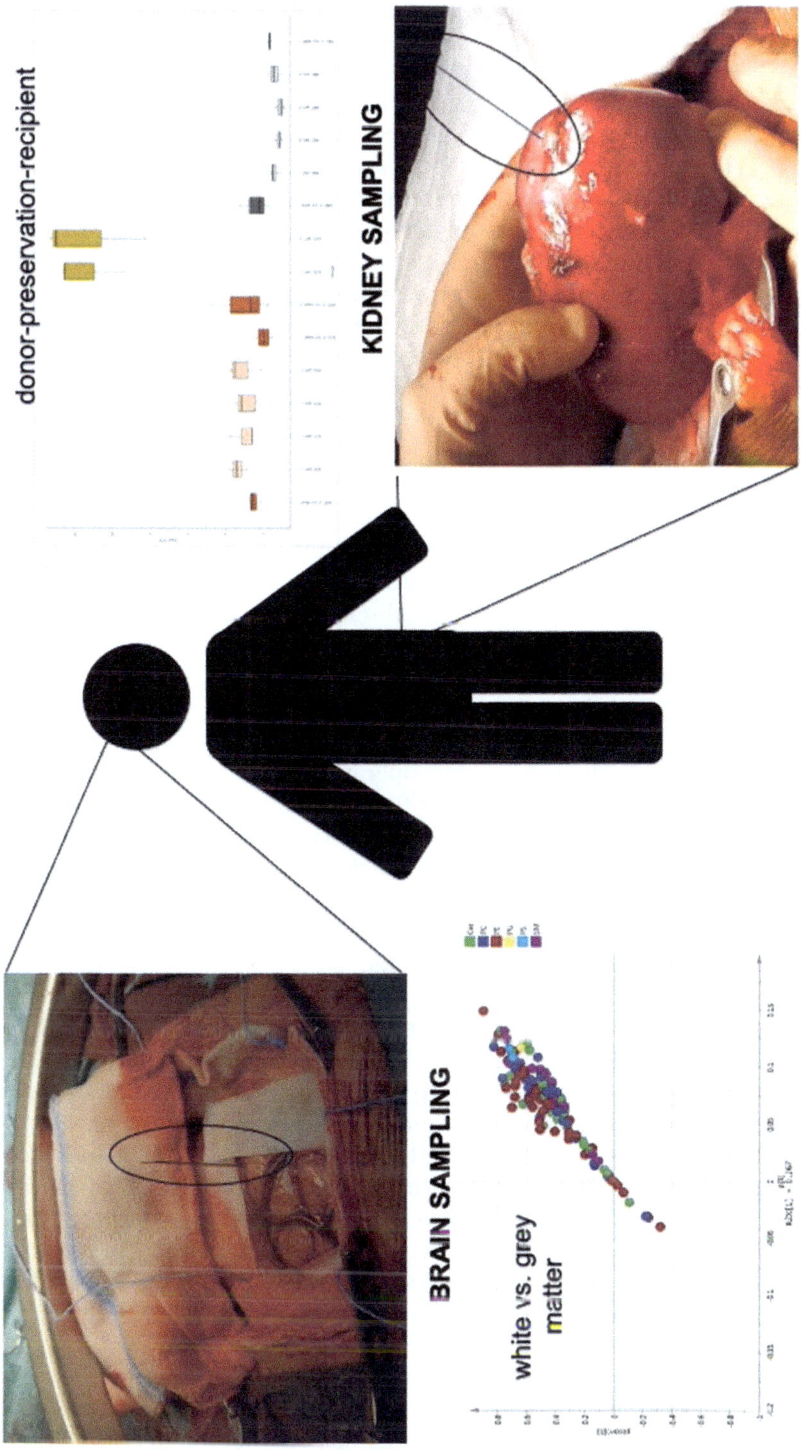

Figure 5.5 Human brain *in vivo* sampling during brain tumor surgery and S-plot representing lipid species differentiating white and gray matter of the brain (left) and human kidney sampling after its harvest from the donor, with box-plot of hypoxantine showing changes of the metabolite over the transplantation procedure timespan ("from donor to recipient") (right).

preconditioning of the graft, and later ones (6 and 21 h) with the effect of prolonged ischemia and oxidative stress.

SPME probes can also be safely used in a hospital environment, including in emergency wards or operating rooms. The first reported *in vivo* human SPME study was associated with a clinical trial for *in vivo* lung chemoperfusion (IVLP) with doxorubicin, demonstrating the capability of SPME to determine changes in the concentration of the drug over the time of the surgical procedure as well as the biodistribution of doxorubicin.[87] In addition, global metabolomic analysis evidenced the presence of a previously described lung cancer biomarker and dynamic changes of metabolites, indicating lung toxicity and injury induced by the IVLP procedure. Although the instrumental measurements were still conducted in the analytical laboratory outside the hospital facility, thus not providing real-time results, the outcome of the study clearly indicates that SPME is capable of filling the gap as a simple and fast sampling-sample preparation tool that provides very rapid and reliable results when coupled to analytical instrumentation on-site. Another example of such practical application of the technology in the clinical environment is the recent development of a protocol for determination of the phenotype of brain tumors.[35,88] The representative results of the developed protocol for sampling of meningiomas and gliomas performed on-site immediately after tumor resection yielded correlations for metabolomic and lipidomic profiles with the histological signature of both types of tumors and additionally, genotype of gliomas. Given these results and those of another study, profiling human brain sampling *in vivo* (see Figure 5.5), one can easily foresee intra-operative or intra-biopsy evaluation of tumor biomarkers *in vivo* in the near future. When used as a diagnostic, such information could play a critical role in decisions regarding chemo- and radiotherapy following surgery. Based on the data reported thus far, extension of the analysis to larger cohorts followed by validation of potential biomarkers and development of a fast method for targeted analysis based on SPME–MS are indicated as future directions.

SPME has also been used for metabolomic analysis of human muscle biopsies obtained from patients undergoing diagnosis of malignant hyperthermia (MH), a potentially fatal pharmacogenetic disorder of skeletal muscle that is triggered by exposure to certain anesthetics.[89] The current diagnostic procedure is invasive, as it involves an extraction of *ca.* 2 g of muscle, which is subsequently dissected to smaller muscle strips and used for caffeine–halothane contracture testing (CHCT). Although some pathological mechanisms of the disease are already known, many factors and metabolic pathways still remain uncovered. To reveal possible metabolic alterations in muscles of MH-susceptible patients (MHS) the authors performed SPME sampling at the baseline (in the absence of anesthetic triggers), and after caffeine and halothane exposure, then compared the data with results obtained by the traditional tissue metabolomics approach performed by a commercial subcontractor. The results of the two techniques were in agreement, showing a similar pattern of metabolite differences between the MHS cohort and MH-negative patients (MHN). While the studied group was small due to the rarity of MH and the limitations in attaining a cohort large enough to perform studies focused on the search of clinically relevant biomarkers, the proof-of-concept experiments showed the potential of the approach for *in vivo* studies as a good alternative to invasive tissue collection.

5.5 Extraction Phases and SPME Devices for LC-based Metabolomics

Generally speaking, extraction and pre-concentration of polar compounds present challenging tasks for sample preparation in many areas including environmental, food, clinical, and other bioanalytical applications. Although this task is challenging, successful implementation of subsequent critical steps of the analytical workflow is entirely dependent on careful optimization of the extraction and pre-concentration steps of the analytical workflow. For instance, biological matrices such as urine, salvia, sweat, and blood fractions are rich in polar metabolites that provide direct or indirect measures about the state of the biological system. Many endogenous biomarker molecules such as sugars, amino acids, and primary amines are polar; however, the extraction and pre-concentration of such polar metabolites are intricate as these compounds possess significant affinity for aqueous matrices, diminishing the sensitivity of the final analytical method. In spite of great efforts reported in the literature to address the common issue of poor polar analyte detection, in many of these studies analytes that were considered as polar compounds could be only deemed as relatively polar compounds, as most of them have Log *P* values above 0. In some cases, even compounds with Log *P* values of 3 to 5 were judged as polar compounds because of the polar moieties found in the structures, which blur the boundaries for clear classifications. Here, we will only consider molecules with Log *P* values below 1 as polar compounds; as such, the following discussion will be limited to this fraction of metabolites.

In the last ten years, great efforts have been made to show the potential of SPME and its related techniques for both *in vivo* and *in vitro* targeted and untargeted metabolomics. However, to date, only a limited number of commercial devices are capable of addressing or partially addressing the requirements of untargeted metabolomic approaches, the poor extraction of polar metabolites being the major limitation of these devices. This limitation has begun to be addressed with advancements in novel extractive materials, with significant developments in the preparation of new SPME devices with extractive phases applicable to polar metabolites recently reported. With the intention of developing an extractive phase suitable for general metabolome coverage, commercial particles available for SPE have been investigated and used for development of novel SPME devices that provide balanced coverage for both polar and non-polar metabolites.[90,91] In this regard, systematic evaluations have provided vast amounts of information about the capabilities of commercial particles. In early attempts, Vuckovic and Pawliszyn investigated the performance of commercial extractive particles by making 42 different types of SPME fibers and testing them with 36 metabolites possessing diverse physicochemical properties (Log P in a range of -7.9 to 7.4).[90] The study not only defined the extractive phase that provides the best coverage, but also identified extractive phases with improved recoveries for polar metabolites. Mixed-mode coatings such as C18 or C8 particles with benzenesulfonic acid moieties, polar-enhanced polystyrene–divinylbenzene with weak anion exchange moieties, and phenylboronic acid with diol moieties appeared as the best candidates for simultaneous balanced coverage of both polar and non-polar analytes.[90] The general trend observed in the above study suggests a decrease in absolute recoveries as the polarity of the metabolite increases. For instance, absolute recoveries in the range of 0.5–5% were reported for very polar analytes such as sucrose (Log $P - 3.7$) and glutamic acid (Log $P - 3.69$), while for more polar compounds, this value drops to *ca.* 0.1% (choline, Log $P - 4.7$), and in some cases, no extraction occurred. It should be noted that SPME recoveries of several percent or less would be sufficient as long as these amounts are enough for instrumental quantitation. Higher recoveries are possible when the volume of the extractive phase is increased. One of the most practical ways to improve sensitivity toward polar analytes is to use TFME and the abovementioned extractive phases. Such an attempt was reported by Mousavi *et al.* for selection of the best TFME coating for *in vitro* untargeted metabolomic investigations,[91] with the abovementioned Vuckovic study paving the way for this study. Mousavi tested both silica-based particles modified with phenylboronic acid (PBA), C18, mixed mode (C18-benzene sulfonic acid modified), and polymeric particles such as hydrophilic–lipophilic balanced (HLB), PS-DVB-WAX, and mixtures of HLB with PS-DVB-WAX. The results of the study revealed that HLB, PBA, and PS-DVB-WAX provided better extractions for polar analytes compared to the rest of the extractive phases, with the best performance observed when HLB was mixed with PS-DVB-WAX (50/50, w/w). The described coating in TFME format was used for profiling of *E. coli* metabolome following stimulation of the cell line with cinnamaldehyde[92] or eugenol.[93] The results revealed dysregulations in levels of polar metabolites such as aminoacids (*i.e.* phenylalanine, serine, proline, and valine), TCA cycle compounds (*i.e.* fumaric acid, malic acid, and glucose 6-phosphate), pyridines and derivatives, imidazopyrimidines, indoles and derivatives, as well as sugars, showing the potential of the coating for the capture of polar metabolites. Currently, C8-benzene sulfonic acid, C18-benzene sulfonic acid, and HLB coatings are widely used for multi-residue analysis and untargeted metabolomics where both polar and non-polar analytes are of interest. Since the primary aim of the abovementioned studies was to perform untargeted metabolomics, the main focus was to find a coating capable of providing balanced coverage for all possible metabolites, in particular for *in vitro* studies, as the used TFME devices are relatively big and not suitable for *in vivo* applications. For applications in living systems where low invasiveness is a must, miniaturized devices in fiber geometry are required. However, it is predicted that miniaturization of the device would further challenge the sensitivity of SPME fibers for highly polar metabolites. Nonetheless, miniaturized SPME fibers have been developed and used in brain sampling of neurotransmitters with Log *P* values in a range of −4.2 to 0.5.[33,34] Various commercially available polymeric particles including HLB (5 μm), strong cation exchange (MCX, 20 μm), strong anion exchange (MAX, 20 μm), weak cation exchange (WCX, 20 μm), weak anion exchange (WAX, 20 μm), and DSC-MCAX and LC-SCX, both with strong cation exchange properties, were tested. In terms of extraction performance, the MCX particle-coated SPME fibers showed the best recoveries for the targeted neurotransmitters. However, one of the problems with the commercial particles is that the diameter of these particles is too large as they are synthesized mostly for SPE applications; even after grinding the particles, large fiber-to-fiber variations may be observed. For such miniaturized devises, the best solution is to use in-house synthesis aiming to obtain particles with a few or sub-micrometer diameters. In the same study described above, in-house synthesized HLB-SCX particles (2 μm in diameter) provided the second-best performance, with the advantage of yielding less fiber-to-fiber variations.[94]

As stressed several times before, when using extractive phases designed for capture of a wide range of analytes, the extraction recoveries of polar analytes will be low compared to those of non-polar analytes. However, the free concentration (fraction available for the extraction) of polar analytes is generally higher compared to the free concentration of non-polar analytes (showing high binding). This feature compensates for the low affinity of polar analytes, yielding a reasonably balanced coverage of a wide range of compounds.[2] On this note, to get more a realistic picture, the performance of SPME can be compared to microdialysis (MD) which is the 'gold standard' for extraction of polar/small molecules *in vivo*. Two studies have been reported[28,29] to date for in-tandem *in vivo* SPME and MD probes that allow us to compare the performances of the two approaches. Boyaci *et al.* performed sampling from rat hippocampus with SPME (mixed mode and C18) and MD probes.[29] The ion maps obtained in this study (see Figure 5.6), especially in positive ionization mode, show the larger number of features in low retention time and m/z values for MD extracts indicating the advantage of the MD over SPME for coverage of polar metabolome. More specifically, the results suggested that SPME has good coverage for relatively non-polar and non-polar compounds such as lipids and steroids while such compounds were not observed in MD extracts. On the contrary, MD extracts provided very good coverage for more polar and small molecules such as amino acids and neurotransmitters, but yielded little or no information regarding non-polar compounds. Considering this, in untargeted studies, utilization of the two methods would provide complementary and thus more comprehensive information, while MD would provide more sensitive results for targeted analysis of polar analytes. Although similar results in terms of general coverage were obtained in the study performed by Cudjoe *et al.*,[27] up-regulation

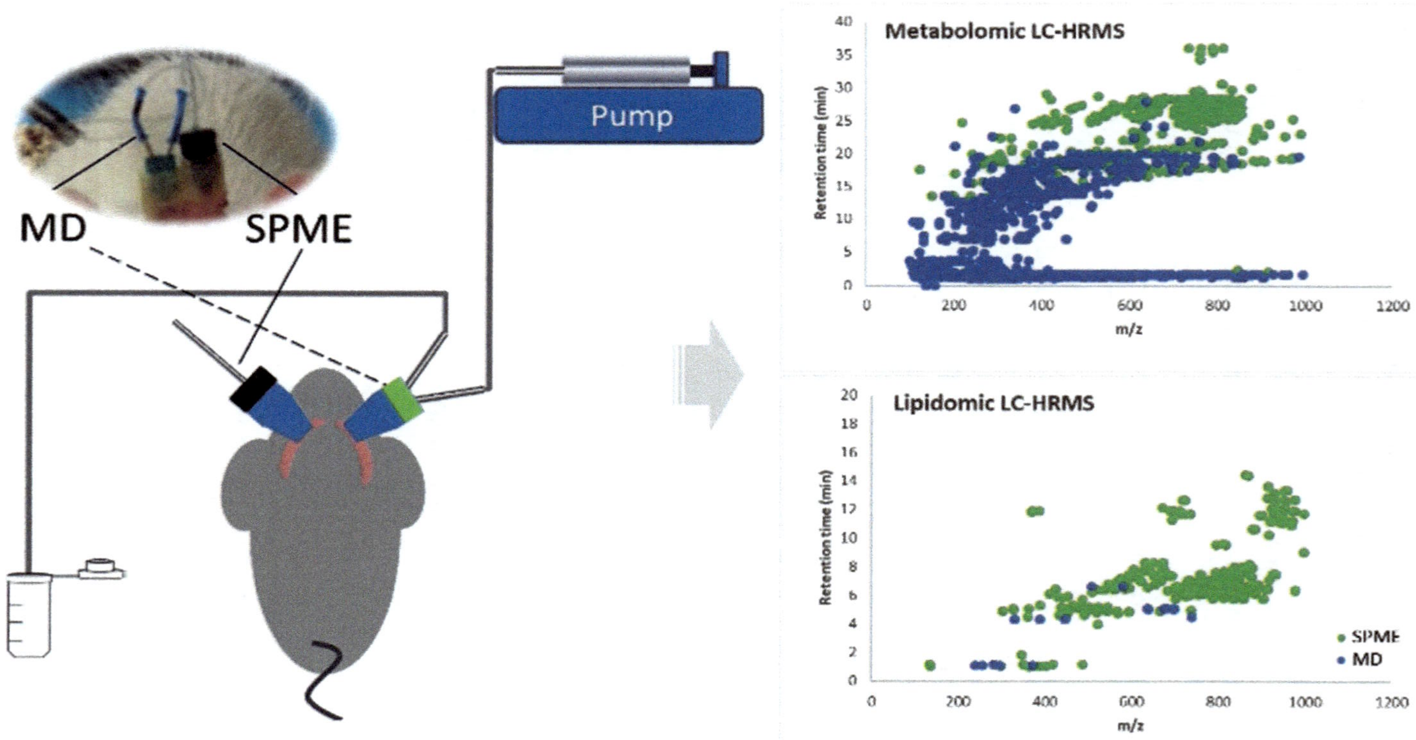

Figure 5.6 Feature coverage of SPME and MD in rat brains, using a reversed phase-general metabolomic method with PFP column (MM-SPME and MD) and lipidomic method with C18 column (C18-SPME and MD), both in the positive ionization mode. Lipidomic LC-HRMS:
Adapted from ref. 29 with permission from American Chemical Society, Copyright 2020.

of selected polar neurotransmitters after and selective serotonin reuptake inhibitor (SSRI) administration were successfully monitored and quantified with both methods. This clearly suggests that in targeted metabolomic approaches, it is feasible to tune SPME chemistry toward particular metabolite(s) and obtain sufficient sensitivity even for polar analytes.

As summarized above, most of the attention with respect to the capture of polar analytes has been directed at SPME fibers (for *in vivo*) and TFME blades (for *in vitro*). However, in-tube SPME, which is mostly used in combination with LC, also provides a useful platform for extraction of polar metabolites. Many polar commercial GC capillary columns have been utilized for in-tube SPME.[94] In this technique, the extractive phase volume is relatively large; together with the abovementioned polar phases, the technique affords exclusive solutions for very polar metabolites. For instance, Kataoka *et al.* used an amine styrene divinylbenzene coated capillary column (60 cm$\times$0.32 mm i.d., 10 μm film thickness) for in-tube SPME to extract nicotine and its metabolite cotinine along with other tobacco related alkaloids from saliva and urine samples prior to their LC–MS detection. In this study, several other polar phases were also investigated including PDMS, 14% cyanopropyl phenylmethylsiloxane, polyethylene glycol (PEG), carbon molecular sieves, and a DVB polymer. The results revealed superior performance for the amine stylene divinylbenzene polymer in comparison to the other tested in-tube SPME columns, providing LODs in a range of 15–40 pg mL^{-1}. However, one of the major problems with in-tube SPME devices is the leaching of the extractive phase,[95] which decreases their reusability and repeatability.

In conclusion, SPME is less effective for extraction of very polar analytes as it has traditionally been developed for GC applications, where relatively nonpolar analytes are analyzed. Owing to their larger extractive phases when compared to traditional fiber SPME, TFME and in-tube SPME can in many cases provide sensitive solutions for polar metabolites with collected samples, whereas for *in vivo* applications, tuning of the coating chemistry is a more reasonable solution for sensitive determinations. Generally speaking, the flexibility of SPME and its related techniques enable a variety of customizations, enabling case-specific solutions to address common issues of sensitivity and range.

5.6 Concluding Remarks

Solid phase microextraction is a relatively new method for investigations of global metabolome. The unique features of SPME, in particular its suitability for sampling of living systems with the ability to extract representative metabolome and capture the labile portion of the metabolome, offer new opportunities for scientists to better characterize investigated systems. The great variety of geometries of SPME sampling devices, coating chemistries, sampling modes, as well as its versatility in coupling to different analytical instrumentation broaden the range of metabolites that can be extracted *via* the technique. Furthermore, the simplicity offered by SPME devices and sampling protocols makes the method ideal for on-site use, including field sampling at hospital facilities. Although SPME is constantly being improved to increase sensitivity without compromising the analysis time and reversely, decreasing the analysis time without losing sensitivity, broaden the range of analytes extracted during one experiment, and furthermore, enable faster and cheaper analyses on-site by new coupling strategies, the method cannot be considered as a replacement for currently used approaches. Rather, SPME is proposed as a tool for applications where other methods cannot fulfill the expectations of the desired research, or as a complementary tool to traditional methods as a way of broadening the scope of knowledge attainable in investigations. The main limitation of the technique at the present time is the limited number of commercially available devices, especially those dedicated to polar metabolome compounds. However, great expansion in this field is expected in the next few years, as evidenced by the growth in literature reports on new metabolomic-related applications of SPME, and particularly with respect to the increasing number of fundamental works dedicated to the stability of metabolites in a variety of matrices extracted in different conditions, such as *in vivo vs. ex vivo* applications. Current investigations are focused on the influence of conditions and storage time of SPME devices, particularly on the non-enzymatic conversion of metabolites. In summary, up-to-date studies on the utilization of SPME in metabolomics and related studies indicate that the method has great potential to fill the gap in the current portfolio of analytical approaches used routinely in research labs as well as industrial or clinical practice.

Abbreviations

5,6-DHET	5,6-dihydroxy-8*Z*,11*Z*,14*Z*-eicosatrienoic acid
AF-PTFE	amorphous fluoropolymers
BaP	benzo[a]pyrene
Bio-SPME	biocompatible solid phase microextraction
Car	carboxen
CBS	coated blade spray
C_e^{eq}	equilibrium concentration of analyte in the extractant
C_s^{eq}	equilibrium concentration of analyte in the sample matrix
DART	direct analysis in real time
DBS	deep brain stimulation
DBS	dried blood spot
DESI	desorption electrospray ionization
DI	direct immersion
DSC-MCAX	mixed mode with strong cation exchange
DVB	divinylbenzene
EBS	extracted blood spot
EWAS	exposome-wide associated studies

GC	gas chromatography
HETE	hydroxyeicozatetraenoic acids
HLB	hydrophilic–lipophilic balanced
HODE	hydroxyoctadecadiene acids
HS	headspace
IL	ionic liquid
IVLP	*in vivo* lung chemo-perfusion
K_{es}	partition coefficient or distribution coefficient between the SPME extraction phase and the sample
LC	liquid chromatography
LC-SCX	silica-based strong cation exchange
LOD	limit of detection
MAHs	monocyclic aromatic hydrocarbons
MALDI	matrix-assisted laser desorption ionizatoin
MAX	strong anion exchange
MCX	strong cation exchange
MD	microdialysis
MH	malignant hyperthermia
MHN	MH-negative patients
MHS	MH-susceptible patients
MOI	microfluidic open interface
MS/MS	tandem mass spectrometry
nano-ESI	nano-electrospray ionization
OPP	open port probe
PA	polyacrylate
PAHs	polycyclic aromatic hydrocarbons
PAN	polyacrylonitrile
PBA	phenylboronic acid
PDMS	polydimethylsiloxane
PEG	polyethylene glycol
PIL	polymeric ionic liquid
poly([DDMGlu][MTFSI])	*N*,*N*-didecyl-*N*-methyl-D-glucaminium poly(2-methyl-acrylic acid 2-[1-(3-{2-[2-(3-trifluoromethanesulfonylamino-propoxy)-ethoxy]-ethoxy}-propylamino)-vinylamino]-ethyl ester)
poly([VBHDIM][NTf2])	poly(1–4-vinylbenzyl-3-hexadecylimidazolium) bis[(trifluoromethyl)sulfonyl] imide
POPs	persistent organic pollutants
PP	protein precipitation
PPCPs	personal care products
PS-DVB	cross-linked polystyrene–divinylbenzene
RAM	restricted access material
ROV	remotely operated vehicle
RSD	relative standard deviation
SCX	strong cation exchange

SPE	solid phase extraction
SPME	solid phase microextraction
SSRI	selective serotonin reuptake inhibitor
TCA cycle	tricarboxylic acid cycle
TFME	thin film microextraction
ToF	time of flight
UF	ultrafiltration
Vac-HS-SPME	vacuum-assisted headspace solid phase microextraction
V_e	volume of the extraction phase
VOCs	volatile organic compounds
V_s	sample volume
WAX	weak anion exchange
β-NAD	β-nicotinamide adenine dinucleotide

Acknowledgements

The authors would like to thank Professor Janusz Pawliszyn for his valuable comments during preparation of the chapter. Barbara Bojko and Ezel Boyaci would like to acknowledge the COST Action CA 16215 for the opportunity of collaboration.

References

1. J. Pawliszyn, *Handbook of Solid Phase Microextraction*, Chemical Industry Press, Beijing, China, 1st edn, 2009.
2. N. Reyes-Garcés, E. Gionfriddo, G. A. Gómez-Ríos, M. N. Alam, E. Boyacı, B. Bojko, V. Singh, J. Grandy and J. Pawliszyn, *Anal. Chem.*, 2018, **90**, 302.
3. W. Filipiak and B. Bojko, *TrAC, Trends Anal. Chem.*, 2019, **115**, 203.
4. N. Reyes-Garcés and E. Gionfriddo, *TrAC, Trends Anal. Chem.*, 2019, **113**, 172.
5. B. Bojko, N. Reyes-Garcés, V. Bessonneau, K. Goryński, F. Mousavi, E. A. Souza Silva and J. Pawliszyn, *TrAC, Trends Anal. Chem.*, 2014, **61**, 168.
6. A. Roszkowska, N. Miękus and T. Bączek, *J. Sep. Sci.*, 2019, **42**, 285.
7. M. Hemmati, C. Nix, J. Crommen, A.-C. Servais and M. Fillet, *TrAC, Trends Anal. Chem.*, 2020, **127**, 115899.
8. F. S. Mirnaghi, K. Goryński, A. Rodriguez-Lafuente, E. Boyaci, B. Bojko and J. Pawliszyn, *J. Chromatogr. A*, 2013, **1316**, 37.
9. M. F. Musteata, M. Walles and J. Pawliszyn, *Anal. Chim. Acta*, 2005, **537**, 231.
10. S. Perera, A. Berthod, E. Dodbiba and D. W. Armstrong, *Rapid Commun. Mass Spectrom.*, 2012, **26**, 853.
11. B. Onat, H. Rosales-Solano and J. Pawliszyn, *Anal. Chem.*, 2020, **92**, 9379.
12. X. Guo, T. Ye, L. Liu and X. Hu, *J. Sep. Sci.*, 2016, **39**, 1533.

13. H. Hashemi-Moghaddam and M. Hagigatgoo, *Chromatographia*, 2015, **78**, 1263.
14. F. Mousavi, B. Bojko, V. Bessonneau and J. Pawliszyn, *J. Proteome Res.*, 2016, **15**, 963.
15. S. M. Z. Hossain, B. Bojko and J. Pawliszyn, *Anal. Chim. Acta*, 2013, **776**, 41.
16. K. Jaroch, E. Boyaci, J. Pawliszyn and B. Bojko, *Sci. Rep.*, 2019, **9**, 402.
17. T. Moriya, Y. Satomi and H. Kobayashi, *Metabolomics*, 2016, **12**, 179.
18. D. Vuckovic, I. de Lannoy, B. Gien, R. E. Shirey, L. M. Sidisky, S. Dutta and J. Pawliszyn, *Angew. Chem., Int. Ed.*, 2011, **50**, 5344.
19. G. Ouyang and J. Pawliszyn, *Anal. Chim. Acta*, 2008, **627**, 184.
20. S. Risticevic, E. A. Souza-Silva, E. Gionfriddo, J. R. DeEll, J. Cochran, W. S. Hopkins and J. Pawliszyn, *Sci. Rep.*, 2020, **10**, 6724.
21. S. Risticevic, E. A. Souza-Silva, J. R. DeEll, J. Cochran and J. Pawliszyn, *Anal. Chem.*, 2016, **88**, 1266.
22. A. Roszkowska, M. Yu, V. Bessonneau, J. Ings, M. McMaster, R. Smith, L. Bragg, M. Servos and J. Pawliszyn, *Environ. Pollut.*, 2019, **249**, 109.
23. A. Roszkowska, M. Yu, V. Bessonneau, L. Bragg, M. Servos and J. Pawliszyn, *J. Environ. Sci. Technol. Lett.*, 2018, **5**, 431.
24. A. Roszkowska, M. Yu, V. Bessonneau, L. Bragg, M. Servos and J. Pawliszyn, *Sci. Rep.*, 2018, **8**, 6980.
25. A. Napylov, N. Reyes-Garces, G. Gomez-Rios, M. Olkowicz, S. Lendor, C. Monnin, B. Bojko, C. Hamani, J. Pawliszyn and D. Vuckovic, *Angew. Chem., Int. Ed.*, 2020, **59**, 2392.
26. J. J. Grandy, B. Onat, V. Tunnicliffe, D. A. Butterfield and J. Pawliszyn, *Sci. Rep.*, 2020, **10**, 1360.
27. M. Kim, J. Lee, C. H. Yang and S. Lee, *Anal. Chim. Acta*, 2016, **923**, 55.
28. E. Cudjoe, B. Bojko, I. Delannoy, V. Saldivia and J. Pawliszyn, *Angew. Chem., Int. Ed.*, 2013, **52**, 12124.
29. E. Boyaci, S. Lendor, B. Bojko, N. Reyes-Garces, G. Gomez-Rios, M. Olkowicz, M. Diwan, M. Palmer, C. Hamani and J. Pawliszyn, *ACS Chem. Neurosci.*, 2020, **11**(22), 3749–3760.
30. H. Piri-Moghadam, F. Ahmadi, G. A. Gómez-Ríos, E. Boyacı, N. Reyes-Garcés, A. Aghakhani, B. Bojko and J. Pawliszyn, *Angew. Chem., Int. Ed.*, 2016, **55**, 7510.
31. T. Vasiljevic, V. Singh and J. Pawliszyn, *Talanta*, 2019, **199**, 689.
32. J. Deng, Y. Yang, M. Xu, X. Wang, L. Lin, Z.-P. Yao and T. Luan, *Anal. Chem.*, 2015, **87**, 9923.
33. S. Lendor, S.-A. Hassani, E. Boyaci, V. Singh, T. Womelsdorf and J. Pawliszyn, *Anal. Chem.*, 2019, **91**, 4896.
34. S.-A. Hassani, S. Lendor, E. Boyaci, J. Pawliszyn and T. Womelsdorf, *J. Neurophysiol.*, 2019, **122**, 1649.
35. J. Bogusiewicz, P. Z. Goryńska, K. Łuczykowski, K. Burlikowska, K. Jaroch, K. Chmara, K. Goryński, D. Paczkowski, J. Furtak, M. Harat, J. Pawliszyn and B. Bojko, *Presented in Part at XX Euroanalysis, Istanbul, Turkey*, September, 2020.

36. J. Pawliszyn, *Presented in Part at 21st International Symposium on Advances in Extraction Technologies ExTech 2–19, Guangzhou, China*, November, 2019.
37. W. Li and F. L. S. Tse, *Biomed. Chromatogr.*, 2010, **24**, 49.
38. F. S. Mirnaghi and J. Pawliszyn, *Anal. Chem.*, 2012, **84**, 8301.
39. G. A. Gómez-Ríos, N. Reyes-Garcés, B. Bojko and J. Pawliszyn, *Anal. Chem.*, 2016, **88**, 1259.
40. G. A. Gómez-Ríos and M. F. Mirabelli, *TrAC, Trends Anal. Chem.*, 2019, **112**, 201.
41. B. Hu, B. Zheng, D. Rickert, G. A. Gómez-Ríos, B. Bojko, J. Pawliszyn and Z.-P. Yao, *Anal. Chim. Acta*, 2019, **1075**, 112.
42. G. A. Gómez-Ríos, C. Liu, M. Tascon, N. Reyes-Garcés, D. W. Arnold, T. R. Covey and J. Pawliszyn, *Anal. Chem.*, 2017, **89**, 3805.
43. M. Tascon, M. N. Alam, G. A. Gómez-Ríos and J. Pawliszyn, *Anal. Chem.*, 2018, **90**, 2631.
44. G. A. Gómez-Ríos and J. Pawliszyn, *Angew. Chem., Int. Ed.*, 2014, **53**, 14503.
45. G. A. Gómez-Ríos and J. Pawliszyn, *Chem. Commun.*, 2014, **50**, 12937.
46. P. A. D'Agostino, J. R. Hancock, C. L. Chenier and C. R. J. Lepage, *J. Chromatogr. A*, 2006, **1110**, 86.
47. S. Lendor, G. A. Gómez-Ríos, E. Boyaci, H. Vander Heide and J. Pawliszyn, *Anal. Chem.*, 2019, **91**, 10141.
48. J. Deng, W. Li, Q. Yang, Y. Liu, L. Fang, Y. Guo, P. Guo, L. Lin, Y. Yang and T. Luan, *Anal. Chem.*, 2018, **90**, 6936.
49. Z. Takáts, J. M. Wiseman and R. G. Cooks, *J. Mass Spectrom.*, 2005, **40**, 1261.
50. É. A. Souza-Silva, E. Gionfriddo and J. Pawliszyn, *TrAC, Trends Anal. Chem.*, 2015, **71**, 236.
51. H. Xu, G. S. Chen, Z. H. Xiong, Y. X. Fan, X. C. Wang and Y. Liu, *TrAC, Trends Anal. Chem.*, 2016, **80**, 12.
52. L. Setkova, S. Risticevic and J. Pawliszyn, *J. Chromatogr. A*, 2007, **1147**, 224.
53. S. Risticevic, J. R. DeEll and J. Pawliszyn, *J. Chromatogr. A*, 2012, **1251**, 208.
54. S. Risticevic and J. Pawliszyn, *Anal. Chem.*, 2013, **85**, 8987.
55. E. Gionfriddo, E. A. Souza-Silva and J. Pawliszyn, *Anal. Chem.*, 2015, **87**, 8448.
56. M. J. Trujillo-Rodríguez, H. Nan, M. Varona, M. N. Emaus, I. D. Souza and J. L. Anderson, *Anal. Chem.*, 2019, **91**, 505.
57. E. Gionfriddo, É. A. Souza-Silva, T. D. Ho and J. L. Anderson, *Talanta*, 2018, **188**, 522.
58. Z. Tang and Y. Duan, *Talanta*, 2017, **172**, 45.
59. N. P. Brunton, D. A. Cronin and F. J. Monahan, *Flavour Fragrance J.*, 2001, **16**, 294.
60. E. Psillakis, *Anal. Chim. Acta*, 2017, **986**, 1.

61. M. Sýkora, E. Vítová and H. H. Jeleń, *Eur. Food Res. Technol.*, 2020, **246**, 573.
62. É. A. Souza Silva and J. Pawliszyn, *Anal. Chem.*, 2012, **84**, 6933.
63. É. A. Souza-Silva and J. Pawliszyn, *J. Agric. Food Chem.*, 2015, **63**, 4464.
64. É. A. Souza-Silva, E. Gionfriddo, R. Shirey, L. Sidisky and J. Pawliszyn, *Anal. Chim. Acta*, 2016, **920**, 54.
65. Z. Bai, A. Pilote, P. K. Sarker, G. Vandenberg and J. Pawliszyn, *Anal. Chem.*, 2013, **85**, 2328.
66. S. Mirnaghi, F. Mousavi, S. M. Rocha and J. Pawliszyn, *J. Chromatogr. A*, 2013, **1276**, 12.
67. G. Chen, J. Qiu, J. Xu, X. Fang, Y. Liu, S. Liu, S. Wei, R. Jiang, T. Luan, F. Zeng, F. Zhu and G. Ouyang, *Chem. Sci.*, 2016, **7**, 1487.
68. E. Gionfriddo, E. Boyacl and J. Pawliszyn, *Anal. Chem.*, 2017, **89**, 4046.
69. A. Roszkowska, M. Yu and J. Pawliszyn, *A New Paradigm for Environmental Chemistry and Toxicology: From Concepts to Insights*, Springer, Singapore, 2020, vol. 1.
70. C. Wild, *Cancer Epidemiol., Biomarkers Prev.*, 2005, **14**, 1847.
71. C. S. Bloszies and O. Fiehn, *Curr. Opin. Toxicol.*, 2018, **8**, 87.
72. T. J. Athersuch and H. C. Keun, *Mutagenesis*, 2015, **30**, 755.
73. R. Barouki, E. Melén, Z. Herceg, J. Beckers, J. Chen, M. Karagas, A. Puga, Y. Xia, L. Chadwick, W. Yan, K. Audouze, R. Slama, J. Heindel, P. Grandjean, T. Kawamoto and K. Nohara, *Environ. Int.*, 2018, **114**, 77.
74. M. Llompart, M. Celeiro, C. García-Jares and T. Dagnac, *TrAC, Trends Anal. Chem.*, 2019, **112**, 1.
75. V. Bessonneau, B. Bojko and J. Pawliszyn, *J. Chromatogr. A*, 2014, **1328**, 113.
76. V. Bessonneau, J. Ings, M. McMaster, R. Smith, L. Bragg, M. Servos and J. Pawliszyn, *Sci. Rep.*, 2017, **7**, 44038.
77. B. Bojko, B. Onat, E. Boyaci, E. Psillakis, T. Dailianis and J. Pawliszyn, *Front. Mar. Sci.*, 2019, **6**, 632.
78. V. Bessonneau, J. Ings, M. McMaster, R. Smith, L. Bragg, M. Servos and J. Pawliszyn, *Environ. Res.*, 2016, **151**, 216.
79. J. Xu, G. Chen, S. Huang, J. Qiu, R. Jiang, F. Zhu and G. Ouyang, *TrAC, Trends Anal. Chem.*, 2016, **85**, 26.
80. F. Mousavi, E. Gionfriddo, E. Carasek, E. A. Souza-Silva and J. Pawliszyn, *Metabolomics*, 2016, **12**, 169.
81. T. Abaffy, M. G. Möller, D. D. Riemer, C. Milikowski and R. A. DeFazio, *Metabolomics*, 2013, **9**, 998.
82. N. Reyes-Garcés, M. Diwan, E. Boyaci, G. A. Gómez-Ríos, B. Bojko, J. N. Nobrega, F. R. Bambico, C. Hamani and J. Pawliszyn, *Anal. Chem.*, 2019, **91**, 9875.
83. K. Burlikowska, I. Stryjak, J. Bogusiewicz, B. Kupcewicz, K. Jaroch and B. Bojko, *Metabolites*, 2020, **10**, 255.
84. I. Stryjak, N. Warmuzińska, M. Hamar, M. Selzner and B. Bojko, *Presented in Part at 30th International Symposium on Pharmaceutical and Biomedical Analysis, Tel Aviv, Israel*, September, 2019.

85. I. Stryjak, N. Warmuzińska, K. Łuczykowski, M. Hamar, P. Urbanellis, E. Wojtal, M. Masztalerz, M. Selzner, Z. Włodarczyk and B. Bojko, *J. Visualized Exp.*, 2020, **160**, 1.
86. I. Stryjak, N. Warmuzińska, J. Bogusiewicz, K. Łuczykowski and B. Bojko, *J. Sep. Sci.*, 2020, **43**, 1867.
87. B. Bojko, N. Looby, M. Olkowicz, A. Roszkowska, B. Kupcewicz, P. Reck des Santos, K. Ramadan, S. Keshavjee, T. K. Waddell, G. Gómez-Ríos, M. Tascon, K. Goryński, M. Cypel and J. Pawliszyn, *J. Pharm. Anal.*, 2021, **11**, 37.
88. J. Bogusiewicz, P. Z. Goryńska, M. Gaca, K. Chmara, K. Goryński, K. Jaroch, D. Paczkowski, J. Furtak, M. Harat and B. Bojko, *J. Visualized Exp.*, 2020, **159**, 1.
89. B. Bojko, T. Vasiljevic, E. Boyaci, A. Roszkowska, N. Kraeva, C. A. Ibarra Moreno, A. Koivu, M. Wąsowicz, A. Hanna, S. Hamilton, S. Riazi and J. Pawliszyn, *Can. J. Anaesth.*, 2021, DOI: 10.1007/s12630-020-01895-y.
90. D. Vuckovic and J. Pawliszyn, *Anal. Chem.*, 2011, **83**, 1944.
91. F. Mousavi, B. Bojko and J. Pawliszyn, *Anal. Chim. Acta*, 2015, **892**, 95.
92. F. Mousavi, B. Bojko, V. Bessonneau and J. Pawliszyn, *J. Proteome Res.*, 2016, **15**, 963.
93. F. Mousavi, E. Gionfriddo, E. Carasek, E. A. Souza-Silva and J. Pawliszyn, *Metabolomics*, 2016, **12**, 169.
94. H. Kataoka, R. Inoue, K. Yagi and K. Saito, *J. Pharm. Biomed. Anal.*, 2009, **49**, 108.
95. T. Inukai, S. Kaji and H. Kataoka, *J. Pharm. Biomed. Anal.*, 2018, **156**, 272.

CHAPTER 6

State-of-the-art Capillary Electrophoresis Mass Spectrometry Methods for Analyzing the Polar Metabolome

KYLE D. DUNCAN AND INGELA LANEKOFF*

Department of Chemistry – BMC, Uppsala University, Uppsala, Sweden
*Email: ingela.lanekoff@kemi.uu.se

6.1 Introduction

Metabolites are small biomolecules such as lipids, sugars, and amino acids that are involved in all cellular and biological functions. Small changes in physiological levels of critical metabolites can cause significant health effects and be indicative of disease diagnoses or progression. As a result, capturing a snapshot of an organism's dynamic metabolome can be highly informative and help understand fundamental biological processes or provide objective clinical diagnoses of diseases. Similar to genomics, transcriptomics, and proteomics, profiling and characterizing an organism's metabolome has become a prominent field at the interface between chemistry and biology. However, when compared to the four nucleotide bases sequenced in genomics, or the 20 essential amino acid building blocks in proteomics, the structural diversity of metabolite molecules is vast and can range from highly charged polar species to bulky hydrophobic structures.[1] Moreover, in addition to the high structural complexity, cellular metabolites can range in concentration over several orders

New Developments in Mass Spectrometry No. 10
Advanced Mass Spectrometry-based Analytical Separation Techniques
for Probing the Polar Metabolome
Edited by Rawi Ramautar

Published by the Royal Society of Chemistry, www.rsc.org

of magnitude (*e.g.*, <1 nM to >1 μM) and their abundances are highly dynamic (*e.g.*, metabolic changes in seconds).[2,3]

Given these known challenges in measuring metabolites in biological samples, analytical techniques built to profile an organism's metabolome must be capable of detecting a wide range of structurally diverse metabolites over a wide range of concentrations from ultra-trace to highly abundant species. Several nicely written recent reviews cover the full spectrum of metabolomics workflows and the important information that can be obtained from different biological samples.[4–8]

Of the available techniques used to screen large numbers of metabolites from biological samples, mass spectrometry (MS) has proliferated into a primary method of choice. This is due to the capability for MS to provide label-free simultaneous quantification of hundreds to thousands of metabolites at ultra-trace levels with a high degree of accuracy and precision. Additionally, at high MS mass accuracy and resolution, molecular formulas can be assigned to detected peaks, which can ultimately simplify molecular annotation of unknown features in untargeted metabolomic studies. However, even with the high selectivity of MS, separation techniques such as chromatography, hydrophobic interaction chromatography, or capillary electrophoresis are usually implemented in metabolomic workflows due to metabolite structural complexity, ranges in metabolite abundance, and for reduction in sample matrix effects. In particular, certain separation techniques are best suited for targeting subsections of the comprehensive metabolome. For example, reversed phase liquid chromatography is advantageous for separating hydrophobic metabolites and lipids, and not ideal for separating highly polar and charged metabolites. Conversely, capillary electrophoresis is an ideal method for separating polar and charged metabolites, as CE separation is based on the variable electrophoretic mobilities of charged species.

The polar subset of the known metabolome includes small and hydrophilic amino acids, nucleic acids, sugars, and other small organic acids that are heavily involved in primary and secondary metabolism, and many other fundamental biological pathways. This makes the polar metabolome a highly interesting and important metabolite class to study and characterize. Henceforth, this chapter discusses current CE–MS methodology used to screen the polar metabolome and highlights recent biological studies (2013 and onward) that employ CE–MS to probe the polar metabolome. We cover traditional CE–MS biological sample analysis such as biofluids (*e.g.*, urine, plasma, and serum), cell and tissue extracts, and place particular emphasis on probing the polar metabolites of minute biological samples. These minute samples include limited biofluid and tissue extracts (*e.g.*, <10 μL or 2 mg), CE–MS analysis of single cells, and spatially defined CE–MS analysis directly from cellular unique regions in tissues. Throughout this chapter, particular emphasis is placed on the sample preparation steps required, CE–MS instrumentation used, types of metabolites detected, and biological insights garnered from the polar metabolite analyses.

6.1.1 Capillary Electrophoresis Mass Spectrometry

Capillary electrophoresis is an analytical separation technique that has undergone rapid growth since the first modern CE applications in the early 1980s. In CE analysis, a small volume of sample (*e.g.*, a few nL) is loaded into a separation capillary containing background electrolyte (BGE), and a high potential is applied between the sample inlet and outlet end of the separation capillary. Analytes migrate through the capillary based on their electrophoretic mobility, which is generally correlated to their molecular size and charge. For example, smaller molecules with a higher charge density will migrate faster than larger molecules with a lower charge density. Key figures of merit for CE separation include very narrow migration bands with high amounts of theoretical plates and the capacity to inject very low sample volumes. CE separation can therefore generate high quality baseline separated peaks for many analytes in limited sample volumes. Moreover, CE–MS is well suited for separation of polar charged molecules, rather than non-polar neutral species often separated by reversed phase liquid chromatography. For general background and current state-of-the-art on the fundamentals of CE separation, several reviews nicely summarize current progress and recent trends over the recent years.[9,10] While CE has been coupled to many different detectors, including UV–vis absorbance, fluorescence, and electrochemical detection, coupling CE to mass spectrometry (CE–MS) has gained considerable attention due to the specificity and selectivity of modern mass spectrometers.

MS allows for the simultaneous label-free quantification of hundreds of molecules with detection limits often in the pM to nM range. Furthermore, co-migrating electrophoretic peaks can be distinguished by MS as long as the mass resolution of the mass spectrometer is adequate to differentiate co-migrating mass-to-charge ratios. Additionally, when high resolution and high mass accuracy MS instrumentation is available, molecular formulas can be assigned to electrophoretic peaks, which can significantly help with peak annotation in untargeted CE–MS analyses. For background information in pairing CE to MS, readers are directed to several recent review articles that cover CE–MS interface designs, CE–MS based metabolomics, and CE–MS technology developments.[11–15]

Despite the growing interest and high potential for CE–MS to be implemented in clinical and routine measurements of polar metabolites, there has been resistance to replacing known, reliable, and routine techniques commonly employed, such as reversed phase liquid chromatography. This is partly due to the perceived complexity in interfacing CE to MS, where there are limited commercially available systems that are thought to require highly trained personnel for routine monitoring scenarios. Therefore, in the hope to further expand the user base of CE–MS, this section of the chapter will cover common CE–MS parameters, including sample injection strategies, common CE–MS interface designs, and considerations for mass spectrometer selection. Specifically, these points will be discussed through the lens of using CE–MS to measure and screen polar metabolites from biological samples.

6.1.2 Sample Injection Techniques

6.1.2.1 Hydrodynamic and Electrokinetic Injection

Sample loading in CE is generally carried out *via* one of two primary strategies: hydrodynamic or electrokinetic injection. In both of these approaches, the CE capillary is first flushed with clean background electrolyte (BGE), then moved to a separate sample vial for analyte loading, and finally returned to the BGE vial to facilitate CE separation. While hydrodynamic and electrokinetic sample injection approaches are both effective methods to load sample into the CE capillary prior to electrophoretic separation, several factors must be taken into account to ensure the correct strategy is selected.

Hydrodynamic loading can be implemented by applying a positive pressure to the inlet, a negative pressure to the outlet, or by raising the sample inlet height in relation to the capillary outlet. Accordingly, hydrodynamic sample loading volumes are dependent on injection time, sample height displacement, and/or the pressure applied to the sample. In general, hydrodynamic loading offers a robust and reproducible means for sample injection and often a higher degree of certainty for injected amount when compared to electrokinetic injection, with the injection RSD value typically below 2%.[16,17] One key feature of hydrodynamic loading is that the sample injected into the CE capillary remains in the same condition as the original sample. This ensures uniform injection for all analytes that is not biased to molecules with high or low electrophoretic mobility.

Electrokinetic injection uses the electroosmotic flow and electrophoretic mobility of analytes to load the CE capillary with sample. For electrokinetic injection, a high voltage is applied to the sample, and molecules are actively drawn into the CE capillary by electromigration and electroosmotic flow of the sample into the CE capillary. Moreover, the injection voltage can be optimized for specific applications, where low voltage electrokinetic injections over a longer time period have been shown to improve injection reproducibility compared to high voltage injections over a shorter time period.[17] However, as analytes are actively drawn into the CE capillary, electrokinetic injection is biased to molecules with high electrophoretic mobilities. Specifically, high mobility species are injected to a greater extent when compared to species with low mobilities. Electrokinetic loading is therefore very susceptible to matrix effects and changes in sample ionic strength, which will directly affect the amount of analyte loaded into the CE capillary. As a result, high ionic strength biological samples need significant sample pre-treatment, such as liquid–liquid extraction for reliable electrokinetic injection. To further minimize bias, users must make sure to matrix match samples and standards and/or include several internal standards over the migration window to account for variable analyte injections. However, while electrokinetic sample loading introduces a known bias, it can drastically improve sensitivity compared to hydrodynamic loading. For example, with hydrodynamic loading, increasing the percentage of injected sample into the capillary volume directly increases the sensitivity (*e.g.*, 10× more sample corresponds to 10× more signal). Conversely, with electrokinetic injection, it is possible to yield up to 1000× improvement in sensitivity with longer injection times.[16] Additional gains in sensitivity for electrokinetic injection are possible by implementing field amplified sample stacking, which exploits differences in electric field between the sample and electrolyte.[18,19] Field amplified sample stacking from low conductivity samples can provide greater than 1000× enhancement in sensitivity.

Ultimately, the selection of the appropriate CE sample loading method depends on the analytical demands for the application, the composition of the sample, and the abundance of the analytes of interest. For robust and reproducible sample injection over many different sample types, hydrodynamic loading is often preferred. However, for studying high mobility metabolites at low biological abundance, electrokinetic sample loading could be an ideal choice.

6.1.2.2 Multi-segment Injection for Higher Throughput Analysis

One current bottleneck in CE–MS metabolomic workflows is the relatively low sample throughput. Often measurement duty cycles for samples can be over 40 minutes, including capillary flushing with BGE between each sample. However, CE separations run over isocratic conditions with a uniform electric field, unlike gradient chromatographic methods that rely on changing the mobile phase composition to reduce analyte–column interactions. Taking advantage of the static CE separation conditions, Kuehnbaum and co-workers increased CE–MS sample throughput by developing multi-segment injection CE–MS (MSI-CE–MS), where several sample plugs are loaded hydrodynamically prior to CE separation.[20] This concept is illustrated in Figure 6.1, where each sample segment is sandwiched between a BGE spacer. Upon CE zonal separation, individual analyte electrophoretic peaks are detected with equal spacing using a high-resolution time-of-flight mass analyzer. Moreover, online sample dilutions or standard additions can be carried out to determine method linearity and unambiguously distinguish endogenous metabolite features from matrix and background peaks. In summary, implementation MSI-CE–MS notably improves CE–MS sample throughput by using a single CE–MS separation event to provide robust sample replicate measurements for improved biological assessments. The authors further state that MSI-CE–MS provides similar sample throughputs to methods without any separation and provides additional bioinformatic merits without isotopic labeling, hardware modifications, or infrastructure investments beyond a commercially available CE–MS instrument.

6.1.3 Capillary Electrophoresis Mass Spectrometry Interface Designs

When coupling CE to MS, the most popular ionization source employed is electrospray ionization (ESI). Electrospray is well-suited for transferring

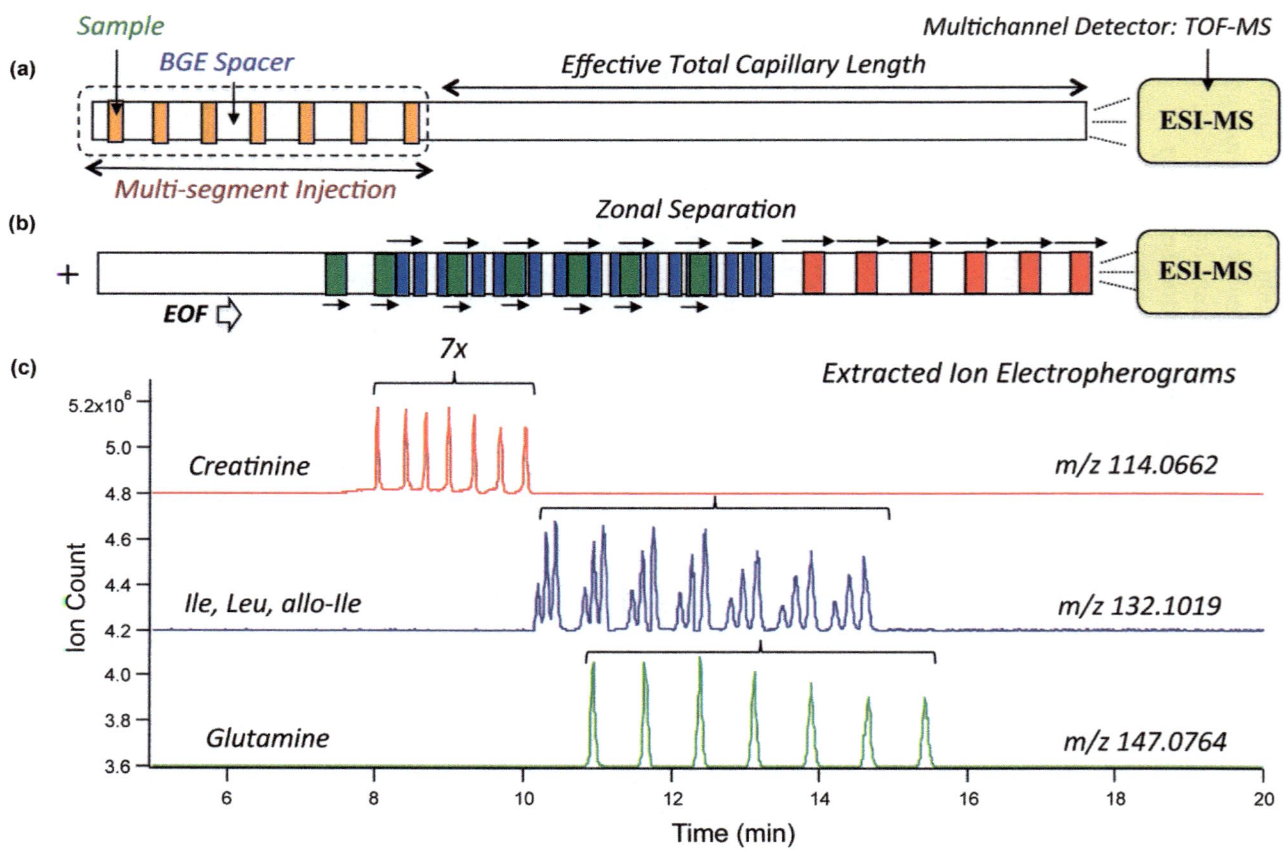

Figure 6.1 (a) Multiplexed separation based on serial injection of seven discrete sample segments within a single capillary by MSI-CE–MS, where (b) ions migrate as a series of zones in free solution prior to ionization. (c) This format enables reliable quantification of polar metabolites and their isomers in different samples since ionization occurs within a short time interval (≈2–6 min) under steady-state conditions when using a high mass resolution TOF-MS.
Reproduced from ref. 20 with permission from American Chemical Society, Copyright 2013.

charged analytes from the liquid phase directly to the gas phase for MS analysis. Moreover, due to the low BGE flowrates used in CE separation (low nL min^{-1}), ionization suppression effects often observed in ESI are typically less apparent than for higher flow rate chromatographic separation methods. However, the coupling of CE to ESI-MS remains a challenging endeavor. Rather than simply placing the CE capillary outlet in a waste vial as in conventional CE separation coupled to a UV detector, the capillary outlet must be set up to consistently electrospray toward the MS while keeping the CE separation circuit closed. This often entails adding a supplemental ESI voltage to the ESI emitter in addition to the high separation voltage added to the sample. Furthermore, the magnitude of the supplemental ESI voltage must be tuned to match the solvent flowrate out of the CE–MS interface. In general, two strategies have been adopted to interface CE to ESI-MS, including a co-axial sheath flow type interface and a sheathless emitter interface design; both discussed in more detail below.

6.1.3.1 Sheath Flow

The co-axial sheath flow CE–MS interface was one of the first developed methods to interface CE to MS and remains to be the standard interface for many users that is commercially available.[21] In this interface design, the high separation potential is supplied to the inlet BGE vial and connected to a conductive sheath liquid (*e.g.*, 1 : 1 methanol : water with 0.1% acetic acid) that co-axially flows on the outside of the CE capillary. At the emitter tip, the co-axial sheath liquid mixes with the CE capillary eluent to complete the CE circuit, and migrating analytes, BGE, and sheath liquid are simultaneously electrosprayed toward the MS for analysis (Figure 6.2). The sheath liquid can either be pneumatically driven, or delivered with a syringe pump. Further, the CE separation capillary can either protrude out of the outer sheath flow interface where the Taylor cone forms as in Figure 6.2, or inserted into a tapering emitter tip as described by Lindenburg and others.[11] The continuously flowing conductive sheath liquid provides a simple way to complete the CE circuit that is stable for many consecutive sample injections.

Since the early development of the sheath flow CE–MS interface, several advancements have been published. A study by Lindenburg *et al.* implemented a microvial interface to lower the sheath liquid flow rate for metabolome analysis.[22] It was found that by lowering the sheath liquid flow rate to 100–500 nL min^{-1} metabolite detection limits were significantly reduced, and the number of molecular features identified from single zebra fish embryos nearly doubled compared to the conventional sheath flow interface. Another study by Sasaki *et al.* developed a CE-ESI adapter for coupling a sheath flow CE interface to ThermoFisher Orbitrap instruments.[23] Specifically, a grounded nebulizer was used to apply an independent ESI voltage to minimize droplet formation at the capillary tip and provide a more stable electrospray. Using their sheath flow CE–MS adapter, high resolution CE separations were achieved for polar metabolites extracted from plasma, and

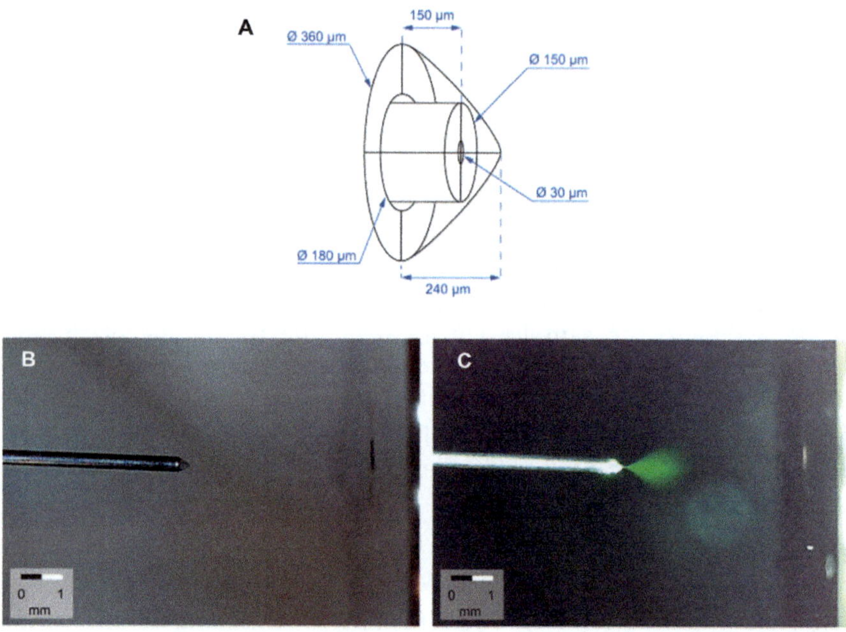

Figure 6.2 Diagram showing the measurements employed to define the cone geometry for the numerical simulations (A). Microphotographies of the Taylor cones and the electrospray droplets offspring with white cenital (B) and laser (C) lighting.
Reproduced from ref. 21 with permission from John Wiley & Sons, Copyright © 2017 WILEY-VCH Verlag GmbH & Co. KGaA, Weinheim.

the authors claimed to identify up to 270 metabolites. Although the added liquid of the sheath flow dilutes the analytes prior to detection, sheath flow CE–MS interfaces provide a robust way to pair CE with ESI–MS.

6.1.3.2 *Sheath-less Interfaces*

To avoid dilution of migrating analyte electrophoretic peaks in sheath liquids, another major class of CE–MS interfaces employ a sheath-less design. Sheath-less interfaces do not utilize a sheath liquid to complete the CE separation circuit, rather, the circuit between the sample and CE capillary outlet is connected directly to a conductive emitter tip for ESI. The most simplistic sheath-less CE–MS interface designs include coating a pulled CE capillary emitter tip with a conductive material such as gold foil,[24] vapor deposited metal,[25] and graphite.[26] In these sheath-less designs the CE circuit is completed by a thin film of BGE that flows out from the capillary and coats the conductive emitter tip. Although these sheath-less interfaces have been successful within particular research groups, none have been widely adopted or commercialized. One reason may be that the stability of sheath-less interfaces often deteriorates over time, and the quality of CE–MS runs may not be uniform for large numbers of consecutive sample analyses.

To improve sheath-less device robustness, a creative approach developed by Hirayama *et al.* constructed a sheath-less CE–MS interface based on a small microcrack in the capillary wall of a pulled emitter tip.[27] Here, the CE circuit between the sample and emitter tip was completed by a small amount of conductive buffer mixing with the BGE within the microcrack of the capillary emitter tip. In addition, a small dialysis membrane was added to reduce migrating metabolite loss into the conductive buffer. The authors argue that this cost-effective solution can be adapted to any commercial capillary, however, the reproducibility of microcracks in multiple capillary interfaces has yet to be demonstrated. Further expanding on this idea, a porous tip sheath-less emitter was recently established and developed into a commercial product by Beckman Coulter. The emitter tip of a capillary is etched with hydrofluoric acid for a controlled amount of time to create micro-holes within the fused silica walls of the CE capillary. As a result, the CE circuit is completed by connecting the capillary inlet high voltage to a conductive liquid in contact with the porous emitter tip. Electrical current can then travel between the outer conductive liquid and BGE in the CE capillary *via* small micro-holes in the emitter. Ramautar *et al.* demonstrated the utility of this sheath-less CE–MS interface over more conventional sheath flow setups.[28] Specifically, 300 molecular features were detected with a commercially available sheath flow platform, while the number of detected features was increased to 900 using the Beckman Coulter sheath-less interface. These results are visualized in Figure 6.3, where many more molecular features are observed in the density map corresponding to the sheath-less interface (Figure 6.3A). Some notable urinary metabolites exclusively detected by sheath-less CE–MS include tri-peptides, acylcarnitines, hypoxanthine, adenine, nicotinamide, serine, and proline. Biozard and co-workers further characterized a beveled version of the emitter tip by Beckman Coulter and compared a standard flat tip to their new beveled design.[29] After analyzing 10 consecutive urine sample injections for both beveled and unbeveled designs, they found that over 300 metabolites were detected in all ten injections using each setup. Although the beveled tip did not help identify new metabolite features, the signal intensities for detected metabolites were three-fold higher when compared to data generated with the standard flat tip emitter. The added sensitivity was not at the cost of tip robustness, as both designs were good for 40–50 runs, and the long-term performance of urine screening was observed to be quite stable over four years.

In summary, the stability of sheath-less CE–MS interfaces has come a long way over recent years. However, long-term stability for sheath-less CE–MS interfaces continues to be a challenge that can hinder large sample studies and routine monitoring applications. In addition, sheath-less CE–MS interfaces typically require highly trained personal to troubleshoot problems, which could hinder applications in routine metabolomic screening applications. However, given the demonstrated improvements in sensitivity

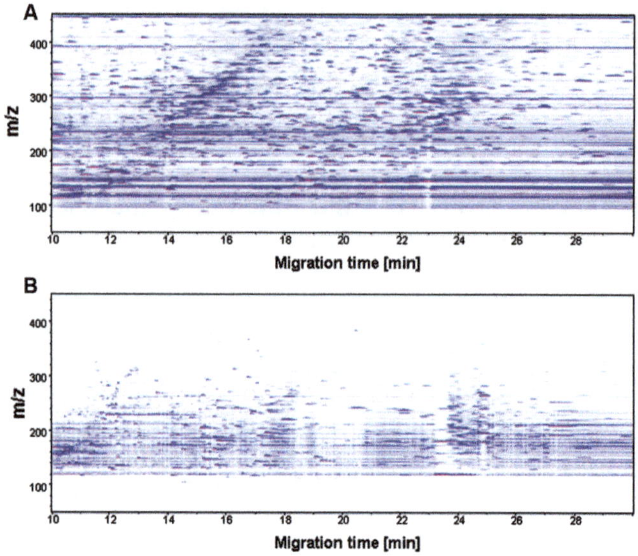

Figure 6.3 Density maps of the molecular features detected in human urine in terms of migration time, *m*/*z*, and ion intensity by sheath-less CE–MS using a porous tip sprayer (A) and sheath-liquid CE–MS (B). Experimental conditions: (A) BGE, 10% acetic acid (pH 2.2); sample injection, 2.0 psi for 30 s (~1% of capillary volume); (B) BGE, 10% acetic acid (pH 2.2); sample injection, 0.5 psi for 30 s (~1% of capillary volume).
Reproduced from ref. 28 with permission from American Chemical Society, Copyright 2011.

for sheath-less designs over the traditional sheath flow interfaces, the development of more robust sheath-less CE–MS designs will likely attract future users. Furthermore, the sensitivity of sheath-less CE–MS interfaces is ideal for studies globally profiling changes in an organism's metabolome, because less biologically abundant species can be detected alongside highly concentrated central metabolites such as amino acids and sugars.

6.1.4 Mass Spectrometer Selection

In addition to selecting a suitable CE–MS interface design, the choice of MS analyzer is equally important for establishing effective polar metabolite CE screening methods. Although CE separation capacity is very high with many theoretical plates, the migration bands in CE are typically quite narrow with electrophoretic peaks normally migrating over a few seconds. As a result, the mass spectrometer scan duty cycle must be fast enough to resolve the migrating electrophoretic peaks. Thus, mass spectrometers with fast duty cycles in the order of microseconds, such as time-of-flight (TOF) analyzers, are often paired with CE separation. The mass accuracy of TOF systems is typically quite high (<2 ppm), which can allow for molecular formulas to be assigned to migrating CE peaks. However, the mass resolving power of TOF analyzers is lower than competing Fourier transfer MS analyzers such as Orbitrap and Fourier Transform Ion Cyclotron Resonance (FTICR) instruments. As a result, CE migration peaks may represent the sum of several isobaric analytes. Therefore, for a higher degree of confidence in untargeted CE–MS based polar metabolomics, high mass resolving instruments, such as Orbitrap or FTICR MS, are used at the cost of lower scan duty cycles. Standard resolving powers for Orbitrap and FTICR analyzers are around 100 000 and >200 000, respectively, and typical Orbitrap and FTICR scan duty cycles are ~2 Hz and ~1 Hz; although higher end systems can achieve slightly faster scan times. For additional confidence in molecular assignments, tandem mass spectrometry experiments using TOF, Orbitrap, and FTICR analyzers will further lower the scan rate by approximately two times. Furthermore, each of the previously listed mass analyzers is often limited to MS^2 (MS/MS). For molecular confirmation using MS^n analysis, lower mass resolving ion trap MS systems are needed. Although annotation is improved, MS^n duty cycles are typically >1.5 Hz, which will limit the number of scans that define electrophoretic peaks. Overall, the selection of MS analyzer depends on the application. For a balance between scan duty cycle and mass accuracy, CE is often interfaced to TOF MS systems. If higher specificity for untargeted metabolite screening is desired, CE separation should be paired with Orbitrap or FTICR mass analyzers. Finally, if MS^n specificity is needed for confident molecular annotation, then ion trap mass spectrometers are best suited, provided the scan duty cycle is sufficient to resolve the migrating CE peaks.

6.2 Recent Advances in CE–MS Profiling of the Polar Metabolome

Over recent years there have been many applications of CE–MS based polar metabolomics reported in the literature. This section provides example studies published since the beginning of 2013 that have used CE–MS to measure metabolites in complex chemical samples ranging from biofluids to whole tissue extracts. The CE–MS interface employed, required sample preparation steps, and biological information obtained in each example study are highlighted to showcase the utility of CE–MS in modern polar metabolomics.

6.2.1 Biofluids

6.2.1.1 Urine

Urine analysis is central in clinical diagnostics due to easy and non-invasive sample collection, and the large sample quantities that can be obtained from an individual with minimal effort or undesired effects. Furthermore, urine has a relatively low sample complexity and protein content, making it highly suitable for rapid polar metabolomic workflows. In particular, low urine protein content allows for simplified sample preparation procedures, which often only entail sample dilution in water or BGE prior to CE–MS analysis. In addition, many polar metabolites in urine have been established as biomarkers for health and disease, making it a useful sample for profiling polar urine metabolomes in search of indicators of healthy or diseased states. Urine analysis by CE–MS reflects metabolite changes indicative of the entire organism's metabolism in equilibrium with the environment, including nutrition, drugs, and gut microbiome activities. Therefore, given the importance to profile the polar urine metabolome, several recent studies have highlighted key instrumental modifications and implemented CE–MS routines for urine metabolomics to further understand and identify disease pathologies and diagnostics.

Long-term separation repeatability and reliability are key parameters in developing clinical and biologically relevant CE–MS urine analysis methods. Recently, two studies have evaluated different systems for repeatability of many sequential sample injections over several days of sample analysis. First, an optimized CE–MS setup and data analysis pipeline for urine metabolome profiling over a period for several years was developed by Boizard and others.[29] This platform aimed at providing clinical relevance to urinary metabolome based-detection of obstructive nephropathy in infants. Through new CE–MS interface designs, the urinary polar metabolome was profiled from 15 healthy and 34 newborns with ureteropelvic junction obstruction (UPJ). Urine was diluted with denaturing solution (2 M urea, 0.0125% NH_4OH, 100 mM NaCl, and 0.01% SDS) and ultrafiltered using 20 kDa centrifugal filtering. Samples were then lyophilized and reconstituted in water. Of the 4879 molecular features detected, 92% or 4487 features were found to have an RSD below 10% for 128 QC injections over a three-year period, and nearly 3000 molecular features had an RSD below 5%. These results are highly promising for long-term screening applications and routine CE–MS analysis. Comparison of the urinary metabolome for healthy and UPJ newborns revealed 32 molecular features associated with UPJ, which could potentially serve as UPJ biomarkers. Another study looking at long-term CE–MS method stability was carried out by Yamamoto *et al.*, which aimed to reproducibly expand the range of measurable polar metabolites to anionic species over extended analysis times with CE–MS.[30] The authors employed multi-segment-injection with a sheath flow interface and showed high throughput (<5 min per sample) screening of 30 anionic urinary metabolites for over 230 repeated sample injections over 3 days. It was found that a BGE pH of less than 9 was essential to avoid aminolysis of the polyimide capillary coating and avoid capillary microcracks, which can significantly reduce method repeatability. In their study, urine samples were diluted 5× in deionized water and centrifuged to remove debris. Detected urinary metabolites included small organic acids such as lactic, uric, and hippuric acids, in addition to sulfonated and glucuronide metabolites. Overall, these two studies demonstrate the robust and repeatable separation of polar metabolites in urine over extended time periods, and the potential for CE–MS based clinical diagnostics and biomarker discovery.

Established and commercially available CE–MS systems have recently been used to profile polar urinary metabolome aberrations in many diseases ranging from coronary heart disease to gastric cancer, and search for potential diagnostic biomarkers.[31–33] Isbell *et al.* analyzed morphine and isobaric morphine glucuronide metabolites by CE–MS and optimized BGE conditions to 25 mM ammonium borate buffer at pH 9.5.[31] Sample preparation consisted of a 1 : 10 dilution of urine into BGE prior to analysis. They used a commercial sheath flow interface for analysis, and detection limits were below 3 $\mu g\, mL^{-1}$ for morphine and morphine glucuronide metabolites. Using their optimized method, it was found that a pilot study cohort had a ratio of 5 : 1 for M3G and M6G glucuronide metabolites in their urine. Another study by Cieslarova *et al.* combined CE–MS, dual diode array detection, and tandem mass spectrometry for the quantification of cardiovascular disease biomarkers in urine from healthy and infarcted patients.[32] Pooled urine samples were diluted in isopropanol and centrifuged, then the supernatant was evaporated to dryness and reconstituted with water prior to CE–MS analysis with a sheath liquid electrospray interface. Limits of detection for several cardiovascular disease biomarkers including trimethylamine-*N*-oxide, carnitine, and creatinine were in the low μM range. A pilot study for CE–MS analysis of patient urine from 19 healthy and 8 infarcted people showed significantly higher levels of trimethylamine-*N*-oxide and carnitine in the infarcted patient urine. Finally, Chen *et al.* employed CE–MS to search for possible urinary biomarkers for gastric cancer patients.[33] The urine from patients with and without gastric cancer was diluted 2× in BGE and hydrodynamically injected into the separation capillary interfaced to the MS with a commercially available sheath flow platform. Overall, the authors found 12 metabolites significantly different in the gastric cancer patients, including several amino acids and small organic acids (*e.g.*, citric, malic, and lactic acids). From these data, they conclude that amino acid levels could serve as potential biomarkers for gastric cancer detection. These studies nicely demonstrate the potential for CE–MS based polar metabolomics to provide biological insight into disease pathology or diagnosis.

Rather than explore disease pathology or diagnosis, a recent ecotoxicology study by Zeng and co-workers profiled the polar urinary metabolome of rats fed with Bisphenol-A to determine the metabolic pathways effected by Bisphenol-A exposure.[34] Using a commercially available sheath flow interface, rat urine samples diluted 5× were screening for cationic and anionic metabolites, and multivariate pattern recognition was employed to uncover the differences between rats fed with a control diet and rats fed with varying doses of Bisphenol-A. The authors found perturbations in 42 urinary metabolites in the rats exposed to Bisphenol-A including, amino acids, polyamines, nucleosides, organic acids, carbohydrates, pterins, polyphenols, and sugar phosphates (Figure 6.4). Overall, the authors relate the metabolite alterations from Bisphenol-A exposure to amino acid biosynthesis and neurotransmitter signaling pathways and provide evidence of specific toxicological effects of Bisphenol-A exposure. Additionally, this study suggests that toxic effects are present even at small Bisphenol-A doses of under 50 mg per day.

In summary, untargeted urinary metabolite profiling provides a snapshot into an organism's metabolome, including important nutritional, primary, and secondary metabolite excretory information. Many studies have used commercially available CE–MS instrumentation to measure cationic and anionic metabolites in urine to understand metabolic alterations in disease state and toxicology. Furthermore, sample preparation procedures for urine prior to CE–MS analysis are often simple due to low total amounts of protein and cell debris. As such, typically simple dilution and centrifugation steps to remove debris are required prior to CE–MS. Structurally diverse polar metabolites have been observed in many different disease models and clinical samples ranging from small organic acids to cyclic nucleotides. In the end, the high separation capacity of CE–MS for polar metabolites demonstrates clear potential for CE–MS in clinical urinalysis pipelines.

6.2.1.2 *Blood Plasma and Serum*

Blood is one of the most commonly used biofluids in metabolomics studies, where rich metabolic information can be used to investigate the health and disease of biological systems.[35] In particular, plasma and serum are two blood surrogates with reduced sample complexity that differ by their preparation. Plasma is the liquid fraction collected after removal of red blood cells and is stabilized with an anticlotting agent, and serum is the liquid fraction of blood after the natural clotting process occurs. While these two blood surrogates have been perceived to be similar in composition, metabolome differences between plasma and serum can be large, and the suitable sample must be selected for each application.[36,37] When compared to metabolite analysis in urine discussed above, plasma and serum are slightly more complex biofluids that pose additional challenges for CE–MS analysis of polar metabolites. One particular challenge is the significantly higher protein content in plasma and serum. Therefore, to selectively extract polar small molecule metabolites, proteins and insoluble cellular debris must be removed prior to CE–MS analysis. As a result, sample preparation steps for plasma and serum analysis are generally more substantial when compared to urine analysis. A typical polar metabolite extraction from plasma or serum includes protein precipitation, liquid–liquid extraction, ultracentrifugation, freeze drying, and reconstitution of extracted metabolites in water or BGE. Nonetheless, blood metabolites can yield substantive and useful information regarding an organism's metabolome, and many biomarkers for health and disease have been discovered in plasma and serum. The following sections discuss some of the recent studies and advancements in CE–MS for exploring the polar metabolome of blood plasma and serum.

6.2.1.2.1 Plasma. Due to the high separation capacity of CE and the specificity of MS detection, untargeted polar metabolome profiling of plasma is highly advantageous for monitoring changes in the blood metabolome after treatment. For example, a recent study by Liu *et al.* employed untargeted CE–MS to profile the plasma metabolome of patients undergoing hemodialysis and high flux hemodialysis.[38] The authors highlight the

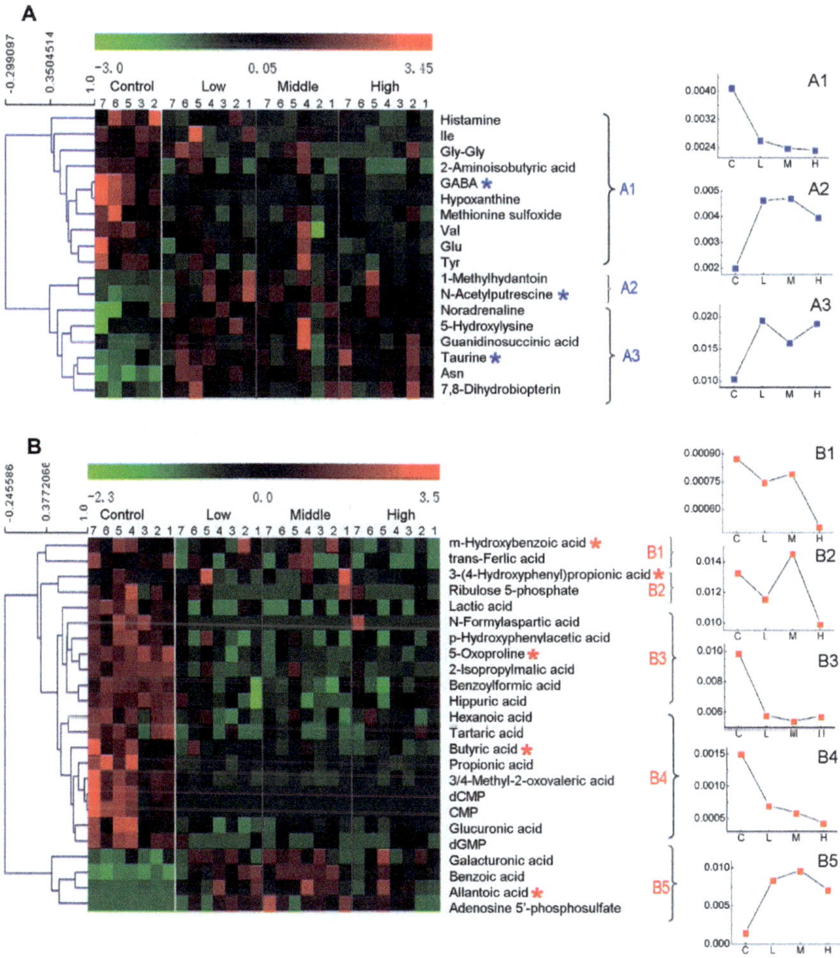

Figure 6.4 Heat maps produced by hierarchical clustering of the most significantly different metabolites (high-dose group *vs.* control group, $p<0.05$) obtained from (A) the cationic mode and (B) the anionic mode. The variation trends of representative metabolites with * in the heat maps presented on the right panel, in which A1–A3 are from the cationic mode and B1–B5 are from the anionic mode. C, L, M, and H mean the control, low-dose, middle-dose, and high-dose groups, respectively.
Reproduced from ref. 34 with permission from American Chemical Society, Copyright 2013.

importance of plasma metabolism for uncovering therapeutic effects of dialysis for patients suffering uremia due to chronic kidney disease. Plasma samples from three groups, control, pre/post hemodialysis, and pre/post high flux hemodialysis were prepared and screened for differences in polar metabolite abundances. After polar metabolite extraction, a sheath flow CE–MS interface was used to profile cationic and anionic metabolites. In total, 144 metabolites were detected, and many metabolites were found to be significantly lower after one hemodialysis treatment. In addition, 11 metabolites were found to be more efficiently removed using high flux hemodialysis. These results demonstrate the efficacy of regular and high flux hemodialysis therapies for treating uremia, as evidenced by the removal of a large number of uremic retention solutes after treatment. Furthermore, this study highlights the applicability for CE–MS to profile polar metabolite changes before and after treatment in order to determine therapeutic outcomes/success.

Although much work has gone into improving the stability and reliability of untargeted CE–MS methodology, the utility and dependability for CE–MS to discover robust biomarkers is still an underexplored area. Therefore, a recent study by Zhang *et al.* evaluated CE–MS analysis and untargeted data workflows (*e.g.*, MetaboAnalyst) to reliably discover biomarkers.[39] To show biomarker discovery in a controlled experiment, synthetic biomarkers were added to two groups of pooled human plasma. Specifically, several non-endogenous isotopically labeled amino acids were spiked into pool human plasma in two groups, and two untargeted data analysis methods were evaluated for their potential for biomarker discovery. After protein precipitation, liquid–liquid extraction, evaporation to dryness, and reconstitution in de-ionized water, metabolites from the spiked pooled human plasma were analyzed using a co-axial sheath flow CE–MS platform. It was found that the two groups of spiked plasma samples were separated by the synthetic biomarkers, indicating the potential utility of CE–MS based biomarker discovery. Moreover, there was little difference in the results generated from both untargeted data analysis strategies. These results demonstrate the robust capacity for CE–MS coupled to untargeted data workflows to discover meaningful biomarkers for human health and disease. Another example of applying untargeted CE–MS to investigate potential plasma biomarkers was reported by Kume and co-workers for investigation of metabolic responses to chronic fatigue.[40] Polar metabolites were extracted from plasma from rats subjected to chronic fatigue conditions and compared to control plasma from specimens. A commercially available sheath flow CE–MS interface was employed for non-targeted analysis of the plasma polar metabolome in both positive and negative mode, and supervised hierarchical clustering and principle component analysis was employed to discern metabolite aberrations in the fatigued rats. In this study, many polar plasma metabolites were revealed to be significantly different between fatigued and control rats, including perturbations to branched-chain amino acid metabolism, the urea cycle, and proline metabolism (Figure 6.5). The authors conclude that CE–MS analysis based polar metabolomics is a useful technique for determining potential biomarkers for diagnosing patients with chronic fatigue.

Diagnosis of mental health disorders such as schizophrenia and autism spectrum disorder currently relies on experienced medical professional patient assessments, which do not incorporate objective physiological biomarkers. Therefore, recent attention has been paid to CE–MS based metabolomics

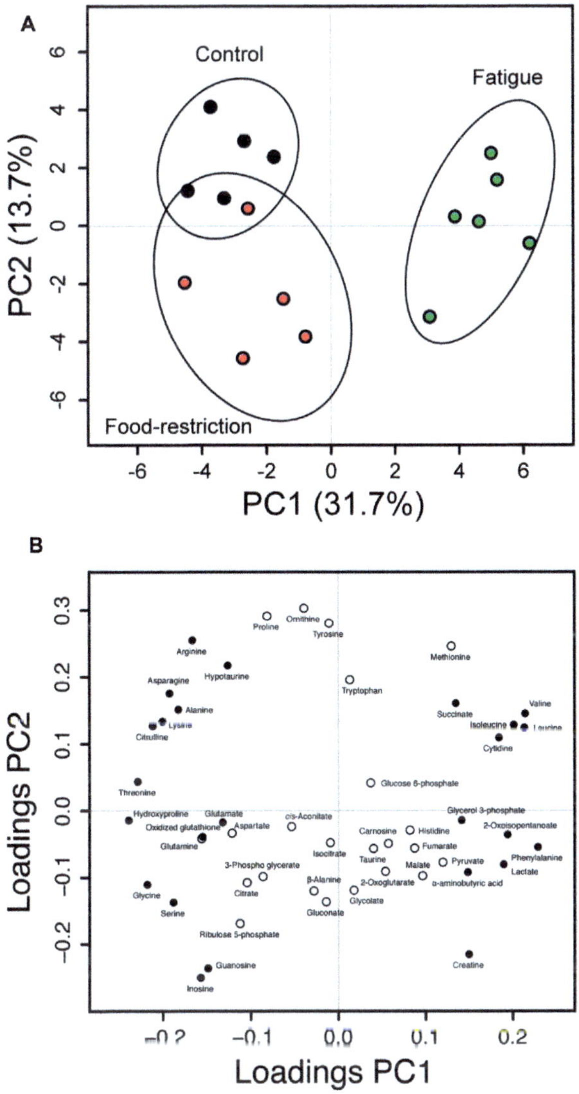

Figure 6.5 Score and loading plots of principal component analysis of plasma metabolome CE–MS data. (A) PCA score plot of PC2 *versus* PC1. The control group ($n = 5$), food-restricted group ($n = 5$), and fatigued group ($n = 5$) are shown as black, green, and red circles, respectively. Black ellipses represent the 90% confidence intervals for each group. (B) PCA loading plot of PC2 *versus* PC1. The data were analyzed after being mean-centered and variance scaled.

methods to identify potential putative biomolecular markers for objective diagnosis. For example, a study by Koike *et al.* employed a commercially available CE–MS device to screen for potential plasma biomarkers indicative of early stages of schizophrenia.[41] Briefly, plasma from 30 patients with first-episode schizophrenia was compared with plasma from 38 healthy individuals. Following liquid–liquid extraction, spin-filtration, freeze drying, and reconstitution in water, five metabolites, including creatine, glutamate, and betaine, were identified to be significantly different between control and first-episode schizophrenic patients. While the authors state that alteration of these metabolites in schizophrenia correlates with well-known metabolic hypotheses, the fact that most patients suffering from their first-episode of schizophrenia receive medication could potentially bias the identification of potential biomarkers through unknown drug metabolism pathways. Another study by Kuwarbara *et al.* searched the polar plasma metabolome for potential objective biomarkers for autism spectrum disorder.[42] They analyzed plasma from 25 individuals with high-functioning autism spectrum disorder and compared the results with plasma obtained from 28 age-matched control patients. Using a commercially available sheath flow CE–MS instrument 17 polar metabolites were found to be significantly different between autism spectrum disorder and control patients. In particular, higher levels of arginine and taurine were identified in autistic patient plasma, in addition to lower levels of oxoproline and lactic acid. These results demonstrate the potential for objective diagnosis of patients with a spectral disease normally diagnosed through observed behaviors by trained clinicians.

Another neurodegenerative disease that was recently studied by CE–MS based polar metabolomics was Huntington's disease. Although genotyping can determine whether an individual has the potential to develop Huntington's disease, biomarkers for determining the progression of the disease are not yet fully established. Therefore, Pont *et al.* developed a CE–MS method utilizing on-line SPE to explore the plasma metabolome of rats with one Huntington's disease allele.[43] Extracted plasma from control and diseased rats was pooled and subjected to protein precipitation, liquid–liquid extraction, and centrifugal filtration. Metabolite extracts were loaded onto an on-line C_{18} SPE column, into a CE separation capillary, and introduced to the MS using a sheath flow interface. The authors profiled plasma from mice 8, 12, and 30 weeks after birth and found 29 metabolites relevant in distinguishing Huntington's disease rats from controls. The tentatively identified relevant metabolites included dipeptides, carnitines, and several eicosanoids, although the latter are detected as sodium adducts with fairly large mass errors. Perturbations in these metabolomic pathways were most apparent after 12 weeks. Nonetheless, the authors demonstrate preliminary evidence of plasma-based biomarkers for following the early progression of Huntington's disease.

6.2.1.2.2 Serum. The previous section highlights studies that employ CE–MS to profile the plasma polar metabolome, however, recent studies have also demonstrated the utility of CE–MS based serum polar metabolomics. When selecting plasma or serum for diagnostic tests or biomarker discovery it is important to consider both blood surrogates, as the clotting factors in serum or platelets and stabilizing agents in plasma may generate bias or misleading results.[44] Sample preparation procedures to extract polar metabolites from serum are very similar to plasma and often entail protein precipitation, liquid–liquid extraction, ultracentrifugation, and lyophilization. Below, several recent examples of studies employing CE–MS to profile the serum polar metabolome are discussed.

Untargeted analysis of serum metabolites offers the potential to screen serum from diseased patients and compare molecular differences from serum obtained from healthy control patients. As a result, metabolite signatures or fingerprints can be correlated to disease states that are difficult to diagnose without specific targeted tests. For example, hepatocellular carcinoma remains a malignant cancer with a high mortality rate that can be difficult to diagnose at early stages of disease progression. Therefore, work by Zeng and others employed untargeted CE–MS metabolomic profiling of serum from hepatocellular carcinoma patients to search for potential molecular biomarkers for early diagnosis.[45] Following the sample preparation procedure described above, extracted serum metabolites were analyzed by a commercially available sheath flow CE–MS device. Polar cationic and anionic metabolites were screened, the data was combined, and both single and multivariate and statistical models were implemented to search for key molecular differences between the cancer patient serum and healthy control individuals. Results from the univariate statistical analysis performed are shown in Figure 6.6. From these data, the authors developed a serum biomarker discovery model that included tryptophan, glutamine, and 2-hydroxybutyric acid, which had high potential for early diagnosis of hepatocellular carcinoma. Another recent study employing untargeted CE–MS serum metabolite screening uncovered potential biomarkers useful in following the progression of insulin resistance to impaired fasting glucose or impaired glucose tolerance, and finally to type 2 diabetes.[46] Specifically, Ciborowski *et al.* profiled the serum metabolome of a total of 197 people that included sample sets of overweight, obese, and normal weight people. Polar metabolites were extracted from serum samples and analyzed by a sheath flow CE–MS interface coupled to a TOF mass analyzer. The data showed numerous metabolites with the potential to track type 2 diabetes mellitus evolution and progression, including branched chain amino acids, lysine, acetylcarnitines, and methionine. However, the authors note that a larger patient cohort is needed to further validate their findings. Finally, a recent untargeted application of CE–MS polar metabolite screening was carried out by Naz and co-workers, where they employed a commercially available sheath flow CE–MS system to profile polar metabolites in serum from rats with ventilator-induced lung injury, an experimental model for acute lung injury.[47] The sample preparation for extraction of polar metabolites from serum samples in this study was slightly different and involved solid phase extraction and

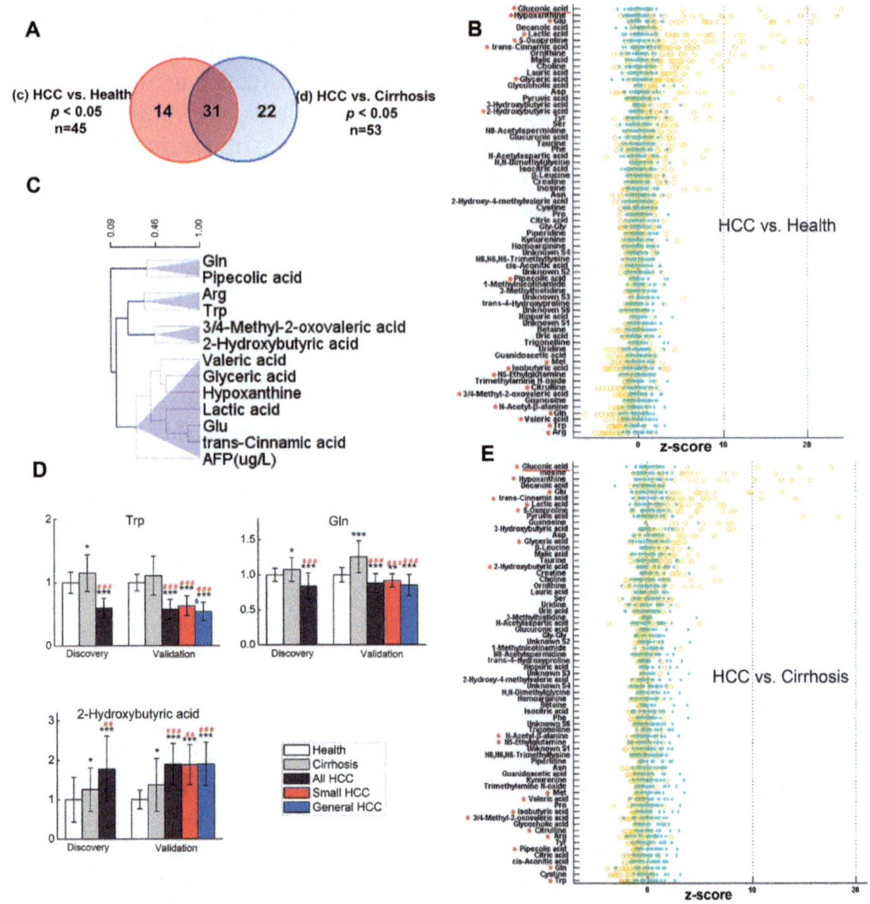

Figure 6.6 Univariate statistical analysis of hepatocellular carcinoma (HCC), cirrhosis, and healthy cells. (A) Venn diagram of the Wilcoxon Mann–Whitney test results. On the basis of all significant differential metabolites, (B) the *z*-score plot for the comparison between HCC subjects (yellow) and healthy controls (blue) and (E) the *z*-score plot for the comparison between HCC subjects (yellow) and cirrhosis subjects (blue). Each point represents one metabolite in one sample. A total of 19 metabolites were cross-selected (*i.e.*, the intersection) based on the preliminary screening result of multivariate analysis and univariate Wilcoxon Mann–Whitney test, which were plotted with * in *z*-score plots. For HCC subjects, the *z*-score ranges of gluconic acid are −0.38 to 1718.68 (HCC *vs* Health) and −0.27 to 1256.77 (HCC *vs* cirrhosis), respectively. (C) Dendrogram of biomarker candidates and AFP based on the Pearson correlation coefficients. (D) Histogram of potential biomarkers. The first set of columns comprises health, cirrhosis, and HCC groups (discovery set), and the second set of columns comprises health, cirrhosis, all HCC, small HCC, and general HCC groups (validation set). The black * means the statistical significance between the health group and other groups. The red # means the statistical significance between the cirrhosis group and other groups. The blue & means the statistical significance between the small HCC and general HCC groups.* (# or &): $0.01 < p < 0.05$, ** (##): $0.001 < p < 0.01$, and *** (###): $p < 0.001$. All data are presented as mean ± SD. Reproduced from ref. 45 with permission from American Chemical Society, Copyright 2014.

ultracentrifugation with a 30 kDa cut-off. From this study, the authors identified over 1100 molecular features from the prepared rat serum. Furthermore, by using univariate and multivariate statistical methods they revealed 18 compounds that were significantly different between rats with acute lung injury and controls. Overall, accumulation of asymmetric dimethyl arginine and ornithine, and a decrease in arginine were found to agree with previous results from the literature.

To search for objective molecular biomarkers to follow the progression of Alzheimer's disease, González-Domínguez *et al.* screened serum polar metabolites from patients with Alzheimer's disease, mild cognitive impairment, and healthy controls.[48] Polar metabolites were extracted from serum into 5% acetonitrile, large proteins (>30 kDa) were removed by ultracentrifugation, and supernatant from the samples was analyzed by a commercially available sheath flow CE–MS interface. Using partial least squares discriminate analysis to classify patients according to disease stage, several potential polar metabolite biomarkers were discovered. These include accumulation of choline, creatinine, dimethyl-arginine, homocysteine–cysteine disulfide, phenylalanine–phenylalanine, and medium chain acylcarnitines in Alzheimer's patients. Additionally, several metabolites were found to decrease when compared to healthy individuals. Overall, the polar metabolite aberrations indicated Alzheimer's disease effects underlying metabolic states including oxidative stress and energy metabolism. Furthermore, the establishment of several potential objective biomarkers for mapping disease progression was discovered with untargeted polar metabolite profiling with CE–MS.

6.2.2 Cell and Tissue Extracts

Direct analysis of tissue extracts can provide a snapshot into metabolic processes related to specific physiological functions. Additionally, cell cultures can be treated with various stimulants or pharmaceuticals prior to profiling of the polar metabolome with CE–MS to determine metabolome responses to chemical treatment. However, when compared to biofluid analysis (*e.g.*, urine, plasma, or serum), greater attention to sample preparation must be taken to quantitatively extract polar analytes from solid tissue or cell pellets into solution for CE–MS analysis. In particular, highly abundant phospholipids, proteins, and other chemical components in these samples can negatively affect the analysis of polar metabolites, especially for analytes present in trace amounts. Nonetheless, polar metabolite extraction from cells or tissue are typically similar to the procedure described above for plasma and serum. In general, polar metabolite extraction sample preparation steps include tissue homogenization, cell lysis and protein precipitation with organic solvents, liquid–liquid extraction, ultracentrifugation, lyophilization, and reconstitution in water or BGE. As an alternative to tissue homogenization, the tissue of interest can be snap frozen in liquid nitrogen and pulverized prior to metabolite extraction.[49] This section highlights and discusses several recent examples employing CE–MS to explore the polar metabolome directly from cell and tissue extracts.

A recent study by Osanai and co-workers employed CE–MS for polar metabolomics to investigate the metabolic response of cyanobacteria under nitrogen starvation.[50] Metabolites from cultured cyanobacteria cells were suspended in methanol, centrifuged through a 5 kDa cut-off filter, evaporated to dryness, and reconstituted in water. The CE–MS system employed was a commercially available sheath flow interface coupled to a TOF mass analyzer. It was found that intermediates in the tricarboxylic acid cycle were increased by nitrogen deprivation, while purine and pyrimidine nucleotides were significantly decreased. This paper represents the utility of CE–MS to probe polar metabolite changes directly from cyanobacterial cultures. Another study demonstrating polar metabolomic analysis from cell culture looked at the effect of vitamin C-induced oxidative stress in human breast MCF7 and colon cancer HC29 cells.[51] Cancer cells were exposed to cytotoxic levels of vitamin C, rinsed, and polar metabolites were extracted. The authors found that levels of metabolites involved in glycolysis and the tricarboxylic acid cycle were significantly increased following treatment with vitamin C, and ATP levels were found to decrease. In the end it was concluded that vitamin C inhibited energy metabolism through NAD depletion, which ultimately induced cancer cell death.

The extraction of polar metabolites from adherent cells poses additional challenges compared to dissected tissue or suspended cells, as adherent cells must be detached prior to cell counting and metabolite extraction. Therefore, Ibanez and co-workers developed a robust sample preparation protocol for polar metabolite extraction from adherent cells.[52] Colon cancer HT-29 cells were detached from culture plates with trypsin and counted to ensure 10 million cells were present, and the cells were combined with water and small abrasive glass beads. Cells were then subjected to three cycles of snap freezing, rapid thawing in an ultrasonic bath, and disruption using a mixer mill. Following this intensive procedure, all cell debris was removed through ultracentrifugation with a 3 kDa cut-off filter, and the filtrate was subsequently injected onto the CE capillary coupled to a sheath flow ESI interface and TOF MS. The authors point out that their selection of optimal extraction conditions enabled higher volumes of sample to be injected, ultimately reducing detection limits compared to conventional sample preparation methods. Their polar metabolite extraction method was demonstrated on HT-29 cells treated with a potent ornithine decarboxylase inhibitor, difluoromethylornithine (DFMO). Using a non-targeted approach, 10 metabolites were found to be significantly altered in cancer cells treated with DFMO, signifying the effect of DFMO on polyamine metabolism and intracellular thiol redox state. Another recent effort aiming to improve metabolite coverage and enhance method sensitivity was published by Huang *et al.*, where they developed a method for a one pot two stage derivatization procedure for cell culture metabolite screening.[53] Their developed method was subsequently applied to profile polar metabolites from bovine aortic endothelial cells. Polar metabolites were isolated with similar procedures as above, and the derivatization that added a charged quaternary amine on all derivatized species was performed in ~100 minutes. Using this approach, the authors found 15 metabolites, including alpha-keto-gluterate, pyruvate, succinate, fumarate, lactate, and many amino acids altered in bovine aortic cells treated with glucose.

Profiling metabolites from cells or tissue provides a snapshot of the cell or organism metabolome, where annotated peaks ideally represent a single metabolite. However, another common branch of metabolome research introduces stable isotope labeled metabolites and monitors the exchange, anabolism, and catabolism of labeled metabolites.[54] A recent study by Zeng *et al.* developed a new protocol for combining conventional untargeted metabolomics with stable isotope tracing using CE–MS.[55] Their four step process involved treating human prostate cancer cells with uniformly labeled ^{13}C glucose for several time intervals, untargeted CE–MS data acquisition in positive and negative polarity, feature alignment and detection, and integrated metabolomics and pathway analysis of metabolite flux. Polar metabolites were extracted from the cancer cells by protein precipitation, liquid–liquid extraction, ultra-centrifugation with a 5 kDa cut-off, freeze drying, and finally reconstitution in de-ionized water. Detected CE–MS molecular features were matched with heavy isotope labeled features, and each detected metabolite was deconvoluted to account for natural heavy isotope abundance and incorporated ^{13}C from catabolism (Figure 6.7). Using their approach, 76 polar metabolites were identified including amino acids, nucleosides, sugar phosphates, amines, and organic acids. In addition to aberrations in glycolysis for cancer cells detected by conventional untargeted metabolomics, they found that metabolites in the glycolysis pathway also had a significantly different flux, measured as incorporation of ^{13}C into metabolite structures. Overall, the authors demonstrated that their methodology is widely applicable in biological studies and provides both classical metabolome 'snapshot' and dynamic metabolite flux information in a unified workflow.

6.3 CE–MS Profiling of the Polar Metabolome from Minute Sample Amounts

CE is an optimal technique for low volume samples, since only nanoliter volumes need to be injected onto the capillary for analysis. In combination with the high separation efficiency, high sensitivity, and minimal injection volumes required, CE–MS offers possibilities for analysis of samples with limited quantities. Limited sample amounts can be an issue for studies where the material availability is low due to unique samples, small model organisms, targeted cellular regions, or limited number of cells. For example, the metabolome of healthy and diseased infants has been compared in ~5 μl samples of infant sweat[56] and ~3 μl samples of whole blood in dried

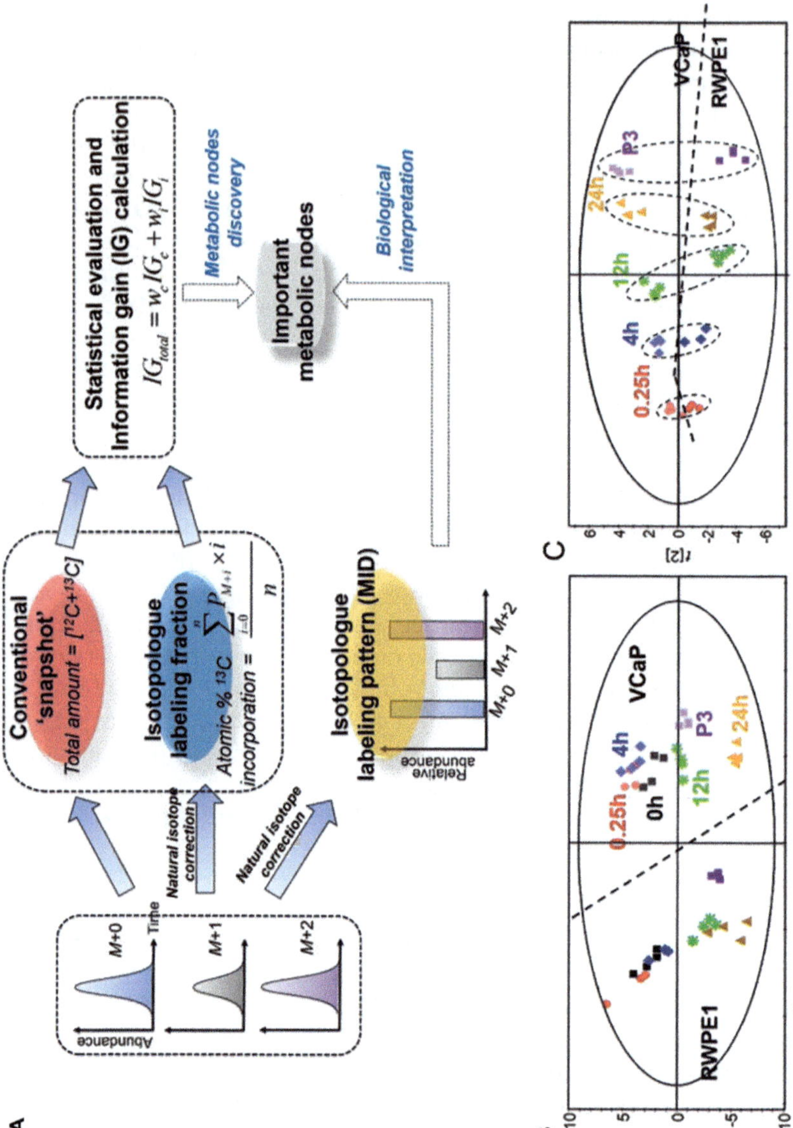
A
Abundance
Time
M+0
M+1
M+2
Natural isotope correction
Natural isotope correction
Conventional 'snapshot'
Total amount = [12C+13C]
Isotopologue labeling fraction
Atomic % 13C incorporation =
Isotopologue labeling pattern (MID)
Relative abundance
M+0
M+1
M+2
Statistical evaluation and Information gain (IG) calculation
Metabolic nodes discovery
Important metabolic nodes
Biological interpretation
B
RWPE1
VCaP
0.25h
0h
4h
12h
P3
24h
t[1]
t[2]
C
0.25h
4h
12h
24h
P3
VCaP
RWPE1
t[1]
t[2]

blood spots.[57] Furthermore, tissue samples down to 2 mg have been used to find metabolic signatures associated with exercise,[58] fetal development,[59] polycystic kidney disease stage,[60] and drug response.[61] In addition, metabolomic characterization of zebra fish embryos[22] and ~10 000 cells[62,63] has been reported after necessary developments to allow for higher signals from low abundance metabolite species. However, a major constraint when analyzing limited samples is that sample preparation steps remain necessary to remove tissue debris and proteins that would otherwise clog the small CE capillary. In addition, if the starting volume is limited, there are limited possibilities to concentrate the sample prior to CE–MS analysis. Therefore, it is essential to have sample preparation strategies that provide high recovery of the endogenous metabolites, minimize oxidation and hydrolysis artifacts, and use small volumes to reconstitute the metabolites in solution after sample clean-up. Furthermore, for samples with low analyte concentrations, it may be necessary to move from a CE–MS interface using a sheath flow interface toward a sheath-less interface that will not dilute the metabolites after CE separation.

6.3.1 Minute Biofluids, Tissue, and Cells

Similar to large volume samples, samples with limited quantities need to be cleaned up to remove cell debris and proteins that otherwise may interfere with the polar metabolite analysis. Sample pretreatment procedures for minute samples are therefore very similar as previously discussed. In short, the sample is often lyophilized or lysed with organic solvent, and the metabolites are extracted using an appropriate liquid–liquid extraction procedure. Subsequently, cell debris and precipitated proteins are removed by centrifugation and ultracentrifugation with a 3–5 kDa cut-off filter whereupon the sample is dried and reconstituted in an appropriate solvent and volume. Generally, minute amounts of tissue are reconstituted in ~2–75 times[22,58–61] their original volume, while biofluids are reconstituted in 4–20 times their original volume.[56,57] Overall, this generates total

Figure 6.7 (A) Scheme illustrating the utilization of the achieved metabolomics data for the discovery and interpretation of polar metabolite profiles including crucial metabolic nodes. After data pretreatment multidimensional datasets including conventional "snapshot" (=total amount of metabolites), isotopologue labeling fraction, and isotopologue labeling pattern are generated. These data are combined for the screening of important biological nodes based on statistical evaluation and information gain (IG) calculation. The isotopologue labeling pattern supports the biological interpretation. (B) PCA score plots of time courses of profiles of conventional metabolite responses (by using total metabolite amounts). (C) Isotopologue labeling fraction of prostate cancer cells (VCaP) and control cells (RWPE-1) grown in [U-^{13}C]-glucose for up to three passages (−P3; that is, 6 days in total).
Reproduced from ref. 55 with permission from John Wiley & Sons,

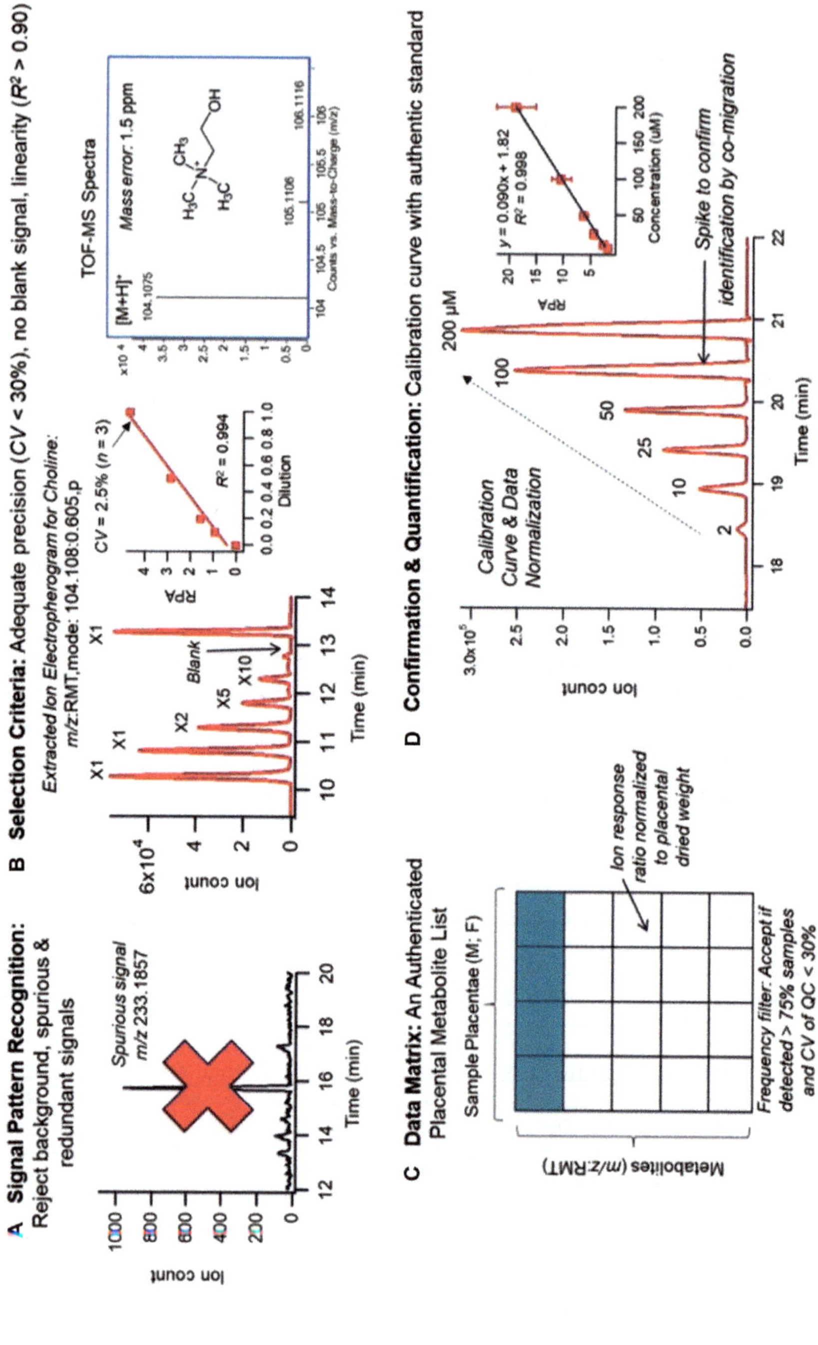
A Signal Pattern Recognition: Reject background, spurious & redundant signals
Spurious signal m/z 233.1857
Ion count
Time (min)
B Selection Criteria: Adequate precision (CV < 30%), no blank signal, linearity (R² > 0.90)
Extracted Ion Electropherogram for Choline: m/z:RMT,mode: 104.108:0.605,p
X1
X1
X2
X5
X10
X1
Blank
Ion count
Time (min)
CV = 2.5% (n = 3)
R² = 0.994
RPA
Dilution
TOF-MS Spectra
[M+H]⁺
Mass error: 1.5 ppm
C Data Matrix: An Authenticated Placental Metabolite List
Sample Placentae (M; F)
Metabolites (m/z:RMT)
Ion response ratio normalized to placental dried weight
Frequency filter: Accept if detected > 75% samples and CV of QC < 30%
D Confirmation & Quantification: Calibration curve with authentic standard
Calibration Curve & Data Normalization
200 µM
Spike to confirm identification by co-migration
Ion count
Time (min)
y = 0.090x + 1.82
R² = 0.998
RPA
Concentration (uM)

prepared sample volumes of 10–100 μl, which is much larger than the minute amounts of sample originally used, and more in line with conventional prepared volumes from larger sample quantities. However, one major difference is that the absolute metabolite concentrations of minute samples are much lower, since the original sample is diluted by sample preparation rather than concentrated. As a result, the recovery of sample preparation and the ionization efficiency of the metabolites during electrospray become much more important for minute samples. To counteract this ultimate dilution of minute sample volumes, several articles report decreased limits of detection and increased number of detected metabolites when using a sheath-less electrospray interface instead of a sheath flow interface.[22,60,62,63] These results are attributed to the notable analyte dilution that occurs post CE separation, as the sheath liquid typically has a much higher flow rate than the background electrolyte. In particular, Hirayama *et al.* reported 4.4 times better detection limits on average for endogenous metabolites in a cell sample,[62] and Lindenburg *et al.* reported an enhanced metabolome coverage using a microvial interface that reduced the sheath flow by 10–100 times.[22] To enable differentiation of low abundance metabolites from noise, Saoi *et al.* employed multi-segment injection CE–MS in their sheath flow system.[59] Figure 6.8 shows a hydrodynamically loaded injection series of 7 samples for untargeted metabolomics of 2 mg placental tissue to study fetal development. By including a dilution series where the same sample was diluted 1×, 2×, 5×, and 10×, in addition to blank samples,

Figure 6.8 Overview of the metabolomics data workflow to authenticate placenta-derived metabolites when using multiplexed separations based on MSI-CE–MS. (A) A dilution trend filter was initially used to filter out background, spurious, and redundant signals generated in ESI–MS to avoid false discoveries. A representative extracted ion electropherogram of a spurious peak (*m*/*z* 233.1857) under positive ion mode is depicted that does not fulfill selection criteria and thus removed from a list of molecular features from a pooled placental extract. (B) Choline (*m*/*z*: RMT 104.108:0.605:p) is an authentic metabolite from placenta that is measured with good precision (CV = 2.5%, $n = 3$), lack of signal in the blank extract, and shows good linearity upon dilution (R^2 >0.90) forming a distinctive temporal signal pattern reflecting serial sample injection. (C) Using these selection criteria, a final data matrix of authentic placental metabolites was reliably detected in a majority of samples in the cohort (>75%) with adequate precision (CV <30%) based on pooled placental extracts used as QCs ($n = 12$) in the study. (D) Absolute quantification for a majority of placental metabolites was performed using MSI-CE–MS where a 6-point calibration curve was performed in one experimental run; all signals were normalized to an internal standard and placental dried weight, and metabolite identification was confirmed by co-migration.

they could distinguish metabolite peaks from matrix peaks and low abundance redundant signals. When only minute samples are available, several features may be on the boarder of the limit of detection, which can complicate confident detection and quantification. Thus, the dilution series is a highly useful approach to differentiate endogenous metabolites from background noise.

6.3.2 Single-cell Capillary Electrophoresis Mass Spectrometry

CE is the most powerful separation technique for single cell metabolomics, since the high separation efficiency generates narrow analyte bands that provide high analyte signal-to-noise values. A single mammalian cell has a diameter of around 10 μm, which corresponds to a volume of approximately 500 fL.[64] Furthermore, within this volume, there are only around 10 pg of carbohydrates and 10 pg of lipids. These low volumes and sample amounts put an extremely high demand on analytical techniques to enable detection. Although CE–MS of 10 μm cells is within reach, cells that have been used for single cell metabolomics with CE–MS are typically larger than 40 μm. Recent literature studies report the use of neurons,[19,65–67] frog embryos,[68–72] and HeLa RBC007 cells.[73]

One major challenge with analysis of single cells is the extraction of the minute material from a single cell. There have been several innovative approaches since 2013 to make this possible, and destructive extraction of metabolites from an entire single cell is the strategy primarily reported.[19,65,67–70,73] In this approach, the cell is dissected from a cluster of cells or from tissue and placed in a vial. Subsequently, metabolites are extracted using several μL of organic solvent, which is typically methanol based. Once the metabolites are in solution, a very small aliquot of several nanoliters is injected into the CE capillary for separation and detection by MS. The benefits of this strategy include the fact that there is no need for substantive sample preparation, and the organic solvent will precipitate proteins to the bottom of the vial, hindering them from being injected onto the column and potentially disturbing the analysis. Furthermore, by keeping all the material in one pot, there is negligible analyte loss due to interactions with the sample vial, digestion container, or pipette surfaces prior to injection. However, the strategy also has several drawbacks. In particular, microliters of solvent are typically used to extract the material from the sub-picoliter volumes of single cells. Thus, the metabolites are heavily diluted, and their final concentrations may end up under the limit of detection. This caveat could be circumvented by using a smaller solution volume, however, the accuracy for delivering low nanoliter volumes is fairly low, which in turn will limit the quality and robustness of the acquired data, and its applicability for quantification. A different strategy to extracting material from the entire cell is to withdraw a minute volume of the material from inside a single cell. A typical procedure is to use a sharp tip made out of glass, insert it into the cell, and physically remove material from the cytoplasm of the cell by aspirating it into the tip.[66,71,72] In studies by Onjiko *et al.* and Portero *et al.*, the cytoplasm extract was ejected from the tip into the vial preloaded with organic solvent.[71,72] Similarly, Aerts *et al.* placed the extract in a vial, but instead of ejecting it they broke off the capillary tip containing the cellular extract and left it in the vial with the added solvent.[66] All three studies used a few μL of organic solvent to dissolve polar metabolites prior to CE–MS analysis. The benefit of extracting only a small volume of the cell is that the cell can be kept alive for subsequent studies. However, this approach also suffers from drawbacks of metabolite dilution. In addition, the repeatability may be challenging since it requires more or less blind sampling of the heterogeneous cellular cytoplasm that is known to have chemical gradients and subcellular structures. Despite the limitations, both strategies have the advantage of diluting with a solvent into which internal standards can easily be incorporated and used as markers for migration time, injection volume, and possibly quantification.

Sample injection into the CE column does not differ much for single cells compared to other types of samples, and most often hydrodynamic loading by raising the sample vial is applied. However, Kawai *et al.* reported hydrodynamic loading using applied pressures, and Liao *et al.* described an electrokinetic injection strategy.[19,73] By using a field amplified sample injection approach Liao *et al.* achieved 100–300 times enhancement in detection limit compared to hydrodynamic loading. Furthermore, they argued the importance of removing salts from the single cells by rinsing the cell prior to injection to minimize ion competition between metabolites and salts, in particular sodium chloride, during electrokinetic injection and enhance metabolite detection.[19] The limited sample pretreatment that often proceeds single cell analysis compared to biofluids or tissue makes these considerations of extra importance.

The low total abundance of metabolites in a single-cell sample puts high demands on the separation efficiency. With a high separation efficiency, it is possible to focus the band of migrating metabolites into a sharp peak where the detected signal of metabolites is as high as possible. Typical separation conditions for single-cell analysis include bare silica columns and 1% formic acid,[19,65,66,68–72] 10% acetic acid,[73] or 20 mM ammonium bicarbonate[67] as BGE. Although a main benefit of CE is that only small amounts of sample need to be injected, analysis of metabolites from single cells can benefit from increasing the injected volume, and thereby the amount of material, onto the capillary followed by sample stacking.[19] To increase the separation efficiency and allow for higher sample volumes, Kawai *et al.* used a dual preconcentration strategy that included both isotachophoresis and sample stacking (Figure 6.9).[73] With this approach, they were able to inject 1.2 μL into the capillary while still ensuring narrow bands for metabolite detection. As a result, they achieved almost 400 times higher sensitivity compared to no preconcentration and an LOD of 450 fM for lysine.[73]

When only minute amounts of sample are available, every factor of the CE–MS analysis becomes increasingly important to optimize. In addition to

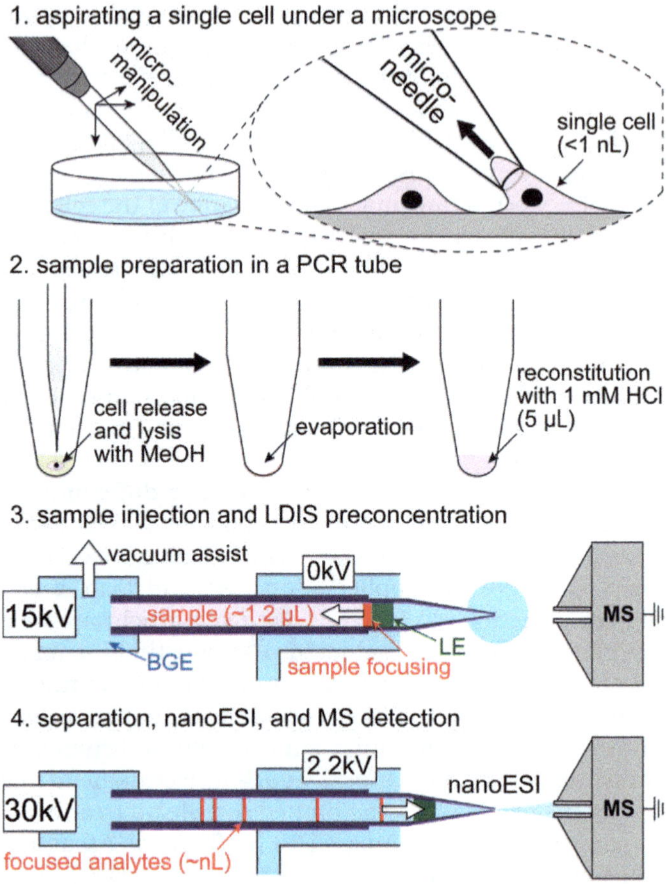

Figure 6.9 Single cell sampling, preparation, injection, and focusing of analyte bands using isotachophoresis and stacking. (1) A single cell was collected by manipulating a microneedle onto the targeted cell and aspirated into the needle by applying negative pressure with a syringe. (2) The collected single cell inside the needle was released and lysed in methanol in a PCR tube. The methanol-extracted metabolites were then further extracted by 1 mM HCl. (3) Preconcentration of the large volume sample inside the capillary: a large-volume sample (*ca.* 1.2 μL) and a short plug of the leading electrolyte (3 M KOH) were introduced into the capillary from the inlet, the samples were concentrated behind the leading electrolyte and transported to the cathode against the electroosmotic flow. (4) The band of analytes was focused by isotachophoresis and separated by capillary zone electrophoresis prior to ESI–MS.
Reproduced from ref. 73 with permission from American Chemical Society, Copyright 2019.

injection and on column preconcentration, it is possible to improve the detection limits by optimizing the conditions for ionization. However, as with CE–MS analyses of larger sample sizes, the main contributing factors to

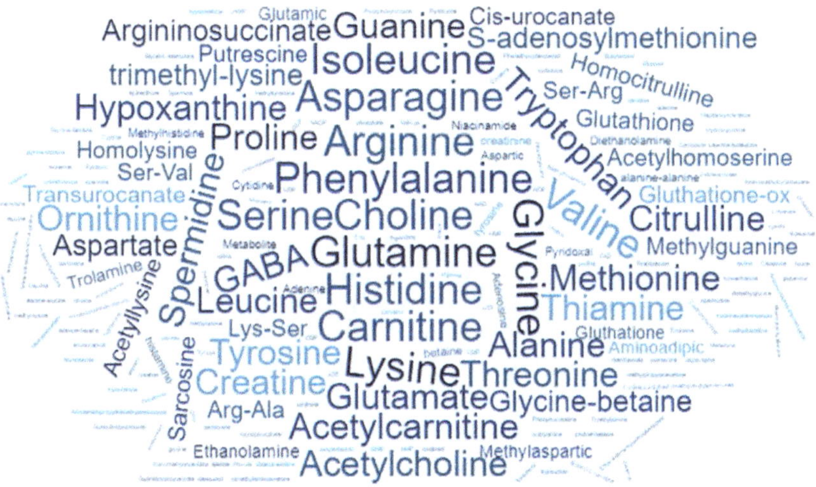

Figure 6.10 Word cloud of metabolites reportedly detected in CEMS analyses of single cells based on ref. 19 and 66–73.

improved detection limits are the use of a sheath-less interface where migrating metabolites are not diluted during ionization, and a tapered electrospray capillary for efficient electrospray ionization.[74]

Single cells contain a wide array of molecules in a large range of concentrations. Overall, 160 metabolites were reported detected from nine scientific articles.[19,66–73] The detected metabolites from these articles are displayed as a word cloud in Figure 6.10. The word cloud reveals that the majority of the metabolites reported in these scientific articles are amino acids and several other polar metabolites. Furthermore, Figure 6.10 displays a high similarity between the detected metabolites, which most likely stem from the similar conditions used for extraction and separation. However, using atypical conditions, Liu *et al.* reported detection and quantification of mono-, di-, and tri-phosphate nucleotides and other molecules of importance in energy metabolism in *Aplysia* R2 neurons. Furthermore, Liao *et al.* reported detection of acyl chain containing carnitines, which are not described in other studies. They speculate that the ability to monitor additional metabolites might be a result of desalting the cell prior to analysis and injecting the sample using field amplification that enhanced the signal.[19] CE–MS of single cells has contributed to measured differences between metabolite abundances in single cell from left and right D1 blastomeres of 8-cell *Xenopus* embryo. In particular, Onjiko *et al.* reported significant changes in spermidine, creatine, putrescine, acetylcarnitine, SAM, ethanolamine, trolamine, isoleucine, leucine, and GABA and recognized that the arginine–proline metabolic pathway was impacted as a result of the physical location of the cells.[69] The field of single-cell CE–MS analysis is still in its infancy with only a limited number of active research groups in the world. However, the field has great potential, and as single-cell analysis using

CE–MS continues to grow and diversify among the scientific community, it is anticipated that the number of reported metabolites and investigated cellular systems will increase.

6.3.3 Surface Sampling Capillary Electrophoresis Mass Spectrometry

Spatial metabolomics is a newly developing field that interfaces mass spectrometry imaging and surface sampling techniques to probe metabolites, lipids, and drugs directly from thin tissue sections.[75] Localizing metabolites to specific morphological regions in tissue can provide new information related to an organism's metabolome, and the involvement of detected metabolites in localized cellular functions. However, the majority of spatial metabolomics methods rely on direct mass spectrometry techniques, which can limit metabolite annotation and biological interpretations if isobaric or isomeric interferences are present. To probe the spatial metabolome of tissue sections more extensively, a recent CE–MS technique was developed to sample, separate, and detect polar metabolites directly from spatially defined regions from the surface of thin tissue sections without previous dissection, extraction or sample preparation. Surface sampling capillary electrophoresis mass spectrometry (SS-CE–MS) couples a stainless steel solvent delivery capillary to a CE capillary, connected by a liquid BGE bridge (Figure 6.11).[76] By electrokinetically loading analytes directly from the tissue surface, over 13 000 deisotoped molecular features were detected from

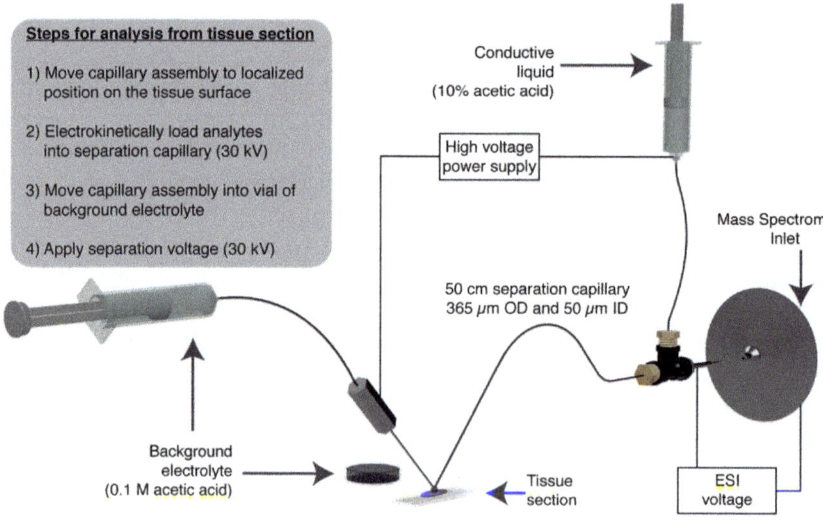

Figure 6.11 Instrumental schematic detailing of the setup and operation of the SS-CE–MS device.
Reproduced from ref. 76 with permission from American Chemical Society, Copyright 2019.

a ~300 µm in diameter spot on the cortex of a rat brain section. Furthermore, replicate sampling of the single cell epidermal layer of *Allium cepa* demonstrated peak area RSD values for measured metabolites of less than 20%. This new SS-CE–MS device will undoubtedly be useful in unambiguous molecular annotation of metabolites sampled directly from cellular unique tissue regions, and further aid in the development of robust untargeted spatial metabolomic techniques.

6.4 Perspective and Future Outlook

The interest of CE–MS based polar metabolomics has grown significantly throughout recent years, largely due to new technological advancements that have improved interfacing CE to ESI–MS. In comparison with other current techniques, CE–MS has the extraordinary power to analyze extremely small volumes of a sample. Although many studies show the analytical stability despite the low injection volume, very few studies make use of this feature. This could be due to minute biological samples still requiring the same sample preparation steps as conventional samples to remove interfering salts, proteins, and lipids, which cause major dilution of low volume samples. A potential solution would be direct injection from the sample into the capillary as in surface sampling CE–MS, however, quantitation remains a challenge. Another solution would be field stacking methods for on-line preconcentration of the sample inside the capillary, however, major sample preparation is required to ensure low ionic strength samples prior to injection. Finally, while sheath flow interfaces remain the most used designs in the literature, recent developments in sheath-less CE–MS interface designs in combination with the achievable enhancement in sensitivity will undoubtedly direct more users toward sheath-less CE–MS in the future. Sheath-less CE–MS developments in combination with efficient sample injection will be important for uncovering low abundance polar metabolites from minute biological samples that play important roles in health and disease.

Acknowledgements

Grants from the Swedish Research Council and the Swedish Foundation for Strategic Research provided the funding for this work.

References

1. O. Yanes, R. Tautenhahn, G. J. Patti and G. Siuzdak, Expanding coverage of the metabolome for global metabolite profiling, *Anal. Chem.*, 2011, **83**, 2152–2161.
2. S. Moco, J. Vervoort, S. Moco, R. J. Bino, R. C. H. De Vos and R. Bino, Metabolomics technologies and metabolite identification, *TrAC, Trends Anal. Chem.*, 2007, **26**, 855–866.

3. F. R. Pinu, S. G. Villas-Boas and R. Aggio, Analysis of intracellular metabolites from microorganisms: Quenching and extraction protocols, *Metabolites*, 2017, DOI: 10.3390/metabo7040053.
4. R. J. Raterink, P. W. Lindenburg, R. J. Vreeken, R. Ramautar and T. Hankemeier, Recent developments in sample-pretreatment techniques for mass spectrometry-based metabolomics, *TrAC, Trends Anal. Chem.*, 2014, **61**, 157–167.
5. A. C. Schrimpe-Rutledge, S. G. Codreanu, S. D. Sherrod and J. A. McLean, Untargeted Metabolomics Strategies—Challenges and Emerging Directions, *J. Am. Soc. Mass Spectrom.*, 2016, **27**, 1897–1905.
6. M. Yan and G. Xu, Current and future perspectives of functional metabolomics in disease studies–A review, *Anal. Chim. Acta*, 2018, **1037**, 41–54.
7. Y. Wang, S. Liu, Y. Hu, P. Li and J. B. Wan, Current state of the art of mass spectrometry-based metabolomics studies - a review focusing on wide coverage, high throughput and easy identification, *RSC Adv.*, 2015, **5**, 78728–78737.
8. C. González-Riano, D. Dudzik, A. Garcia, A. Gil-De-La-Fuente, A. Gradillas, J. Godzien, Á. López-Gonzálvez, F. Rey-Stolle, D. Rojo, F. J. Ruperez, J. Saiz and C. Barbas, Recent developments along the analytical process for metabolomics workflows, *Anal. Chem.*, 2020, **92**, 203–226.
9. R. L. C. Voeten, I. K. Ventouri, R. Haselberg and G. W. Somsen, Capillary Electrophoresis: Trends and Recent Advances, *Anal. Chem.*, 2018, **90**, 1464–1481.
10. R. K. Harstad, A. C. Johnson, M. M. Weisenberger and M. T. Bowser, Capillary Electrophoresis, *Anal. Chem.*, 2016, **88**, 299–319.
11. P. W. Lindenburg, R. Haselberg, G. Rozing and R. Ramautar, Developments in Interfacing Designs for CE–MS: Towards Enabling Tools for Proteomics and Metabolomics, *Chromatographia*, 2014, **78**, 367–377.
12. A. Hirayama, M. Wakayama and T. Soga, Metabolome analysis based on capillary electrophoresis-mass spectrometry, *TrAC, Trends Anal. Chem.*, 2014, **61**, 215–222.
13. R. Ramautar, G. W. Somsen and G. J. de Jong, CE–MS for metabolomics: Developments and applications in the period 2014–2016, *Electrophoresis*, 2017, **38**, 190–202.
14. R. Ramautar, G. W. Somsen and G. J. de Jong, CE-MS for metabolomics: Developments and applications in the period 2016–2018, *Electrophoresis*, 2019, **40**, 165–179.
15. A. Stolz, K. Jooß, O. Höcker, J. Römer, J. Schlecht and C. Neusüß, Recent advances in capillary electrophoresis-mass spectrometry: Instrumentation, methodology and applications, *Electrophoresis*, 2019, **40**, 79–112.
16. M. C. Breadmore, Electrokinetic and hydrodynamic injection: Making the right choice for capillary electrophoresis, *Bioanalysis*, 2009, **1**, 889–894.

17. J. P. Schaeper and M. J. Sepaniak, Parameters affecting reproducibility in capillary electrophoresis, *Electrophoresis*, 2000, **21**, 1421–1429.
18. A. Šlampová, Z. Malá and P. Gebauer, Recent progress of sample stacking in capillary electrophoresis (2016–2018), *Electrophoresis*, 2019, **40**, 40–54.
19. H.-W. W. Liao, S. S. Rubakhin, M. C. Philip and J. V. Sweedler, Enhanced single-cell metabolomics by capillary electrophoresis electrospray ionization-mass spectrometry with field amplified sample injection, *Anal. Chim. Acta*, 2020, **1118**, 36–43.
20. N. L. Kuehnbaum, A. Kormendi and P. Britz-Mckibbin, Multisegment injection-capillary electrophoresis-mass spectrometry: A high-throughput platform for metabolomics with high data fidelity, *Anal. Chem.*, 2013, **85**, 10664–10669.
21. V. González-Ruiz, S. Codesido, S. Rudaz and J. Schappler, Evolution in the design of a low sheath-flow interface for CE-MS and application to biological samples, *Electrophoresis*, 2018, **39**, 853–861.
22. P. W. Lindenburg, R. Ramautar, R. G. Jayo, D. D. Y. Chen and T. Hankemeier, Capillary electrophoresis-mass spectrometry using a flow-through microvial interface for cationic metabolome analysis, *Electrophoresis*, 2014, **35**, 1308–1314.
23. K. Sasaki, H. Sagawa, M. Suzuki, H. Yamamoto, M. Tomita, T. Soga and Y. Ohashi, Metabolomics Platform with Capillary Electrophoresis Coupled with High-Resolution Mass Spectrometry for Plasma Analysis, *Anal. Chem.*, 2019, **91**, 1295–1301.
24. H. Zhang, C. Lou, J. Li and J. Kang, A gold foil covered fused silica capillary tip as a sheathless interface for coupling capillary electrophoresis-mass spectrometry, *J. Chromatogr. A*, 2020, **1624**, 461215.
25. X. Guo, T. L. Fillmore, Y. Gao and K. Tang, Capillary Electrophoresis-Nanoelectrospray Ionization-Selected Reaction Monitoring Mass Spectrometry via a True Sheathless Metal-Coated Emitter Interface for Robust and High-Sensitivity Sample Quantification, *Anal. Chem.*, 2016, **88**, 4418–4425.
26. S. Nilsson, M. Wetterhall, J. Bergquist, L. Nyholm and K. E. Markides, A simple and robust conductive graphite coating for sheathless electrospray emitters used in capillary electrophoresis/mass spectrometry, *Rapid Commun. Mass Spectrom.*, 2001, **15**, 1997–2000.
27. A. Hirayama, H. Abe, N. Yamaguchi, S. Tabata, M. Tomita and T. Soga, Development of a sheathless CE-ESI-MS interface, *Electrophoresis*, 2018, **39**, 1382–1389.
28. R. Ramautar, J. M. Busnel, A. M. Deelder and O. A. Mayboroda, Enhancing the coverage of the urinary metabolome by sheathless capillary electrophoresis-mass spectrometry, *Anal. Chem.*, 2012, **84**, 885–892.
29. F. Boizard, V. Brunchault, P. Moulos, B. Breuil, J. Klein, N. Lounis, C. Caubet, S. Tellier, J. L. Bascands, S. Decramer, J. P. Schanstra and B. Buffin-Meyer, A capillary electrophoresis coupled to mass

spectrometry pipeline for long term comparable assessment of the urinary metabolome, *Sci. Rep.*, 2016, **6**, 1–17.
30. M. Yamamoto, R. Ly, B. Gill, Y. Zhu, J. Moran-Mirabal and P. Britz-McKibbin, Robust and High-Throughput Method for Anionic Metabolite Profiling: Preventing Polyimide Aminolysis and Capillary Breakages under Alkaline Conditions in Capillary Electrophoresis-Mass Spectrometry, *Anal. Chem.*, 2016, **88**, 10710–10719.
31. T. A. Isbell, E. C. Strickland, J. Hitchcock, G. McIntire and C. L. Colyer, Capillary electrophoresis-mass spectrometry determination of morphine and its isobaric glucuronide metabolites, *J. Chromatogr. B: Anal. Technol. Biomed. Life Sci.*, 2015, **980**, 65–71.
32. Z. Cieslarova, M. Magaldi, L. A. Barros, C. L. do Lago, D. R. Oliveira, F. A. H. Fonseca, M. C. Izar, A. S. Lopes, M. F. M. Tavares and A. Klassen, Capillary electrophoresis with dual diode array detection and tandem mass spectrometry to access cardiovascular biomarkers candidates in human urine: Trimethylamine-N-Oxide and L-carnitine, *J. Chromatogr. A*, 2019, **1583**, 136–142.
33. J. L. Chen, J. Fan and X. J. Lu, CE-MS based on moving reaction boundary method for urinary metabolomic analysis of gastric cancer patients, *Electrophoresis*, 2014, **35**, 1032–1039.
34. J. Zeng, H. Kuang, C. Hu, X. Shi, M. Yan, L. Xu, L. Wang, C. Xu and G. Xu, Effect of bisphenol A on rat metabolic profiling studied by using capillary electrophoresis time-of-flight mass spectrometry, *Environ. Sci. Technol.*, 2013, **47**, 7457–7465.
35. M. Suarez-Diez, J. Adam, J. Adamski, S. A. Chasapi, C. Luchinat, A. Peters, C. Prehn, C. Santucci, A. Spyridonidis, G. A. Spyroulias, L. Tenori, R. Wang-Sattler and E. Saccenti, Plasma and Serum Metabolite Association Networks: Comparability within and between Studies Using NMR and MS Profiling, *J. Proteome Res.*, 2017, **16**, 2547–2559.
36. L. Liu, J. Aa, G. Wang, B. Yan, Y. Zhang, X. Wang, C. Zhao, B. Cao, J. Shi, M. Li, T. Zheng, Y. Zheng, G. Hao, F. Zhou, J. Sun and Z. Wu, Differences in metabolite profile between blood plasma and serum, *Anal. Biochem.*, 2010, **406**, 105–112.
37. Z. Yu, G. Kastenmüller, Y. He, P. Belcredi, G. Möller, C. Prehn, J. Mendes, S. Wahl, W. Roemisch-Margl, U. Ceglarek, A. Polonikov, N. Dahmen, H. Prokisch, L. Xie, Y. Li, H. E. Wichmann, A. Peters, F. Kronenberg, K. Suhre, J. Adamski, T. Illig and R. Wang-Sattler, Differences between human plasma and serum metabolite profiles, *PLoS One*, 2011, **6**, 1–6.
38. S. Liu, L. Wang, C. Hu, X. Huang, H. Liu, Q. Xuan, X. Lin, X. Peng, X. Lu, M. Chang and G. Xu, Plasma metabolomics profiling of maintenance hemodialysis based on capillary electrophoresis-time of flight mass spectrometry, *Sci. Rep.*, 2017, **7**, 1–11.
39. W. Zhang, K. Segers, D. Mangelings, A. Van Eeckhaut, T. Hankemeier, Y. Vander Heyden and R. Ramautar, Assessing the suitability of capillary

electrophoresis-mass spectrometry for biomarker discovery in plasma-based metabolomics, *Electrophoresis*, 2019, **40**, 2309–2320.
40. S. Kume, M. Yamato, Y. Tamura, G. Jin, M. Nakano, Y. Miyashige, A. Eguchi, Y. Ogata, N. Goda, K. Iwai, E. Yamano, Y. Watanabe, T. Soga and Y. Kataoka, Potential biomarkers of fatigue identified by plasma metabolome analysis in rats, *PLoS One*, 2015, **10**, 1–17.
41. S. Koike, M. Bundo, K. Iwamoto, M. Suga, H. Kuwabara, Y. Ohashi, K. Shinoda, Y. Takano, N. Iwashiro, Y. Satomura, T. Nagai, T. Natsubori, M. Tada, H. Yamasue and K. Kasai, A snapshot of plasma metabolites in first-episode schizophrenia: A capillary electrophoresis time-of-flight mass spectrometry study, *Transl. Psychiatry*, 2014, **4**, 1–8.
42. H. Kuwabara, H. Yamasue, S. Koike, H. Inoue, Y. Kawakubo, M. Kuroda, Y. Takano, N. Iwashiro, T. Natsubori, Y. Aoki, Y. Kano and K. Kasai, Altered Metabolites in the Plasma of Autism Spectrum Disorder: A Capillary Electrophoresis Time-of-Flight Mass Spectroscopy Study, *PLoS One*, 2013, DOI: 10.1371/journal.pone.0073814.
43. L. Pont, F. Benavente, J. Jaumot, R. Tauler, J. Alberch, S. Ginés, J. Barbosa and V. Sanz-Nebot, Metabolic profiling for the identification of Huntington biomarkers by on-line solid-phase extraction capillary electrophoresis mass spectrometry combined with advanced data analysis tools, *Electrophoresis*, 2016, **37**, 795–808.
44. W. B. Dunn, D. Broadhurst, P. Begley, E. Zelena, S. Francis-McIntyre, N. Anderson, M. Brown, J. D. Knowles, A. Halsall, J. N. Haselden, A. W. Nicholls, I. D. Wilson, D. B. Kell and R. Goodacre, Procedures for large-scale metabolic profiling of serum and plasma using gas chromatography and liquid chromatography coupled to mass spectrometry, *Nat. Protoc.*, 2011, **6**, 1060–1083.
45. J. Zeng, P. Yin, Y. Tan, L. Dong, C. Hu, Q. Huang, X. Lu, H. Wang and G. Xu, Metabolomics study of hepatocellular carcinoma: Discovery and validation of serum potential biomarkers by using capillary electrophoresis-mass spectrometry, *J. Proteome Res.*, 2014, **13**, 3420–3431.
46. M. Ciborowski, E. Adamska, M. Rusak, J. Godzien, J. Wilk, A. Citko, W. Bauer, M. Gorska and A. Kretowski, CE-MS-based serum fingerprinting to track evolution of type 2 diabetes mellitus, *Electrophoresis*, 2015, **36**, 2286–2293.
47. S. Naz, A. Garcia, M. Rusak and C. Barbas, Method development and validation for rat serum fingerprinting with CE-MS: Application to ventilator-induced-lung-injury study, *Anal. Bioanal. Chem.*, 2013, **405**, 4849–4858.
48. R. González-Domínguez, A. García, T. García-Barrera, C. Barbas and J. L. Gómez-Ariza, Metabolomic profiling of serum in the progression of Alzheimer's disease by capillary electrophoresis-mass spectrometry, *Electrophoresis*, 2014, **35**, 3321–3330.
49. P. Gao, M. Ji, X. Fang, Y. Liu, Z. Yu, Y. Cao, A. Sun, L. Zhao and Y. Zhang, Capillary electrophoresis – Mass spectrometry metabolomics analysis

revealed enrichment of hypotaurine in rat glioma tissues, *Anal. Biochem.*, 2017, **537**, 1–7.
50. T. Osanai, A. Oikawa, T. Shirai, A. Kuwahara, H. Iijima, K. Tanaka, M. Ikeuchi, A. Kondo, K. Saito and M. Y. Hirai, Capillary electrophoresis-mass spectrometry reveals the distribution of carbon metabolites during nitrogen starvation in Synechocystis sp. PCC 6803, *Environ. Microbiol.*, 2014, **16**, 512–524.
51. M. Uetaki, S. Tabata, F. Nakasuka, T. Soga and M. Tomita, Metabolomic alterations in human cancer cells by Vitamin C-induced oxidative stress, *Sci. Rep.*, 2015, **5**, 1–9.
52. C. Ibáñez, C. Simó, A. Valdés, L. Campone, A. L. Piccinelli, V. García-Cañas and A. Cifuentes, Metabolomics of adherent mammalian cells by capillary electrophoresis-mass spectrometry: HT-29 cells as case study, *J. Pharm. Biomed. Anal.*, 2015, **110**, 83–92.
53. T. Huang, M. Armbruster, R. Lee, D. S. Hui and J. L. Edwards, Metabolomic analysis of mammalian cells and human tissue through one-pot two stage derivatizations using sheathless capillary electrophoresis-electrospray ionization-mass spectrometry, *J. Chromatogr. A*, 2018, **1567**, 219–225.
54. A. Chokkathukalam, D. H. Kim, M. P. Barrett, R. Breitling and D. J. Creek, Stable isotope-labeling studies in metabolomics: New insights into structure and dynamics of metabolic networks, *Bioanalysis*, 2014, **6**, 511–524.
55. J. Zeng, Z. Wang, X. Huang, S. S. Eckstein, X. Lin, H. Piao, C. Weigert, P. Yin, R. Lehmann and G. Xu, Comprehensive Profiling by Non-targeted Stable Isotope Tracing Capillary Electrophoresis-Mass Spectrometry: A New Tool Complementing Metabolomic Analyses of Polar Metabolites, *Chem. -Eur. J.*, 2019, **25**, 5427–5432.
56. A. N. MacEdo, S. Mathiaparanam, L. Brick, K. Keenan, T. Gonska, L. Pedder, S. Hill and P. Britz-McKibbin, The Sweat Metabolome of Screen-Positive Cystic Fibrosis Infants: Revealing Mechanisms beyond Impaired Chloride Transport, *ACS Cent. Sci.*, 2017, **3**, 904–913.
57. A. Dibattista, N. McIntosh, M. Lamoureux, O. Y. Al-Dirbashi, P. Chakraborty and P. Britz-McKibbin, Temporal Signal Pattern Recognition in Mass Spectrometry: A Method for Rapid Identification and Accurate Quantification of Biomarkers for Inborn Errors of Metabolism with Quality Assurance, *Anal. Chem.*, 2017, **89**, 8112–8121.
58. M. Saoi, M. Percival, C. Nemr, A. Li, M. Gibala and P. Britz-Mckibbin, Characterization of the Human Skeletal Muscle Metabolome for Elucidating the Mechanisms of Bicarbonate Ingestion on Strenuous Interval Exercise, *Anal. Chem.*, 2019, **91**, 4709–4718.
59. M. Saoi, K. M. Kennedy, W. Gohir, D. M. Sloboda and P. Britz-McKibbin, Placental Metabolomics for Assessment of Sex-specific Differences in Fetal Development During Normal Gestation, *Sci. Rep.*, 2020, **10**, 9399.

60. E. Sánchez-López, G. S. M. Kammeijer, A. L. Crego, M. L. Marina, R. Ramautar, D. J. M. Peters and O. A. Mayboroda, Sheathless CE-MS based metabolic profiling of kidney tissue section samples from a mouse model of Polycystic Kidney Disease, *Sci. Rep.*, 2019, **9**, 1–9.
61. N. T. N. Phan, J. Hanrieder, E. C. Berglund and A. G. Ewing, Capillary electrophoresis-mass spectrometry-based detection of drugs and neurotransmitters in drosophila brain, *Anal. Chem.*, 2013, **85**, 8448–8454.
62. A. Hirayama, H. Abe, N. Yamaguchi, S. Tabata, M. Tomita and T. Soga, Development of a sheathless CE-ESI-MS interface, *Electrophoresis*, 2018, **39**, 1382–1389.
63. W. Zhang, F. Guled, T. Hankemeier and R. Ramautar, Utility of sheathless capillary electrophoresis-mass spectrometry for metabolic profiling of limited sample amounts, *J. Chromatogr. B: Anal. Technol. Biomed. Life Sci.*, 2019, **1105**, 10–14.
64. A. Schmid, H. Kortmann, P. S. Dittrich and L. M. Blank, Chemical and biological single cell analysis, *Curr. Opin. Biotechnol.*, 2010, **21**, 12–20.
65. P. Nemes, S. S. Rubakhin, J. T. Aerts and J. V. Sweedler, Qualitative and quantitative metabolomic investigation of single neurons by capillary electrophoresis electrospray ionization mass spectrometry, *Nat. Protoc.*, 2013, **8**, 783–799.
66. J. T. Aerts, K. R. Louis, S. R. Crandall, G. Govindaiah, C. L. Cox and J. V. Sweedler, Patch clamp electrophysiology and capillary electrophoresis-mass spectrometry metabolomics for single cell characterization, *Anal. Chem.*, 2014, **86**, 3203–3208.
67. J. X. Liu, J. T. Aerts, S. S. Rubakhin, X. X. Zhang and J. V. Sweedler, Analysis of endogenous nucleotides by single cell capillary electrophoresis-mass spectrometry, *Analyst*, 2014, **139**, 5835–5842.
68. R. M. Onjiko, S. A. Moody and P. Nemes, Single-cell mass spectrometry reveals small molecules that affect cell fates in the 16-cell embryo, *Proc. Natl. Acad. Sci. U. S. A.*, 2015, **112**, 6545–6550.
69. R. M. Onjiko, S. E. Morris, S. A. Moody and P. Nemes, Single-cell mass spectrometry with multi-solvent extraction identifies metabolic differences between left and right blastomeres in the 8-cell frog (: Xenopus) embryo, *Analyst*, 2016, **141**, 3648–3656.
70. R. M. Onjiko, E. P. Portero, S. A. Moody and P. Nemes, In Situ Microprobe Single-Cell Capillary Electrophoresis Mass Spectrometry: Metabolic Reorganization in Single Differentiating Cells in the Live Vertebrate (Xenopus laevis) Embryo, *Anal. Chem.*, 2017, **89**, 7069–7076.
71. E. P. Portero and P. Nemes, Dual cationic-anionic profiling of metabolites in a single identified cell in a live: Xenopus laevis embryo by microprobe CE-ESI-MS, *Analyst*, 2019, **144**, 892–900.
72. R. M. Onjiko, D. O. Plotnick, S. A. Moody and P. Nemes, Metabolic comparison of dorsal: Versus ventral cells directly in the live 8-cell frog embryo by microprobe single-cell CE-ESI-MS, *Anal. Methods*, 2017, **9**, 4964–4970.

73. T. Kawai, N. Ota, K. Okada, A. Imasato, Y. Owa, M. Morita, M. Tada and Y. Tanaka, Ultrasensitive Single Cell Metabolomics by Capillary Electrophoresis-Mass Spectrometry with a Thin-Walled Tapered Emitter and Large-Volume Dual Sample Preconcentration, *Anal. Chem.*, 2019, **91**, 10564–10572.
74. R. Ramautar, *Capillary Electrophoresis–Mass Spectrometry for Metabolomics*, The Royal Society of Chemistry, 2018.
75. T. Alexandrov, Spatial Metabolomics and Imaging Mass Spectrometry in the Age of Artificial Intelligence, *Annu. Rev. Biomed. Data Sci.*, 2020, **3**, 61–87.
76. K. D. Duncan and I. Lanekoff, Spatially Defined Surface Sampling Capillary Electrophoresis Mass Spectrometry, *Anal. Chem.*, 2019, **91**, 7819–7827.

CHAPTER 7

Ion Mobility–Time-of-flight Mass Spectrometry and Applications for Metabolomics

MAX L. FEUERSTEIN, STEPHAN HANN AND TIM CAUSON*

Institute of Analytical Chemistry, Department of Chemistry, University of Natural Resources and Life Sciences, Vienna (BOKU), Muthgasse 18,1190, Vienna, Austria
*Email: Tim.Causon@boku.ac.at

7.1 Introduction

Metabolomics aims to comprehensively map levels of small molecules originating from endogenous and exogenous sources and relate changes to altered functions in the observed biological system. However, metabolites encompass a high degree of structural diversity, and dealing with the large differences in physicochemical properties is one of the major challenges in surveying the metabolome. Thus, combinations of different analytical techniques are necessary to perform a true comprehensive analysis of the metabolome.[1–3] As the effort required for achieving comprehensiveness is often extremely high, popular non-targeted metabolomics strategies aim to maximize the metabolome coverage with an analytical strategy suited for a wide range of analytes without detailed prior knowledge of the exact sample composition.[1,2] For this application, high-resolution two-stage mass spectrometry (MS) coupled to front end separation techniques such as liquid chromatography (LC), gas chromatography or capillary electrophoresis

New Developments in Mass Spectrometry No. 10
Advanced Mass Spectrometry-based Analytical Separation Techniques for Probing the Polar Metabolome
Edited by Rawi Ramautar

Published by the Royal Society of Chemistry, www.rsc.org

constitute a gold standard for metabolomics due to the high selectivity, sensitivity and suitable dynamic range.[1,2,4] As a key technology for this pursuit, accurate MS allows determination of sum formulas of unknown compounds based on monoisotopic mass and isotopologue distribution pattern. However, besides correct formula annotation, the full identification of metabolites is finally necessary for correct biological interpretation of metabolomics data.[1,5] Therefore the determination of accurate mass, isotopic pattern, fragment spectrum and retention information in combination with a growing number of curated and available compound libraries allows characterization and identity confirmation of unknown compounds in non-targeted analysis.[1,5] Within the scope of non-targeted analysis, recent technological developments have brought ion mobility spectrometry (IMS) to the attention of metabolomics researchers. The commercialization of different types of IMS instruments coupled to QTOFMS (*i.e.*, hybrid IM–QTOFMS) has led to the availability of several analytical platforms that have been successfully used in proteomics,[6,7] metabolomics[8–10] and lipidomics experiments.[11,12] Besides selectivity enhancements and increased separation peak capacity, IM–QTOFMS allows use of mobility-derived quantities (*i.e.*, an ion's collision cross section, CCS) as an additional identification point. The fast separation speed of IMS devices in the millisecond time scale allows comprehensive sampling of transient peaks of front-end separation techniques, such as LC. Therefore, nesting IMS in between chromatographic separation and MS detectors offers the possibility of multidimensional analyte separation allowing an increase in selectivity of analytical workflows without losing coverage.[3,8,10,13] However, as a post-ionization separation technique with limited orthogonality to MS,[14] IMS as a single separation dimension prior to MS detection is not able to address some common issues in non-targeted metabolomics, such as ion suppression,[3] in source fragmentation and adduct formation.[15,16] Additionally, data processing and peak picking and alignment can be challenging because different adducts, charge states and protomers or conformers originating from the same metabolite are often separated due to differences in their mobilities, and their re-alignment is more challenging for data processing than for LC–MS.[3,17,18]

7.1.1 Fundamentals: Gas Phase Mobility and Separation Principals

IMS is used to separate ions based on their mobility differences in a gas-filled ion mobility cell. Ions are accelerated by an electric field and reach a steady state velocity due to collisions with an inert buffer gas such as helium or nitrogen. The mobility (K, $m^2 V^{-1} s^{-1}$) of an ion is determined at the experimentally employed electric field.[8,17,19]

$$K = \frac{v_d}{E} = \frac{L}{t_d E} \tag{7.1}$$

whereby, K depends on the applied electric field strength (E, V m^{-1}) and governs the drift velocity (v_d, m s^{-1}) which is experimentally determined by the drift time (t_d, s) required to traverse a drift path of length (L, m).[19] Typically, the gas phase mobility is reported as a reduced mobility (K_0) or as CCS values derived from K_0.[17,19] The reduced mobility simply takes into account the experimental temperature and pressure (normalization to NIST conditions).

$$K_0 = K\frac{N}{N_0} = K\frac{p}{p_0}\frac{T_0}{T} \tag{7.2}$$

Of direct relevance to identity confirmation, the CCS of the ion is a derived property that reflects the ion and the properties of the buffer gas used in the experiment. CCS values can be understood to reflect several molecular properties including the ion mass, ion charge state, ion size, buffer gas identity and ion conformation. Depending on the type of instrumentation used, several methods exist to determine K_0 or CCS values. However, the current primary method to determine an ion's CCS is referred to as the stepped-field method using DTIM–MS.[17] Low-field conditions are defined *via* examining the effect of the E/N ratio on ion mobility and remain an ongoing fundamental question for IMS.[19,20] When experimental evaluation indicates linear behavior (*i.e.*, low-field conditions), ion mobility is considered to not be affected by the electric field applied and CCS can be derived from mobility *via*:

$$\mathrm{CCS} = \frac{3}{16}\sqrt{\frac{2\pi}{\mu k_b T}}\frac{ze}{N_0 K_0} \tag{7.3}$$

with reduced mass of the ion-neutral complex (μ, kg), Boltzmann constant (k_b, J K^{-1}), analyte integer charge z, the elementary charge (e, C) and CCS (Å^2 or nm^2), which is often treated as equivalent to the momentum transfer collision integral Ω.[19] Besides offering the possibility to determine CCS values, operating under the low field limit minimizes effects related to ion heating such as ion activation.[17,19] Furthermore, the CCS values determined using DTIM–MS instrumentation are routinely employed for external CCS calibration of other classes of IM–MS instrumentation.

7.2 Instrumentation and Data Acquisition

Current commercial IM–MS instrumentation can be classified as time-dispersive, field-dispersive or trapping IMS instrumentation and is discussed extensively in several recent reviews.[8,10,17] Briefly, in time dispersive instruments (*i.e.*, drift tube ion (DTIMS) and traveling wave (TWIMS) instruments) ions are accelerated using an electric field and are separated based on ion–neutral gas interactions while traversing a drift tube filled with stationary inert gas (*i.e.*, nitrogen or helium).[8,10,17] In contrast, trapped ion mobility (TIMS) allows ions to be accumulated in a trapping cell using a voltage gradient and counter flow of inert buffer gas. This "trap-and-release" concept allows mobility-selective release of ions from the trap into the MS part of the instrument using a carefully optimized voltage gradient (Figure 7.1). As the transport and separation mechanisms in the commercial DTIMS, TWIMS and TIMS devices are very different, direct comparison of derived CCS values of metabolites should be performed carefully. For example, in contrast to DTIMS, higher field strengths in TIMS and TWIMS are purported to lead to ion heating and related effects, such as ion activation, conformational changes or isomerization.[17,21,22] However, especially in TIMS devices, this effect is highly dependent upon the applied method parameters such as gas flow rate, axial potential gradients or ion accumulation times.[22] Although field-dispersive IM methods such as field asymmetric waveform ion mobility spectrometry (FAIMS) display high orthogonality to mass spectrometry and have been successfully applied in the context of metabolomics,[23–25] they are mainly used in targeted analysis or fundamental ion characterization. Furthermore, FAIMS is not a generic ion mobility device (rather a mobility-selection device) and explicitly operates at high-field conditions making it a highly specialized approach within IM–MS. From the other commercially available instruments, it is of note that all are currently equipped with quadrupole time-of-flight (QTOFMS) mass spectrometers, which allow accurate mass determination and suitable acquisition speed to cope with sampling of narrow IMS peaks.

It has been observed that mobilities and CCS values determined with different instrument types can differ on a percent level.[17] Thus, for rugged application of IM–MS in metabolomics, it is important to be aware that CCS

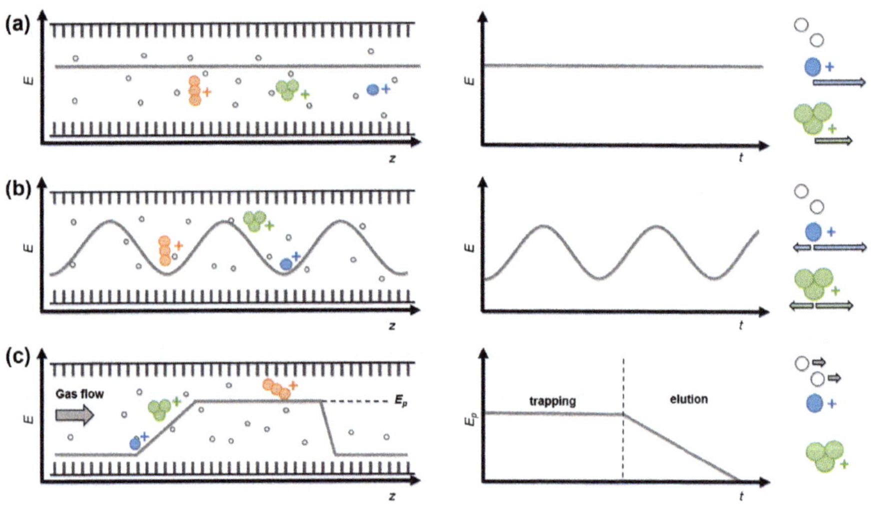

Figure 7.1 Commercially available IM–QTOF instruments perform ion mobility separation based on different operational principles: (a) DTIMS, (b) TWIMS and (c) TIMS.

values of large and small molecular ions are fundamentally dependent on the investigated ion–neutral gas pair and can be further affected by different method parameters, *i.e.*, instrumental platform used, applied electric field strengths and the choice of the buffer gas.[19,26] Finally, K is proportionally related to the charge state of an ion. Hence, ions of the same molecule with different charge states show different mobilities as they are separated in the IM cell.[14,26] These factors lead to important reporting issues for IM–MS datasets as a single metabolite measured on a particular class of IM–MS instrument in a given buffer gas can yield several different ions with respective CCS values. CCS values are thus reported per ion species (including conformers) of the metabolite using a notation that indicates the experimental measurement made (*e.g.*, $^{DT}CCS_{N2}$ for DTIM–MS and nitrogen buffer gas, $^{TW}CCS_{He}$ for TWIM–MS and helium buffer gas).[17,27]

7.2.1 DTIM–QTOFMS

Low-field drift tube IM–MS (DTIM–MS) uses a uniform field to accelerate the analyte ions, while collisions with a buffer gas in a drift tube lead to a reduced gas phase velocity of the ion from which a collision cross section (^{DT}CCS) can be directly derived.[10] Two commercially available instruments are available: a high pressure DTIM–QTOFMS system with a 20 cm drift tube (TOFWerk);[28] and a low pressure DTIM–QTOFMS system with a 78 cm drift cell (Agilent Technologies). In order to perform DTIM separation, short ion injection pulses are necessary to perform high-resolution IM separation. High pressure instrumentation requires use of an ion gate, but trapping funnels can be used to trap and sequentially release ion packages prior to mobility separation in low pressure DTIMS in order to increase the ion utilization efficiency.[29,30] As a point of weakness for non-targeted metabolomics applications, the pulsed operation mode generally leads to higher local ion concentrations in the instrument and can result in earlier onset of system saturation and hence, a reduction in the practical dynamic range. Additionally, space-charge effects can lead to discrimination of low *m/z* ions in the Agilent trapping funnel making careful choice of trapping times and suitable sample dilution mandatory. Employed for all measurements on high pressure DTIM–MS systems,[28] multiplexed operation can also be used on low pressure DTIM–MS systems and, in the latter case, offers a possibility to overcome issues related to space-charge effects in the trapping funnel, improve the sensitivity for metabolomics and retain use of up to ~50% of the ion beam[10,30] (Figure 7.2). In the case of the Agilent system, ions are trapped and released in a modified Hadamard pseudo-random sequence, while the TOFWerk system uses a dedicated Hadamard-type multiplexing. In both examples, the overlapping mobility spectra are deconvoluted as part of the data pre-processing.[3,31] QTOFMS operation with CID is also a feature of both commercial instruments enabling DIA acquisition to be performed (*i.e.*, full DTIM–QTOFMS operation).

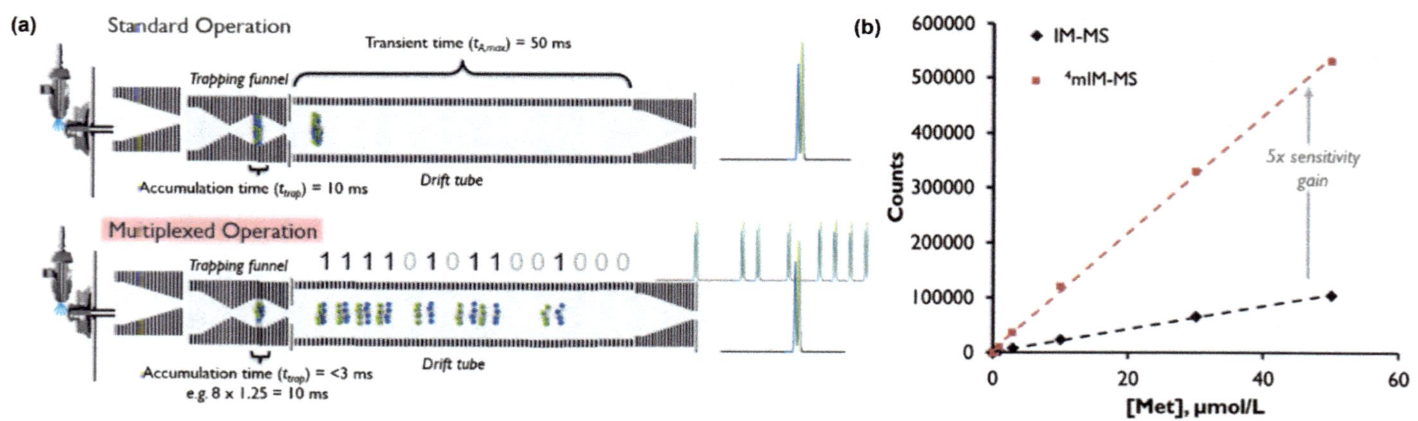

Figure 7.2 (a) Schematic representation of the multiplexing and demultiplexed result in comparison to standard IM–MS operation. In this example, a 20% ion utilization efficiency is realized in both examples, but the 4-bit multiplexing provides reduced system saturation and improved analytical performance for metabolomics as shown by (b) gains in the sensitivity for low molecular mass metabolites (shown here is methionine, $[M+H]^+ = 150.0583$) assessed *via* quantitative spiking experiments.

Of critical importance for the use of IM–MS for metabolomics, the primary method to determine ^{DT}CCS (stepped-field method) with DTIM–MS involves measurement of arrival times (t_A) consisting of the drift time (t_d) and total time spent outside of the drift region (*i.e.*, ion optics, collision cell and the flight tube) referred to as t_0. To be able to determine ^{DT}CCS, a linear calibration approach is used, whereby drift tube field strengths are varied in a stepwise fashion. ^{DT}CCS is then determined by correction of the arrival times using the t_0 (intercept) of the linear model (*i.e.*, $t_d = t_A - t_0$).[32] Commercial low pressure DTIM–QTOF instrumentation can operate with an interlaboratory RSD of <0.3% using this method for $^{DT}CCS_{N2}$ determination.[32] This is important as other external IM–MS calibration methods require accurate ^{DT}CCS values.[17] However, this method is time-consuming and cannot be used for LC–DTIM–QTOFMS methods. Instead, an external single-field calibration is performed using a linear model based on a set of calibrant ions with consensus $^{DT}CCS_{N2}$ values. Single-field calibration is used routinely for DTIM–MS metabolomics[3,32] and allows $^{DT}CCS_{N2}$ determination with <0.6% bias compared to the primary method.[32] In combination with a growing number of curated CCS databases, this is now an established way to use DTIM–MS in identity confirmation workflows for metabolomics.[33–35] Furthermore, the alignment of precursor and fragment ions in the mobility dimension offers the possibility to re-establish precursor–product ion relationships in data independent acquisition (DIA) strategies, such as all ions (AI) fragmentation using the Agilent instrumentation.[36] Additionally, the collision energy can be programmed to be ramped in a drift time-dependent manner.[37]

7.2.2 TWIM–QTOFMS

Traveling wave ion mobility technology was commercialized by Waters in 2006 and allows ions to be separated while traversing a gas-filled IM cell in a time-dispersive setup (Figure 7.1). However, instead of a linear DC potential, the stacked-ring electrode ion guide (SRIG) is used to apply traveling potential waves characterized by a "wave height" (*V*) and traveling speed ($m\,s^{-1}$). Additional RF voltages are used to constrain ion packages radially and increase transmission efficiency. Ions are transported through the buffer gas using patterns of repeating voltage pulses superimposed on the RF-confining field.[38,39] The pulses are then applied to the next pair of electrodes in the SRIG resulting in a traveling potential wave, and ions tend to "surf" on approaching potential waves. Low mobility ions are overtaken by the traveling wave resulting in negative axial traveling velocities, or "roll-over effects", due to the negative acceleration potential gradient exerted on an ion before steady state drift velocity can be re-established.[38,39] During method development, different parameters can be tuned in TWIMS including shape, height and velocity.[17] However, ions can experience higher field strengths in TWIMS resulting in higher effective ion temperatures and can undergo ion activation, fragmentation and isomerization.[21] Since the introduction of the first commercial TWIM–MS instrument, six generations of instruments have been released. This includes different instrumental geometries in the TWIM–QTOF (Vion) and Q-TWIM–TOF Synapt series (G1, G2, G2-S and G2-Si) and cyclic TWIM (cIM) configurations. These variants offer flexible data acquisition strategies to be employed including IM separation of precursor ions prior to CID, IM separation of fragment ions after CID and IM–MS of quadrupole-selected ions.[38] Using TWIMS, ion mobility resolving power is limited (*e.g.*, $R_p \sim 40$ CCS/ΔCCS using Synapt G2-Si) and can be increased using cIM, whereby a single pass offers resolving power of $R_p \sim 65$ CCS/ΔCCS.[40]

In terms of metabolomics applications, a seminal interlaboratory and interplatform comparison in 2014 demonstrated that $^{TW}CCS_{N2}$ can be determined with mean RSD lower than 2% for 97% of ions using different types of TWIM–MS instruments.[9] Furthermore, addition of mobility separation to DIA workflows (HDMS^E) can be employed on a routine basis. Additionally, tailored drift time-dependent collision energy programs allow additional flexibility to cope with insufficient fragmentation in DIA.[6]

From a fundamental perspective, ion transport and separation using TWIMS are demonstrably more complex than separation in DTIMS. In TWIMS, non-static electric fields are used, and non-linear contributions to the determined mobilities can be observed.[19,26] ^{TW}CCS determination thus heavily relies on the quality of external calibration particularly as non-linear contributions to ^{TW}CCS and related uncertainties are not fully understood.[17,41] Careful choice of calibration strategies including the calibration function and choice of calibrant ions is necessary to reduce compound-specific bias on determined mobilities. Some experimental studies have suggested that the used calibrant ions should possess similar chemical properties as the targeted class of analyte molecules to minimize errors caused by the calibration approach.[17,26,38] Additionally, $^{DT}CCS_{He}$ values have often been used for CCS calibration of TWIM–MS systems, which can introduce additional compound-dependent differences in determined mobilities due to the different polarizability of He and N_2.[17] Depending on the used calibration method, linear, polynomial or power functions are used to calibrate for ^{TW}CCS determination, which should be reported alongside the determined CCS values and mobilities.[17,38] Libraries for experimentally determined ^{TW}CCS values for both metabolomics and natural products are commercially available while several other related databases have been published for related applications including pesticides,[42] steroids[34] and several others within the scope of food analysis.[43]

7.2.3 TIMS–QTOFMS

In TIMS technology (commercialized by Bruker in 2016), an axial electric field gradient (EFG) holds ions stationary against a gas flow while an RF field is used for radial confinement of ion packages (Figure 7.1). By decreasing or "ramping" of the plateau level of the EFG, ions are sequentially released based on their mobility differences.[44,45] Prior to release, ions are trapped in the TIMS funnel for a user-defined time period. Typically, trapping times of a few milliseconds are used, but trapping times of up to one second can be applied if required.[44] As a considerable advantage over DTIM–MS and TWIM–MS, TIMS can achieve native resolving powers of up to 250 with a relatively small separation device. However, it is noted that this level of resolution can only be achieved across a narrow *m/z* and mobility range. Operation of TIMS–QTOFMS for generic, non-targeted metabolomics across a suitable mass range yields resolving powers that are not superior to either TWIM–MS or DTIM–MS. Other significant technological developments for TIMS–QTOFMS include the introduction of tandem or dual TIMS also allows a high degree of freedom in data acquisition strategies,[46] which have very recently emerged primarily in developments for proteomics. Parallel Accumulation and Sequential Acquisition (PASEF) allows users to perform synchronized DDA isolation and fragmentation using sub-millisecond switching times of the quadrupole isolation window during a single TIMS scan (~50 ms) and an optimized precursor selection strategy, which allows an increased acquisition speed of DDA without loss of sensitivity.[7,47]

Different calibration strategies have been reported for ^{TIMS}CCS determination and were comparable to ^{DT}CCS with differences of 1%,[17] but larger differences have also been reported elsewhere.[48] A sample-independent transferable calibration strategy allowing TIMS-calibration for various instrument settings and compound classes has been reported recently.[49] However, while some experimental studies indicate compatibility of ^{TIM}CCS values with ^{DT}CCS and ^{TW}CCS values, it has also been demonstrated that gas-phase ion mobilities cannot be derived from first principles using TIMS due to changes in *E/N* caused by non-linear contributions to pressure changes during TIMS separation.[50] Within the metabolomics community, the Sumner group recently published a $^{TIMS}CCS_{N2}$ library for 146 natural products for plant metabolomics.[48]

7.3 Data Processing

Considering the changes in data acquisition in comparison to MS only, the use of IM–MS in metabolomics can lead to an overestimation of the "true" number of metabolites in a sample, because different charge states, adducts and in-source fragments are separated from the precursor ion by IMS.[19] Additionally, the occurrence of protomers,[18,51] other conformers and even kinetically trapped conformations[52] can further complicate automation of data processing and interpretation. To support users in dealing with these complex datasets, all providers of commercially available IM–QTOFMS instruments offer vendor-specific software tools supporting the analysis of IM–QTOFMS data. This includes tools for data visualization, CCS calibration, peak picking, statistical evaluation tools and software used for identity confirmation distributed by TOFWerk (Tofware), Agilent (MassHunter Mass Profiler and IM-MS Browser),[53] Waters (UNIFI,

DriftScope/MassLynx and Progenesis QI)[34,41,54,55] and Bruker (Metaboscape and DataAnalysis).[48] Although LC–IM–QTOFMS is an attractive acquisition strategy for metabolomics, data processing remains a challenging task, and multiple software tools are necessary to allow comprehensive data evaluation for demanding applications.[36] Fortunately, open-source software tools capable of pre-processing data[56] and evaluating datasets have also emerged during recent years including prominently Skyline[36] and MS-DIAL.[57,58]

Skyline is an open-source software for targeted data extraction of LC–MS data from various instrument providers and allows importation and batch-analysis of LC–IM–MS data. Initially, the software was developed for treating proteomics data, but analysis of metabolomics ("small molecule") data is now also supported. The use of mobility-based filtering of the extracted data was demonstrated to have a beneficial effect on data quality in terms of limit of detection and linear range of calibration curves.[36,59] Furthermore, several studies have illustrated the practical application of this software for evaluation of LC–IM–MS metabolomics data with the advantage of direct access of raw datafiles making it an ideal choice for non-target screening and retrospective analysis using retention time, CCS, accurate *m/z* and fragment ions as identification points (Figure 7.3).[30,36]

MS-DIAL is a non-vendor data processing software for metabolomics and lipidomics data supporting analysis of data dependent (DDA) and data independent acquisition (DIA) data and includes centroiding, peak picking, feature alignment, chromatographic deconvolution and the possibility to perform identity confirmation using MS^2 libraries.[57] When treating DIA data, chromatographic deconvolution is a suitable approach to cope with mixed fragment spectra and to re-establish the precursor–fragment relationship.[57] In analogy, deconvolution of IM–MS data is a suitable way to clean fragment spectra of unresolved compounds in IM–QTOFMS, especially for IM-assisted DIA (*e.g.*, $HDMS^E$).[60,61] With release of software version 4, MS-DIAL now offers the possibility to analyze and deconvolute LC–IM–QTOFMS data[58] (Figure 7.4). Besides the built-in functions, MS-DIAL can be linked to different software tools, *e.g.*, MS-Finder can be used for *in silico* fragmentation and identity confirmation,[62,63] and processed LC–IM–QTOFMS data can be imported in the GNPS (Global Natural Product Social Molecular Networking) environment to perform feature-based molecular networking.[64] Relying on an initial conversion of raw datafiles, comparison to vendor-specific solutions and experimental validation remain important for validation of data processing workflows. Overall, given the huge capabilities relevant to metabolomics, MS-DIAL is emerging as a very attractive platform for evaluation of IM–QTOFMS data for a wide range of metabolomics applications.

In addition to these two major non-vendor options, other popular software tools have been applied for the evaluation of IM–MS data relevant to metabolomics studies. Some specialized evaluation tools developed for IM–MS proteomics datasets (*e.g.*, Mobi-DIK extension of OpenSWATH)[65] also have the potential to be adapted for metabolomics applications in the future.

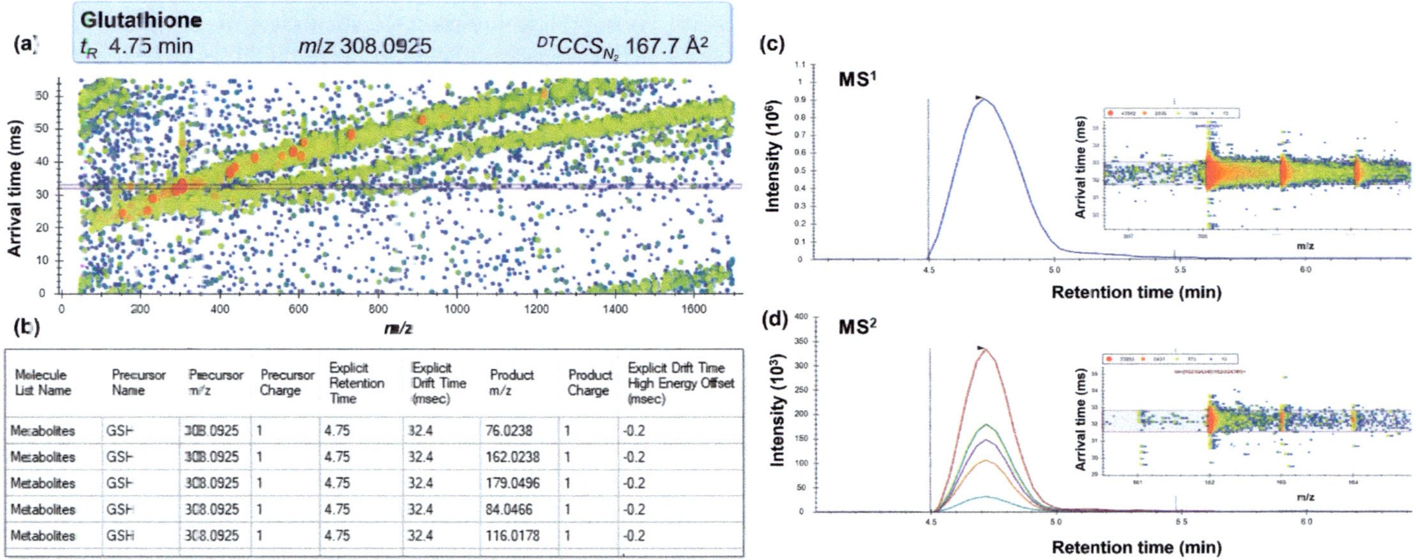

Molecule List Name	Precursor Name	Precursor m/z	Precursor Charge	Explicit Retention Time	Explicit Drift Time (msec)	Product m/z	Product Charge	Explicit Drift Time High Energy Offset (msec)
Metabolites	GSH	308.0925	1	4.75	32.4	76.0238	1	-0.2
Metabolites	GSH	308.0925	1	4.75	32.4	162.0238	1	-0.2
Metabolites	GSH	308.0925	1	4.75	32.4	179.0496	1	-0.2
Metabolites	GSH	308.0925	1	4.75	32.4	84.0466	1	-0.2
Metabolites	GSH	308.0925	1	4.75	32.4	116.0178	1	-0.2

Figure 7.3 (a) Arrival time *vs.* *m*/*z* heat map representation obtained by extracting data from the apex of the chromatographic peak with *m*/*z* 308.0925 and retention time of 4.75 min using Skyline. (b) Transition list used for metabolite screening at TOFMS and QTOFMS levels. (c) Mobility-filtered extracted ion chromatogram (EIC) of precursor ion of glutathione (*m*/*z* 308.0925) and (d) corresponding EICs of IM–QTOFMS fragment ions.

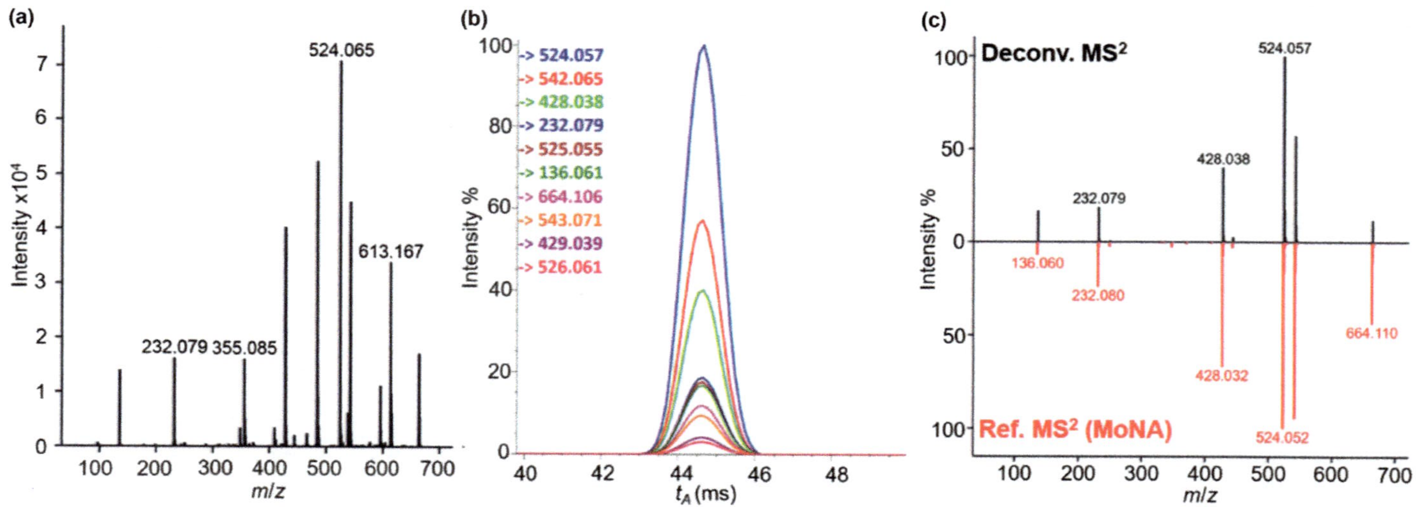

Figure 7.4 (a) Raw high energy spectrum (20 V collision energy) of co-eluting NAD^+ and interfering GSSG with overlapping mobility peaks in cellular yeast extracts acquired using All Ions (AI) fragmentation on a LC–DTIM–QTOF platform. (b) Arrival time spectra of cleaned up high energy spectra after deconvolution using MS-DIAL. (c) Deconvoluted MS^2 spectrum and matching reference spectrum in Mass Bank of North America (MoNA).

7.4 Applications in Metabolomics

The range of metabolomics applications using IM–MS as a key platform reported during recent years is very diverse. Most published applications have focused on the combination of reversed-phase liquid chromatography with IM–MS using electrospray ionization, which provides the broadest coverage for non-targeted metabolomics within a single analysis. As the oldest of the commercial platforms, the use of TWIM integrated in QTOFMS for metabolomics and related applications has been extensively reported since the early work of Paglia[9] and includes studies of plant metabolites,[66,67] drugs of abuse[68] and pesticides.[69] Further applications and relationship to the development of this family of instrumentation can be found in a recent review.[38] Similarly, metabolomics applications using commercial DTIM–QTOFMS platforms encompass a diverse range of sample types over the last 5 years including human plasma[70] and HaCaT cells,[71] characterization of the NIST1950 reference material,[72] environmental studies,[73–75] and also further plant-related applications including medical plants,[76] food[77,78] and non-targeted analysis of different varieties of wine.[53] In comparison, only limited studies relevant to non-targeted metabolomics have been reported with TIMS–MS to date, but, given the rapid developments in related applications,[48,68,79] this is expected to change during the next years. The reader is also referred to a recent review covering metabolomics applications of IM–MS up to 2020 in some detail.[80]

Of importance for the broad acceptance of this technology for metabolomics applications, several studies have reported that no matrix effects affecting CCS determination were observed for complex matrices such as urine,[81] wine[53] and fish feed[69] using different LC–IM–QTOFMS methods and instrument platforms. Additionally, the use of different classes of Q-TWIM–TOFMS and TWIM–QTOFMS instrumentation was evaluated for the analysis of steroids and phase two metabolites in a recent interlaboratory effort which is an exemplary work for establishing IM–QTOFMS within a fully validated method for relevant applications.[34,81] In terms of the metabolome coverage improvement offered by using IM as an additional separation dimension in non-targeted metabolomics, it is noted that using IM–QTOFMS in metabolomics can easily lead to an overestimation of the "true" number of metabolites in a sample due to the separation of different charge states, conformers, adducts and in-source fragments that are separated from the precursor ion by IMS.[19] While this issue has previously led to some erroneous overestimations of the total peak capacity improvement when adding IM separation to metabolomics workflows, the know-how for interpreting higher dimensionality datasets is now available, and a stronger focus on analytical validation can be expected for metabolomics applications in the future.

7.5 Discussion and Outlook

Looking to the technical developments that may support metabolomics applications in the future, some additional instrumental developments are currently emerging to overcome the limited resolving power of current IMS instrumentation, *i.e.*, cyclic IMS and structures for loss-less ion manipulation (SLIM). A recently introduced cyclic Q-TWIM–TOF instrument allows users to perform IMS^n acquisition using multipass ion mobility separation and fragmentation and, at least in single pass mode, determination of ^{TW}CCS values has been demonstrated. However, this technology faces the same challenges and limitations regarding calibration as discussed for TWIM–MS instruments in Section 7.2, and it is noted that the major focus of the development is for characterization of larger molecules (*e.g.*, proteins) and not for metabolomics. Additionally, the decreased acquisition speed with increasing resolving power has to be considered for applying such techniques in combination with chromatographic separations.[82] In the case of SLIM, this technology is based on electrode arrays in long serpentine pathways that allow extended drift paths of >100 m and are offering resolving powers of >400 with cycle times on the second timescale. Using prototype SLIM devices, high resolving power has been demonstrated to allow the discrimination of small differences in the mobilities of isotopologues and isotopomers due to the reduced mass contribution to mobility and different centers of mass, respectively.[83] However, the practical application and suitability for CCS determinations for this technology need to be assessed for LC–IM–MS metabolomics as no commercial product is currently available on the market.[83]

As the key identification point derived from IM–MS, a critical evaluation of the value of CCS values from different instruments must also be part of the future discussion for metabolomics. Together with the proven ruggedness of $^{DT}CCS_{N2}$ as a molecular descriptor, the growing number of publicly available and curated ^{DT}CCS databases is an important step toward routine use of this identification point for non-targeted metabolomics with any of the commercially available platforms. This allows an additional constraint for identity confirmation of unknown compounds as experimental ^{DT}CCS values can be compared against experimental values or, when not available, theoretical CCS values of candidate structures. Although much research effort has been expended toward the experimental generation and curation of a growing number of such CCS repositories,[33–35,48] the coverage of such databases is still limited for routine non-targeted metabolomics, and computationally derived databases based on CCS modeling or machine learning are aiming to fill this gap.[19,66,67,84] Amongst these *in silico* approaches, all offer different benefits and limitations in terms of accuracy or computational effort, but none can currently generate reference CCS values that can be used and applied with equivalently low uncertainty as is the case for *m/z* calibration.[18,19,84,85] It is also important to note that experimental measurement conditions remain important as experimental gas phase structures can differ from predictions, and even small molecule ions are observed to yield more than a single IM peak (*e.g.*, protomeric isomers). Finally, as the current estimates for uncertainty for external calibrant $^{DT}CCS_{N2}$ values lie in the range of 2.7–4.6% (coverage factor = 2),[86] we note that computationally derived CCS values should be used carefully, particularly when cross-platform IM–MS comparisons are considered and experimental biases from calibration procedures cannot be corrected for.[17,19] Inter-platform comparisons have shown a strong correlation of *CCS* values obtained from different instrumental platforms for most small molecule ions, *i.e.*, $^{DT}CCS_{N2}$ and $^{TW}CCS_{N2}$ differences <1%. However, differences can increase to >6% depending on the compound and ion species.[87] Depending on the compound class, average differences between $^{TIMS}CCS_{N2}$ and $^{DT}CCS_{N2}$ can vary between 0.5% and 2.1%.[48] Clearly, comparisons of CCS values obtained from different instrumental IM–QTOFMS platforms need to be performed with caution due to the differences in ion transport, instrumental precision and applied external calibration strategies in order to avoid misinterpretations and misannotations in metabolomics. Unfortunately, the missing agreement on primary calibrants and the lack of common reference materials for different compound classes in different matrices complicate a comprehensive assessment of instrumental performance and subsequent differences in CCS using different IM–QTOFMS platforms,[17] which remains an open issue for future research.

Besides increasing resolution of IMS instruments, strategies to combine the selectivities of IMS separation and quadrupole isolation have been introduced during recent years and show promising results. Such approaches include synchronized application of SONAR technology[88] with IM-enhanced HDMS^E for TWIM–QTOFMS, IM-enhanced DDA (*i.e.*, PASEF for TIMS–QTOFMS)[7] and DIA relying on fast switching of quadrupole isolation windows, which can also be widened beyond 50 *m*/*z* and rapidly ramped as demonstrated using Agilent DTIM–QTOFMS.[89] Recent work has also demonstrated development of a DTIM–QTOFMS DIA strategy using this technology with application to discovery-oriented metabolomics.[90] These emerging technologies offer a practical means to maximize coverage of a molecular class of interest, while excluding unwanted precursors from the DIA fragmentation, which is advantageous for some non-targeted metabolomics applications.

7.6 Conclusions

IM–MS is now a well-recognized technology in the field of metabolomics. Using commercially available platforms, high dimensional metabolomics datasets can be now obtained on a routine basis by experienced users. The addition of a generic IM separation to existing LC–MS metabolomics methods generally allows an increase in selectivity by improving the peak capacity and allows the use of CCS as an additional identification point for all metabolites measured. Furthermore, the use of DIA strategies with IM–QTOFMS workflows allows MS^2 spectra to be deconvoluted with consideration of both chromatographic and IM peaks, which provides high quality datasets for identity confirmation without significant losses in ion utilization efficiency (*e.g.*, compared to DIA with QTOF only). However, some challenges for metabolomics still remain including the resources required for data analysis and interpretation, the limited capacity and resolving power of generic IM analyzers, and finally the inability to predict "true" CCS values for all ions of a given metabolite structure solely from molecular properties. This last issue renders careful external CCS calibration and experimental evaluations as essential aspects for the development of reliable use of this property for validated applications. However, we have seen that IM–QTOFMS has brought and will continue to bring new opportunities for metabolomics studies, particularly when targeted interrogations of large sub-classes of metabolites are a focal point in data evaluation, or when fundamental studies of conformers, protomeric isomers and related ion formation chemistry are of interest.

References

1. A. C. Schrimpe-Rutledge, S. G. Codreanu, S. D. Sherrod and J. A. McLean, *J. Am. Soc. Mass Spectrom.*, 2016, **27**, 1897–1905.
2. F. Fenaille, P. Barbier Saint-Hilaire, K. Rousseau and C. Junot, *J. Chromatogr. A*, 2017, **1526**, 1–12.
3. T. Mairinger, T. J. Causon and S. Hann, *Curr. Opin. Chem. Biol.*, 2018, **42**, 9–15.
4. K. Ortmayr, T. J. Causon, S. Hann and G. Koellensperger, *TrAC, Trends Anal. Chem.*, 2016, **82**, 358–366.
5. I. Blaženović, T. Kind, J. Ji and O. Fiehn, *Metabolites*, 2018, **8**, 31.
6. U. Distler, J. Kuharev, P. Navarro, Y. Levin, H. Schild and S. Tenzer, *Nat. Methods*, 2014, **11**, 167–170.
7. F. Meier, A.-D. Brunner, S. Koch, H. Koch, M. Lubeck, M. Krause, N. Goedecke, J. Decker, T. Kosinski, M. A. Park, N. Bache, O. Hoerning, J. Cox, O. Räther and M. Mann, *Mol. Cell. Proteomics*, 2018, **17**, 2534–2545.
8. J. C. May and J. A. McLean, *Anal. Chem.*, 2015, **87**, 1422–1436.
9. G. Paglia, J. P. Williams, L. Menikarachchi, J. W. Thompson, R. Tyldesley-Worster, S. Halldórsson, O. Rolfsson, A. Moseley, D. Grant, J. Langridge, B. O. Palsson and G. Astarita, *Anal. Chem.*, 2014, **86**, 3985–3993.
10. T. J. Causon, R. T. Kurulugama and S. Hann, in *Ion Mobility-Mass Spectrometry: Methods and Protocols*, ed. G. Paglia and G. Astarita, Springer US, New York, NY, 2020, pp. 79–94.
11. I. Blaženović, T. Shen, S. S. Mehta, T. Kind, J. Ji, M. Piparo, F. Cacciola, L. Mondello and O. Fiehn, *Anal. Chem.*, 2018, **90**, 10758–10764.
12. G. Paglia, P. Angel, J. P. Williams, K. Richardson, H. J. Olivos, J. W. Thompson, L. Menikarachchi, S. Lai, C. Walsh, A. Moseley, R. S. Plumb, D. F. Grant, B. O. Palsson, J. Langridge, S. Geromanos and G. Astarita, *Anal. Chem.*, 2015, **87**, 1137–1144.
13. T. J. Causon and S. Hann, *J. Chromatogr. A*, 2015, **1416**, 47–56.
14. A. Kaufmann, P. Butcher, K. Maden, S. Walker and M. Widmer, *Anal. Chim. Acta*, 2020, **1107**, 113–126.

15. Y.-F. Xu, W. Lu and J. D. Rabinowitz, *Anal. Chem.*, 2015, **87**, 2273–2281.
16. T. De Vijlder, D. Valkenborg, F. Lemière, E. P. Romijn, K. Laukens and F. Cuyckens, *Mass Spectrom. Rev.*, 2018, **37**, 607–629.
17. V. Gabelica, A. A. Shvartsburg, C. Afonso, P. Barran, J. L. P. Benesch, C. Bleiholder, M. T. Bowers, A. Bilbao, M. F. Bush, J. L. Campbell, I. D. G. Campuzano, T. Causon, B. H. Clowers, C. S. Creaser, E. De Pauw, J. Far, F. Fernandez-Lima, J. C. Fjeldsted, K. Giles, M. Groessl, C. J. Hogan Jr, S. Hann, H. I. Kim, R. T. Kurulugama, J. C. May, J. A. McLean, K. Pagel, K. Richardson, M. E. Ridgeway, F. Rosu, F. Sobott, K. Thalassinos, S. J. Valentine and T. Wyttenbach, *Mass Spectrom. Rev.*, 2019, **38**, 291–320.
18. S. Warnke, J. Seo, J. Boschmans, F. Sobott, J. H. Scrivens, C. Bleiholder, M. T. Bowers, S. Gewinner, W. Schöllkopf, K. Pagel and G. von Helden, *J. Am. Chem. Soc.*, 2015, **137**, 4236–4242.
19. V. Gabelica and E. Marklund, *Curr. Opin. Chem. Biol.*, 2018, **42**, 51–59.
20. D. Morsa, E. Hanozin, G. Eppe, L. Quinton, V. Gabelica and E. D. Pauw, *Anal. Chem.*, 2020, **92**, 4573–4582.
21. D. Morsa, V. Gabelica and E. De Pauw, *J. Am. Soc. Mass Spectrom.*, 2014, **25**, 1384–1393.
22. C. Bleiholder, F. C. Liu and M. Chai, *Anal. Chem.*, 2020, **92**, 16329–16333.
23. K. M. Szykuła, J. Meurs, M. A. Turner, C. S. Creaser and J. C. Reynolds, *Anal. Bioanal. Chem.*, 2019, **411**, 6309–6317.
24. J. L. Willems, M. M. Khamis, W. M. Saeid, R. W. Purves, G. Katselis, N. H. Low and A. El-Aneed, *Anal. Chim. Acta*, 2016, **933**, 164–174.
25. M. S. Wei, R. H. J. Kemperman, M. A. Palumbo and R. A. Yost, *J. Am. Soc. Mass Spectrom.*, 2020, **31**, 355–365.
26. M. F. Bush, I. D. G. Campuzano and C. V. Robinson, *Anal. Chem.*, 2012, **84**, 7124–7130.
27. K. J. Pacholarz and P. E. Barran, *Anal. Chem.*, 2015, **87**, 6271–6279.
28. M. Groessl, S. Graf and R. Knochenmuss, *Analyst*, 2015, **140**, 6904–6911.
29. B. H. Clowers, Y. M. Ibrahim, D. C. Prior, W. F. Danielson, M. E. Belov and R. D. Smith, *Anal. Chem.*, 2008, **80**, 612–623.
30. T. J. Causon, L. Si-Hung, K. Newton, R. T. Kurulugama, J. Fjeldsted and S. Hann, *Anal. Bioanal. Chem.*, 2019, **411**, 6265–6274.
31. B. H. Clowers, W. F. Siems, H. H. Hill and S. M. Massick, *Anal. Chem.*, 2006, **78**, 44–51.
32. S. M. Stow, T. J. Causon, X. Zheng, R. T. Kurulugama, T. Mairinger, J. C. May, E. E. Rennie, E. S. Baker, R. D. Smith, J. A. McLean, S. Hann and J. C. Fjeldsted, *Anal. Chem.*, 2017, **89**, 9048–9055.
33. J. A. Picache, B. S. Rose, A. Balinski, K. L. Leaptrot, S. D. Sherrod, J. C. May and J. A. McLean, *Chem. Sci.*, 2018, **10**, 983–993.
34. M. Hernández-Mesa, B. Le Bizec, F. Monteau, A. M. García-Campaña and G. Dervilly-Pinel, *Anal. Chem.*, 2018, **90**, 4616–4625.
35. W. B. Struwe, K. Pagel, J. L. P. Benesch, D. J. Harvey and M. P. Campbell, *Glycoconjugate J.*, 2016, **33**, 399–404.

36. B. X. MacLean, B. S. Pratt, J. D. Egertson, M. J. MacCoss, R. D. Smith and E. S. Baker, *J. Am. Soc. Mass Spectrom.*, 2018, **29**, 2182–2188.
37. E. S. Baker, K. Tang, W. F. Danielson, D. C. Prior and R. D. Smith, *J. Am. Soc. Mass Spectrom.*, 2008, **19**, 411–419.
38. I. D. G. Campuzano and K. Giles, *TrAC, Trends Anal. Chem.*, 2019, **120**, 115620.
39. K. Giles, J. L. Wildgoose, D. J. Langridge and I. Campuzano, *Int. J. Mass Spectrom.*, 2010, **298**, 10–16.
40. M. McCullagh, K. Giles, K. Richardson, S. Stead and M. Palmer, *Rapid Commun. Mass Spectrom.*, 2019, **33**, 11–21.
41. M. Hernández-Mesa, V. D'Atri, G. Barknowitz, M. Fanuel, J. Pezzatti, N. Dreolin, D. Ropartz, F. Monteau, E. Vigneau, S. Rudaz, S. Stead, H. Rogniaux, D. Guillarme, G. Dervilly and B. Le Bizec, *Anal. Chem.*, 2020, **92**, 5013–5022.
42. S. Goscinny, M. McCullagh, J. Far, E. De Pauw and G. Eppe, *Rapid Commun. Mass Spectrom.*, 2019, **33**, 34–48.
43. M. Hernández-Mesa, D. Ropartz, A. M. García-Campaña, H. Rogniaux, G. Dervilly-Pinel and B. Le Bizec, *Molecules*, 2019, **24**, 2706.
44. K. Michelmann, J. A. Silveira, M. E. Ridgeway and M. A. Park, *J. Am. Soc. Mass Spectrom.*, 2015, **26**, 14–24.
45. J. A. Silveira, K. Michelmann, M. E. Ridgeway and M. A. Park, *J. Am. Soc. Mass Spectrom.*, 2016, **27**, 585–595.
46. F. C. Liu, M. E. Ridgeway, M. A. Park and C. Bleiholder, *Analyst*, 2018, **143**, 2249–2258.
47. F. Meier, S. Beck, N. Grassl, M. Lubeck, M. A. Park, O. Raether and M. Mann, *J. Proteome Res.*, 2015, **14**, 5378–5387.
48. M. Schroeder, S. W. Meyer, H. M. Heyman, A. Barsch and L. W. Sumner, *Metabolites*, 2019, **10**, 13.
49. M. Chai, M. N. Young, F. C. Liu and C. Bleiholder, *Anal. Chem.*, 2018, **90**, 9040–9047.
50. C. N. Naylor, T. Reinecke, M. E. Ridgeway, M. A. Park and B. H. Clowers, *J. Am. Soc. Mass Spectrom.*, 2019, **30**, 2152–2162.
51. J. Boschmans, S. Jacobs, J. P. Williams, M. Palmer, K. Richardson, K. Giles, C. Lapthorn, W. A. Herrebout, F. Lemière and F. Sobott, *Analyst*, 2016, **141**, 4044–4054.
52. H. Xia and A. B. Attygalle, *J. Am. Soc. Mass Spectrom.*, 2017, **28**, 2580–2587.
53. T. J. Causon, V. Ivanova-Petropulos, D. Petrusheva, E. Bogeva and S. Hann, *Anal. Chim. Acta*, 2019, **1052**, 179–189.
54. J. Zhang, W. Yang, S. Li, S. Yao, P. Qi, Z. Yang, Z. Feng, J. Hou, L. Cai, M. Yang, W. Wu and D. Guo, *Anal. Bioanal. Chem.*, 2016, **408**, 3881–3890.
55. C. Zhang, T. Zuo, X. Wang, H. Wang, Y. Hu, Z. Li, W. Li, L. Jia, Y. Qian, W. Yang and H. Yu, *Molecules*, 2019, **24**, 2708.
56. A. Bilbao, B. C. Gibbons, G. W. Slysz, K. L. Crowell, M. E. Monroe, Y. M. Ibrahim, R. D. Smith, S. H. Payne and E. S. Baker, *Int. J. Mass Spectrom.*, 2018, **427**, 91–99.

57. H. Tsugawa, T. Cajka, T. Kind, Y. Ma, B. Higgins, K. Ikeda, M. Kanazawa, J. VanderGheynst, O. Fiehn and M. Arita, *Nat. Methods*, 2015, **12**, 523–526.
58. H. Tsugawa, K. Ikeda, M. Takahashi, A. Satoh, Y. Mori, H. Uchino, N. Okahashi, Y. Yamada, I. Tada, P. Bonini, Y. Higashi, Y. Okazaki, Z. Zhou, Z.-J. Zhu, J. Koelmel, T. Cajka, O. Fiehn, K. Saito, M. Arita and M. Arita, *Nat. Biotechnol.*, 2020, **38**, 1159–1163.
59. K. J. Adams, B. Pratt, N. Bose, L. G. Dubois, L. St John-Williams, K. M. Perrott, K. Ky, P. Kapahi, V. Sharma, M. J. MacCoss, M. A. Moseley, C. A. Colton, B. X. MacLean, B. Schilling, J. W. Thompson and Alzheimer's Disease Metabolomics Consortium, *J. Proteome Res.*, 2020, **19**, 1447–1458.
60. M. Brantley, B. Zekavat, B. Harper, R. Mason and T. Solouki, *J. Am. Soc. Mass Spectrom.*, 2014, **25**, 1810–1819.
61. M. E. Pettit, M. R. Brantley, F. Donnarumma, K. K. Murray and T. Solouki, *Analyst*, 2018, **143**, 2574–2586.
62. H. Tsugawa, T. Kind, R. Nakabayashi, D. Yukihira, W. Tanaka, T. Cajka, K. Saito, O. Fiehn and M. Arita, *Anal. Chem.*, 2016, **88**, 7946–7958.
63. A. Vaniya, S. N. Samra, M. Palazoglu, H. Tsugawa and O. Fiehn, *Phytochem. Lett.*, 2017, **21**, 306–312.
64. L.-F. Nothias, D. Petras, R. Schmid, K. Dührkop, J. Rainer, A. Sarvepalli, I. Protsyuk, M. Ernst, H. Tsugawa, M. Fleischauer, F. Aicheler, A. A. Aksenov, O. Alka, P.-M. Allard, A. Barsch, X. Cachet, A. M. Caraballo-Rodriguez, R. R. Da Silva, T. Dang, N. Garg, J. M. Gauglitz, A. Gurevich, G. Isaac, A. K. Jarmusch, Z. Kameník, K. B. Kang, N. Kessler, I. Koester, A. Korf, A. Le Gouellec, M. Ludwig, C. Martin H, L.-I. McCall, J. McSayles, S. W. Meyer, H. Mohimani, M. Morsy, O. Moyne, S. Neumann, H. Neuweger, N. H. Nguyen, M. Nothias-Esposito, J. Paolini, V. V. Phelan, T. Pluskal, R. A. Quinn, S. Rogers, B. Shrestha, A. Tripathi, J. J. J. van der Hooft, F. Vargas, K. C. Weldon, M. Witting, H. Yang, Z. Zhang, F. Zubeil, O. Kohlbacher, S. Böcker, T. Alexandrov, N. Bandeira, M. Wang and P. C. Dorrestein, *Nat. Methods*, 2020, **17**, 905–908.
65. H. L. Röst, G. Rosenberger, P. Navarro, L. Gillet, S. M. Miladinović, O. T. Schubert, W. Wolski, B. C. Collins, J. Malmström, L. Malmström and R. Aebersold, *Nat. Biotechnol.*, 2014, **32**, 219–223.
66. Z. Zhou, M. Luo, X. Chen, Y. Yin, X. Xiong, R. Wang and Z.-J. Zhu, *Nat. Commun.*, 2020, **11**, 4334.
67. Z. Zhou, X. Xiong and Z.-J. Zhu, *Bioinformatics*, 2017, **33**, 2235–2237.
68. K. J. Adams, C. E. Ramirez, N. F. Smith, A. C. Muñoz-Muñoz, L. Andrade and F. Fernandez-Lima, *Talanta*, 2018, **183**, 177–183.
69. J. Regueiro, N. Negreira and M. H. G. Berntssen, *Anal. Chem.*, 2016, **88**, 11169–11177.
70. X. Zhang, R. Knochenmuss, W. F. Siems, W. Liu, S. Graf and H. H. Hill, *Anal. Chem.*, 2014, **86**, 1661–1670.
71. X. Zhang, K. Kew, R. Reisdorph, M. Sartain, R. Powell, M. Armstrong, K. Quinn, C. Cruickshank-Quinn, S. Walmsley, S. Bokatzian, E. Darland, M. Rain, K. Imatani and N. Reisdorph, *Anal. Chem.*, 2017, **89**, 6384–6391.

72. C. M. Nichols, J. N. Dodds, B. S. Rose, J. A. Picache, C. B. Morris, S. G. Codreanu, J. C. May, S. D. Sherrod and J. A. McLean, *Anal. Chem.*, 2018, **90**, 14484–14492.
73. L. Emhofer, M. Himmelsbach, W. Buchberger and C. W. Klampfl, *J. Chromatogr. A*, 2019, **1592**, 122–132.
74. X. Zheng, K. T. Dupuis, N. A. Aly, Y. Zhou, F. B. Smith, K. Tang, R. D. Smith and E. S. Baker, *Anal. Chim. Acta*, 2018, **1037**, 265–273.
75. S. Stephan, J. Hippler, T. Köhler, A. A. Deeb, T. C. Schmidt and O. J. Schmitz, *Anal. Bioanal. Chem.*, 2016, **408**, 6545–6555.
76. M. Fenclova, M. Stranska-Zachariasova, F. Benes, A. Novakova, P. Jonatova, V. Kren, L. Vitek and J. Hajslova, *Anal. Bioanal. Chem.*, 2020, **412**, 819–832.
77. W. Liu, X. Zhang, R. Knochenmuss, W. F. Siems and H. H. Hill, *J. Am. Soc. Mass Spectrom.*, 2016, **27**, 810–821.
78. M. Groessl, A. Azzollini, P. J. Eugster, B. Plet, J.-L. Wolfender and R. Knochenmuss, *CHIMIA Int. J. Chem.*, 2014, **68**, 135–139.
79. K. J. Adams, N. F. Smith, C. E. Ramirez and F. Fernandez-Lima, *Int. J. Mass Spectrom.*, 2018, **427**, 133–140.
80. M.-D. Luo, Z.-W. Zhou and Z.-J. Zhu, *J. Anal. Test.*, 2020, **4**, 163–174.
81. M. Hernández-Mesa, F. Monteau, B. Le Bizec and G. Dervilly-Pinel, *Anal. Chim. Acta: X*, 2019, **1**, 100006.
82. K. Giles, J. Ujma, J. Wildgoose, S. Pringle, K. Richardson, D. Langridge and M. Green, *Anal. Chem.*, 2019, **91**, 8564–8573.
83. R. Wojcik, G. Nagy, I. K. Attah, I. K. Webb, S. V. B. Garimella, K. K. Weitz, A. Hollerbach, M. E. Monroe, M. R. Ligare, F. F. Nielson, R. V. Norheim, R. S. Renslow, T. O. Metz, Y. M. Ibrahim and R. D. Smith, *Anal. Chem.*, 2019, **91**, 11952–11962.
84. C. Bleiholder, T. Wyttenbach and M. T. Bowers, *Int. J. Mass Spectrom.*, 2011, **308**, 1–10.
85. Z. Zhou, X. Shen, J. Tu and Z.-J. Zhu, *Anal. Chem.*, 2016, **88**, 11084–11091.
86. T. J. Causon and S. Hann, *J. Am. Soc. Mass Spectrom.*, 2020, DOI: 10.1021/jasms.0c00233.
87. V. Hinnenkamp, J. Klein, S. W. Meckelmann, P. Balsaa, T. C. Schmidt and O. J. Schmitz, *Anal. Chem.*, 2018, **90**, 12042–12050.
88. M. A. Moseley, C. J. Hughes, P. R. Juvvadi, E. J. Soderblom, S. Lennon, S. R. Perkins, J. W. Thompson, W. J. Steinbach, S. J. Geromanos, J. Wildgoose, J. I. Langridge, K. Richardson and J. P. C. Vissers, *J. Proteome Res.*, 2018, **17**, 770–779.
89. T. Mairinger, R. Kurulugama, T. J. Causon, G. Stafford, J. Fjeldsted and S. Hann, *Rapid Commun. Mass Spectrom.*, 2019, **33**, 66–74.
90. M. L. Feuerstein, R. T. Kurulugama, S. Hann and T. Causon, *Anal. Chim. Acta*, 2021, **1163**, 338508.

CHAPTER 8

NMR Approaches for Probing the Polar Metabolome

PAULO R. RIBEIRO* AND ELISANGELA F. BOFFO

Metabolomics Research Group, Departamento de Química Orgânica, Instituto de Química, Universidade Federal da Bahia, Rua Barão de Jeremoabo s/n, 40170-115, Salvador, Brazil
*Emails: paulodc3@gmail.com; pauloribeiro@ufba.br

8.1 General Metabolomics Aspects

The term "Metabolomics" has raised a great deal of attention and curiosity amongst scientists over the last decade, becoming extremely popular within system biology studies.[1] Metabolomics approaches have a wide range of applications covering plant, micro-organism, animal and human related studies, aiming at providing qualitative and quantitative depiction of the chemical composition of a given organism. This includes assessing small molecules (substrates, intermediates, and products of metabolism) involved in biochemical processes within cells, tissues, organs, or even entire organisms.[2–8] High-resolution nuclear magnetic resonance (NMR) spectroscopy is a highly informative technique for conducting metabolomics studies. Several metabolites can be detected upon one single measurement (about 5 to 8 min) without the requirement of extra steps for sample preparation, such as derivatization. Moreover, it is a robust, non-destructive, non-invasive, reproducible, and fully quantitative method. Nuclear Magnetic Resonance (NMR) is a highly technique that can be used for nonselective (untargeted) and selective (targeted) analysis. Although it is considered to be

New Developments in Mass Spectrometry No. 10
Advanced Mass Spectrometry-based Analytical Separation Techniques for Probing the Polar Metabolome
Edited by Rawi Ramautar

Published by the Royal Society of Chemistry, www.rsc.org

a low-sensitive technique, especially if compared to mass spectroscopy, its sensitivity has improved enormously through the use of higher field strength,[9] cryo- and microprobes,[10–12] and dynamic nuclear polarization.[10] In the following sections, we will discuss the main steps and challenges in assessing polar metabolomes by NMR.

8.2 Sample Preparation

Sample preparation is a crucial step for any empirical study. Errors committed at this step may affect the entire follow-up chain of analysis, leading to undesired or meaningless results. Sample preparation for NMR-based metabolomics studies requires extra attention since it should take into account general aspects related to the NMR acquisition and to the quantitative aspect of the study (metabolomics).

NMR acquisition aspects include the selection of proper solvent and its correct amount to dissolve the samples (extracts) or even to perform extraction, as well as the use of clean and unscratched NMR tubes. Generally, ^{1}H NMR spectra require 5–25 mg of material, whereas other nuclei or 2D experiments may require up to 100 mg of material. Nevertheless, it should be noticed that highly concentrated samples may impose extra difficulties during shimming. Deuterated solvents are used in solution NMR for deuterium lock and shimming purposes. Routine NMR solvents include chloroform-D ($CDCl_3$), acetone-D6 ($(CD_3)_2CO$), benzene-D6 (C_6D_6), deuterium oxide (D_2O), DMSO-D6 ($(CD_3)_2SO$), ethanol-D6 (CD_3CD_2OD), and methanol-D4 (CD_3OD). These solvents might be used individually or even in binary and tertiary mixtures. Additionally, the use of an internal standard is highly recommended, usually 3-(trimethylsilyl)propionic-2,2,3,3-d_4 acid sodium salt (TMSP-d_4 or TSP) or 2,2-dimethyl-2-silapentane-5-sulfonate-d_6 sodium salt (DSS) for polar solvents.[13–18]

The metabolomics optimization (quantitative aspect of the study) ensures the best conditions in which a great number of metabolites can be identified and quantified. It includes the use of quenching protocols, removal of proteins, and the use of extraction protocols either to obtain an enriched fraction of the desired metabolites or to eliminate possible interferents and undesirable compounds, for example lipids and urea.

8.2.1 Quenching

Even though every omics science has their singularities, they all have one aspect in common: the need to use a quenching protocol to protect the integrity of the target molecules. The metabolome of a given organism is easily disturbed by external agents during handling and processing of the sample. Changes in the metabolome might be induced by hypoxia, by degradation resulting from undesired temperature and pH or even by very fast metabolite turnover rates, such as observed for D-glucose (1.0 mM s^{-1}).[16,19,20] Therefore, performing proper quenching is pivotal to prevent metabolome alterations.

For that, the use of fast and uniform sample preparation protocols is imperative, which includes quenching of enzymatic activity. Cryoquenching is defined as the use of freezing temperatures by means of dry ice or liquid nitrogen (N_2) to perform metabolic quenching of biological samples. Nevertheless, liquid nitrogen (N_2) is far more reliable and efficient than dry ice for cryoquenching samples for metabolomics studies. Cryoquenching has been widely applied to animal fluids, cells and tissues,[21–23] human fluids, cells and tissues,[24–27] and plants.[28–30] There is, however, a growing concern regarding the relationship between freezing time and tissue size. Large samples may delay temperature drop and freezing due to low thermal conductivity of the tissue. Clamp freezing tackles this issue increasing the exposed surface area by flattening the tissue between pre-cooled metal blocks, usually aluminum.[31] Therefore, clamp freezing provides an alternative approach to perform reliable assessment of the metabolome of large tissue samples, such as kidney samples, and should be favored over snap freezing (in N_2) for metabolome analysis purposes.[32] Metabolic quenching can also be achieved by quick homogenization of the samples with ice-cold perchloric acid ($HClO_4$), acetonitrile, or methanol,[33–38] although they are more frequently used for microbial metabolomics.

It is astonishing, however, that the majority of the studies omit quenching information in their material and methods section, preventing proper assessment of the robustness of the studies. Since metabolomics approaches are becoming more popular amongst scientists with the most varying backgrounds, we urge the authors to play close attention to the necessary requirements of each step to obtain reliable results and meaningful biological insights.

8.2.2 Protein Removal

Polar extracts usually contain large amounts of carbohydrates and water-soluble proteins. These macromolecules impose serious threats to the detection and identification of polar metabolites by NMR, since they produce broad and low intensity peaks that disrupt the NMR spectrum. As a result, the baseline of the spectrum is compromised, lowering the accuracy of metabolite identification and quantification. The deleterious effect caused by the presence of these macromolecules can be attenuated by using two different approaches: the use of NMR pulse sequences that reduce the impact of macromolecule peaks on a spectrum, such as the Carr–Purcell–Meiboom–Gill (CPMG) pulse sequence,[21] and the removal of proteins from the extract, with subsequent disruption of protein–metabolite interactions.[39] Nevertheless, usually both approaches are applied to ensure proper metabolite identification and quantification in polar NMR metabolomics. One should notice, however, that broad peaks caused by proteins in the sample mainly occur at the aliphatic region of the spectrum, but the use of NMR pulse sequences that reduce the impact of macromolecule peaks on a spectrum might affect peaks within the entire spectrum[40] (Figure 8.1).

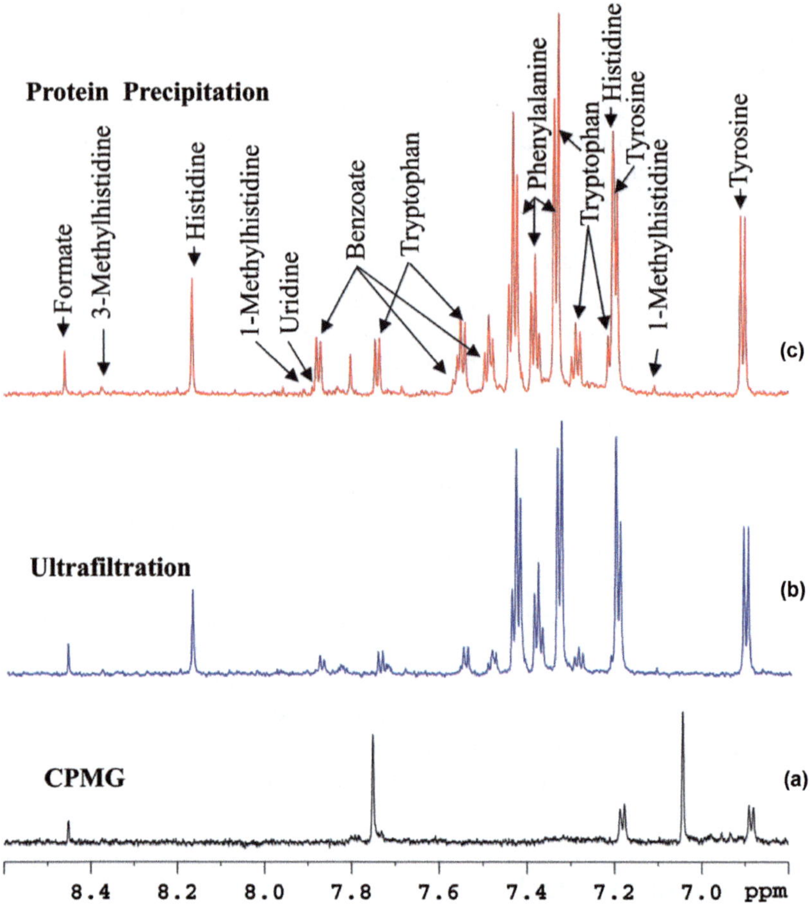

Figure 8.1 Comparison of the aromatic region of 800 MHz, cryo-probe ^{1}H NMR spectra of the same pooled human serum sample obtained by suppressing protein signals (a) by T2 filtering using the CPMG pulse sequence, (b) by ultrafiltration using a 3 kDa molecular weight cutoff filter, and (c) by protein precipitation using methanol (1 : 2). Most of the metabolite signals in the displayed region are missing or significantly attenuated in the T2 filtered spectrum, and many including tryptophan, benzoate, and formate were significantly attenuated in ultrafiltered serum compared to the protein precipitated serum.
Reproduced from ref. 40, https://pubs.acs.org/doi/10.1021/ac5005103, with permission from American Chemical Society, Copyright 2014.

Therefore, protein precipitation seems to provide the most robust approach for routine NMR-based metabolomics of water-soluble extracts with high protein content, and it should be favored over the use NMR pulse sequences that reduce the impact of macromolecule peaks on a spectrum.

Protein removal can be achieved by two main approaches: ultrafiltration using membranes with low molecular-weight cutoff (MWCO, *ca.* 3 kDa) or precipitation mediated by acetonitrile, acetone, methanol, perchloric acid, or trichloroacetic acid. In fact, protein precipitation has been successfully applied for a wide panel of biological samples, such as for fish liver,[41] the model plant *Arabidopsis thaliana*,[12] Jerusalem artichoke (*Helianthus tuberosus* L.),[42] *Rhizostoma pulmo* jellyfish,[43] and mouse liver.[44]

McHugh *et al.*[39] compared different precipitation protocols to assess metabolite recovery upon protein removal in human blood serum. They demonstrated that $CHCl_3$: MeOH : water precipitation allowed faster and more efficient protein removal, including low induced variance caused by sample handling. Another positive aspect of the use of $CHCl_3$: MeOH : water for precipitation purposes is that this approach produced an organic fraction ($CHCl_3$) that can be used for lipidomic analysis by NMR, LC– or GC–MS.[40]

A feasible alternative to the $CHCl_3$: MeOH : water system is to use CD_3OD : D_2O precipitation followed by centrifugation. In this case, blood samples were lyophilized, and the resulting material was resuspended in a CD_3OD : D_2O mixture followed by centrifugation. This approach removed the need to eliminate non-deuterated solvents used during extraction. NMR spectrum obtained using pre-saturation pulses for water suppression (zgpr) with protein precipitation using a CD_3OD : D_2O mixture (5 : 2) produced a flatter spectrum baseline and, therefore, improved the accuracy of metabolite identification and quantification (Figure 8.2a–d). Nevertheless, one should evaluate and assess with approach would provide the best results for a specific type of sample.

8.2.3 Fine Tuning the Polar Metabolome with Extraction Protocols

Extraction protocols for metabolomics studies are highly dependent on the nature of the samples. Plant, animal, and human tissue samples are usually submitted to extraction by grinding them in liquid nitrogen (quenching) followed by addition of solvents. Microwave- and ultrasound-assisted extraction are often applied.[45,46] To assess the polar metabolome these solvents must be able to interact with the desired molecules and, therefore, polar solvents such as water, acetonitrile, ethanol, and methanol, as well as phosphate buffer are the main choice in such approaches. Some studies also apply a solution of strong acids, such as perchloric acid ($HClO_4$), to obtain a selective enriched fraction of the desired metabolites.[47] Nevertheless, addition of a less polar solvent such as chloroform or dichloromethane is frequently applied to remove proteins and lipids present in the samples.[46,48,49] Blood and urine samples do not require extensive extraction protocols. Urine samples can be directly analyzed by NMR, whereas blood samples are usually submitted to protein precipitation prior to the NMR analysis.[46] This is an enormous advantage of NMR metabolomics of urine samples over GC– and LC–MS metabolomics, since GC– and LC–MS require prior depletion of urea by enzymatic treatment.[50–52] However, in some cases it is also possible

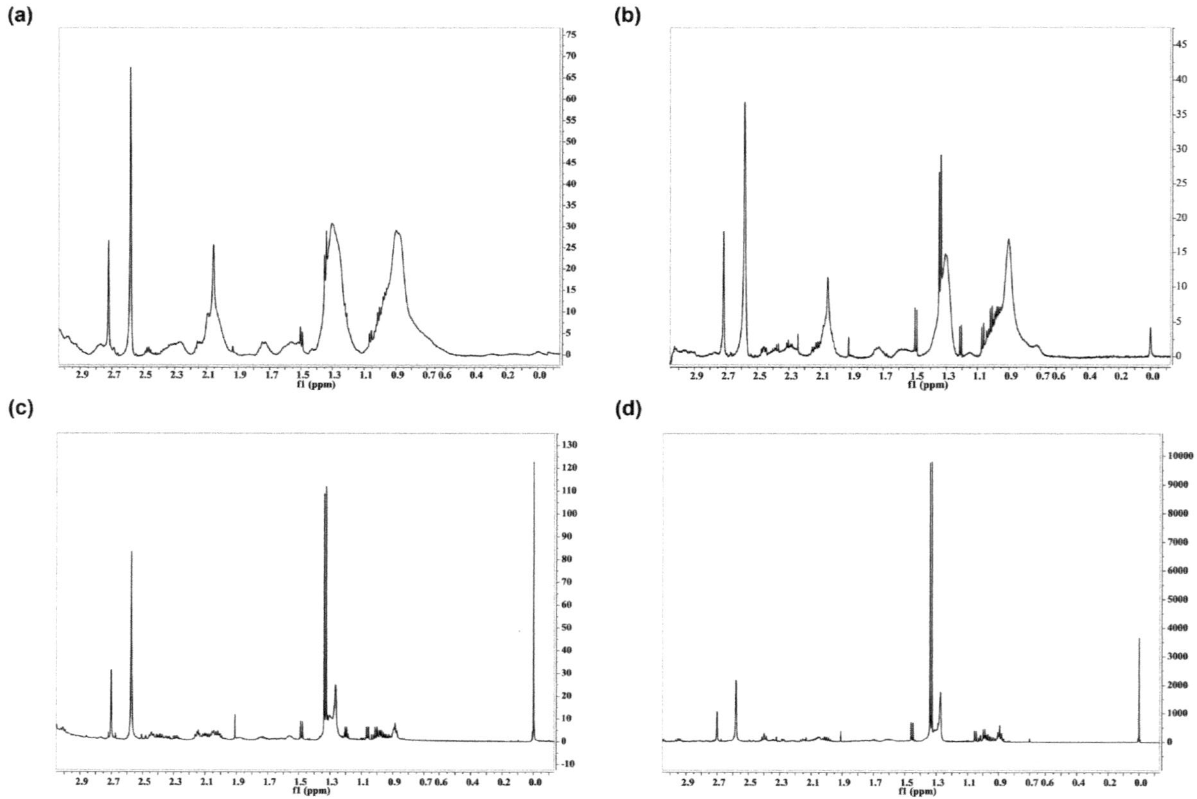

Figure 8.2 Expansion of NMR spectrum (3.0 to 0 ppm) obtained from blood samples of patients with sickle cell disease. NMR spectrum obtained using pre-saturation pulses for water suppression (zgpr) with (a) no protein removal protocol, (b) protein precipitation with a $CD_3OD:D_2O$ mixture (2 : 5), (c) protein precipitation with a $CD_3OD:D_2O$ mixture (4 : 2), and (d) protein precipitation with a $CD_3OD:D_2O$ mixture (5 : 2).
Source: Metabolomics Research Group (data not published).

to proceed with protein precipitation in urine samples prior to the NMR analysis.[53] Assessing the polar metabolome of small samples of tissues and biofluids remains a significant analytical challenge, and, therefore, alternative approaches must be developed in order to assess their metabolome. In these scenarios, microextraction might be an important ally since it allows non-exhaustive sample preparation with a very small volume of extracting phase. Microextraction can be applied as solid-phase microextraction (SPME) or as in liquid-phase microextraction (LPME).[54] Additionally, microextraction allows accurate assessment of the polar metabolome of tissue or blood samples without the necessity to collect a large volume of biological sample for analysis.[55]

8.3 Optimizing Spectra Acquisition and Processing for Polar Metabolomics

8.3.1 Spectra Acquisition

The acquisition of ^{1}H NMR spectra (liquid-state NMR) is fast and allows the analysis of all hydrogens present in the sample in detectable quantities. The main information provided is of chemical shifts, relative signal areas, multiplicity patterns, and coupling constants. A high and homogeneous magnetic field is mandatory for adequate spectral resolution in the metabolomics analysis of complex samples. Most of the studies covered in this review were performed on 11.7 tesla magnetic fields (or greater) (500 MHz to hydrogen frequency), which in part compensates for the low sensitivity of the technique when compared to mass spectrometry.[56]

Parameters related to the acquisition and processing of the experiments can influence the quality of NMR spectra and, therefore, must be adjusted correctly. The probe must be tuned and matched correctly and locked to the signal of the deuterated solvent. Moreover, the adjustment of the shim coils ensures that the field is homogeneous in three dimensions across the sample detection area. The sample temperature must be stabilized to prevent variations of chemical shifts, especially the water peak for water suppression and the radio frequency (RF) pulse length must be calibrated.[57]

The pulse sequence to perform solvent suppression should be chosen carefully to avoid signal distortions or anomalies in the spectrum. Several pulse sequences are used for ^{1}H NMR spectrum acquisition, mainly those that suppress the most intense signal of the spectrum, which is most frequently the solvent signal. The suppression of the solvent signal is important because the presence of this signal impairs the resolution of low intensity peaks and consequently metabolite assignment and quantification.[58,59]

Two pulse sequences are widely used for the acquisition of these spectra: 1D NOESY and 1D CPMG. NOESY (Nuclear Overhauser Effect SpectroscopY) has been among the most often used for pre-saturation in one-dimensional experiments.[56,60–62] This pulse sequence is relatively simple to perform, and it provides reproducible ^{1}H NMR experiments for biological samples with good water suppression. The Carr–Purcell–Meiboom–Gill (CPMG) pulse sequence is used to decrease broad signals from large molecules that present relatively long transverse relaxation times, and the pre-saturation is used together to remove the water signal.[63–65]

Figure 8.3 shows the ^{1}H NMR spectra obtained using the pulse sequences NOESY and CPMG. It is clear that the broad signals present in the ^{1}H NMR spectrum obtained with the NOESY pulse sequence (A) are not observed in the ^{1}H NMR spectrum obtained with the CPMG pulse sequence (B) because of the filter of the signals with high values of T2. Therefore, if protein removal is not performed the CPMG pulse sequence would allow more reliable identification and quantification of polar (or non-polar) metabolites than the NOESY pulse sequence. Nevertheless, the protein removal step is highly recommended.

Le Guennec *et al.*[59] proposed alternative pulse sequences to 1D NOESY presat and CPMG presat. The optimized PURGE (with gradient signs inverted between every scan) and the optimized PROJECT provided results equal to or better than those of the NOESY and CPMG sequences. According to the authors, the optimized PURGE pulse sequence improved the water region when compared with noesypr1d for all samples tested, without changing the other regions. PROJECT's main advantage over CPMG is that it does not cause *J*-coupling distortions when a great improvement when longer T2 filters need to be applied.

Multi-dimensional NMR experiments, mainly 2D NMR, might also be applied to assess the polar metabolome. These experiments provide additional information about chemical structures as they spread overlapping peaks along two orthogonal dimensions (Figure 8.3C). The most used two-dimensional experiments are COrrelation SpectroscopY (COSY), Heteronuclear Single Quantum Coherence (HSQC), Heteronuclear Multiple Bond Correlation (HMBC), and TOtal COrrelation SpectroscopY (TOCSY).[58,66,67] However, other two-dimensional experiments have been used to solve specific problems related to the polar metabolome. ^{1}H–^{15}N HSQC experiments were used to characterize the polar metabolome of human urine samples, whereas a two-dimensional *J*-resolved experiment has been used as an alternative when the overlapping signals in the ^{1}H NMR spectra prevent proper assessment of the polar metabolome. Additionally, the signal splitting and coupling constants provided by *J*-resolved experiment can be used for assigning and confirming the metabolite signals from one-dimensional NMR.[68,69] Diffusion Ordered SpectroscopY (DOSY) was applied to show the autodiffusion coefficients of constituents of Friesian cow milk,[70] whereas spatial molecular-dynamically ordered spectroscopy (SMOOSY) was used to evaluate polar metabolites and their dynamics in intact bodies and heterogeneous systems from NMR data, including spatial *z*-position, chemical shift, and diffusion or relaxation.[71]

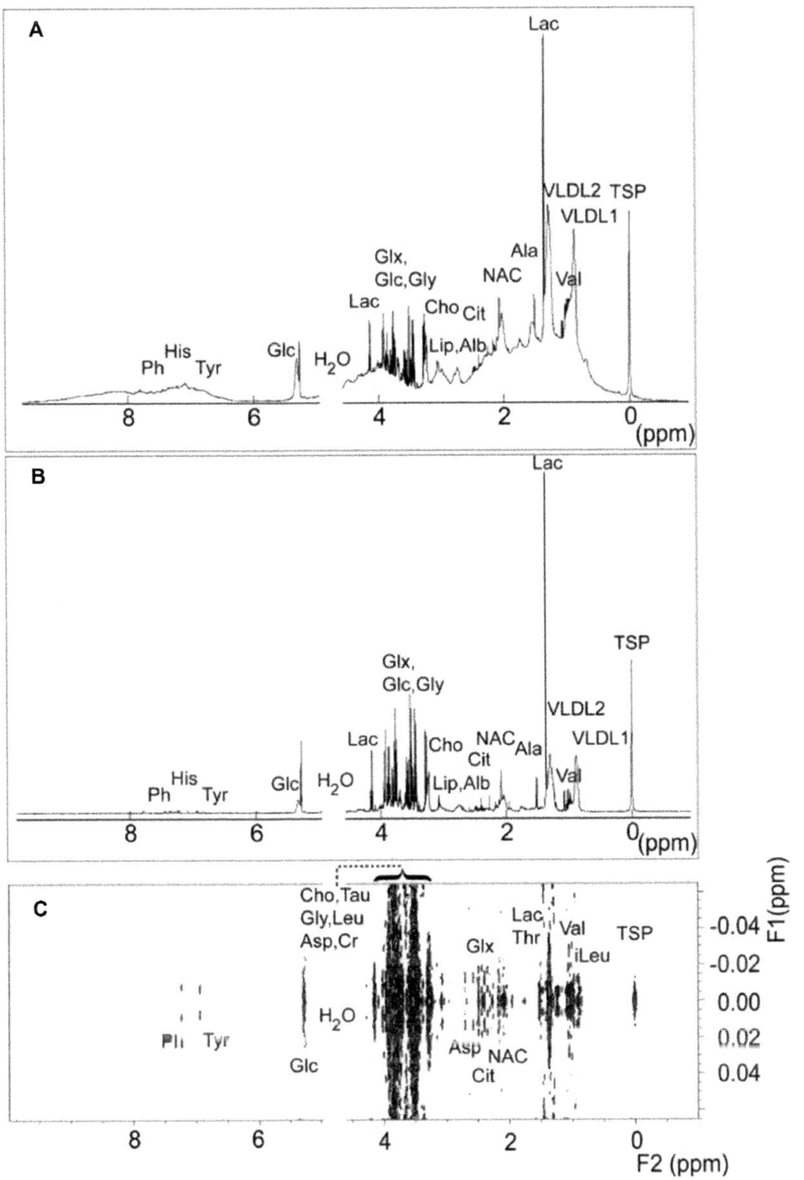

Figure 8.3 Representative spectrum of the 1D 1H J spectrum of a human serum sample acquired on a 600 MHz spectrometer with a (A) noesygppr1d pulse sequence, (B) cpmgpr1d pulse sequence, and (C) jresgpprqf pulse sequence. Ala = alanine; Ace = acetoacetate; Cho = choline; Cit = citrate; Glc = glucose; Glx = glutamine + glutamate; Gly = Glycerol; His = histidine derivatives; iLeu = iso-leucine; Lac = lactate; Leu = leucine; Lip = lipid; Nac = glycoprotein acetyl group; Ph = phenylalanine; Pro = proline; Tau = taurine; Thr = threonine; Tri = trimethylamine; TSP = trimethylsilyl propionic acid; Tyr = tyrosine; Val = valine; VLDL = very low-density lipoprotein. Selected metabolites are labeled for clarity.
Reproduced from ref. 57 with permission from Elsevier, Copyright 2020.

Despite the importance of providing detailed information on the pulse sequence used, many authors limited themselves to reporting that basic sequences for suppressing the residual water signal were used[66,68,72–74] or even did not describe the sequences that were used at all.[18,69] Although these studies probably used basic pre-saturation methods that use low power selective irradiation in the residual water signal, we stress the need to provide complete and detailed information of the pulse sequence used for metabolomics studies, since it may compromise the robustness and credibility of the study.

Other nuclei such as ^{13}C, ^{15}N, and ^{31}P show poor sensitivity due to low natural abundance and low gyromagnetic ratio. However, a broad chemical shift dispersion and relatively narrow line widths make these nuclei interesting for metabolomic analysis of polar compounds.[62,75,76]

8.3.2 Spectra Processing

The signal generated by the experiment acquisition sequences (FID, free induction decay) must be processed correctly, in order to make all spectra comparable among themselves and to extract information from NMR spectral data. The first step is the application of an exponential multiplication to the FID signal to improve the signal-to-noise ratio. However, high values cannot be used, because this operation tends to result in a broadening of the signals (line broadening). The Fourier Transformation converts raw data in the time domain to frequency domain spectrum.[57,77] After that, the spectrum generated must be phased, baselined, and calibrated with respect to the chemical shift of internal reference or solvent signals. Phase adjustments aim at maximizing the absorption character and the symmetry of all signals in an NMR spectrum. Small phasing errors can affect the signals integration, spectral alignment, and spectral binning. Proper adjustment of the baseline allows spectrum signals to be integrated and values to be used in quantitative analysis. These processes can be done manually[74] or through fully automated methods.[60]

Post-processing steps prepare the data for the chemometric analysis, preserving the variance relative to the NMR signals, while reducing other types of variance induced by different sources of biases. They include spectral alignment, binning, and normalization. Spectral alignment is needed to ensure that the signal of a compound can be compared and quantified across multiple NMR spectra. In this process the positions of the signal iteratively shift in various spectra so that the signals from the same compounds can be directly overlapped or aligned.[77] Binning reduces the data dimensionality to minimize the instrumental errors that reflect in small misalignments and to make the analysis easier. In this process, NMR spectra are divided into small spectral regions (or buckets, normally spanning 0.04–0.05 ppm), which are sufficiently wide to include one or more NMR signals. Normalization is a process that makes all samples comparable, as it standardizes parameters such as the use of different numbers of scans, differences in sample concentration, or when more than one NMR spectrometer is used to acquire the spectra. Therefore, it is essential to adjust these parameters properly.[78] After spectra processing, the NMR data are converted into an adequate format to be processed in multivariate statistical analysis.

8.3.3 Spectral Resolution Associated to Magnetic Field Strength and Probe Specificity

The ^{1}H NMR-based metabolomics approach has two main limitations: the relatively small chemical shift window, which can cause signal overlap, imposing difficulties for compound identification and quantification, and the low sensitivity of the NMR technique. However, the increase in the magnetic field strength provides sensitivity gain, decreased experimental time, and an improvement in spectral dispersion that can help in better resolution of overlapping signals.

Figure 8.4 shows a comparison between the ^{1}H NMR spectra acquired in three different fields (400, 500, and 600 MHz). Clearly, signal resolution improves with the increasing magnetic field strength. Nevertheless, similar discrimination pattern in the PCA plot was observed for all three different fields (400, 500, and 600 MHz).

A disadvantage associated with the increase in magnetic field strength is the exponentially higher price for the acquisition and maintenance of the NMR instrument. An alternative to improve sensitivity is the use of more efficient probes, such as the cryogenically cooled probes, that maximize the signal-to-noise ratio. ^{1}H and ^{13}C NMR spectra were acquired using two spectrometers equipped with different hardware configurations: a 400 MHz spectrometer equipped with a room temperature probe and a 700 MHz spectrometer equipped with a cryogenically cooled probe. The increase in the field led to a resolution enhancement, and the use of cryogenic probes led to a sensitivity gain (Figure 8.5).

Another hardware alternative is the use of microprobes that have been designed to maximize the sensitivity for mass-limited samples. Fathi *et al.*[81] simultaneously characterized 52 polar and nine non-polar metabolites in the liver tissue of nestling birds using high-throughput NMR-based analysis from a small quantity of liver tissue. The ^{1}H NMR water signal from the polar fraction was suppressed by means of excitation sculpting (ES), and the two-dimensional ^{1}H–^{13}C heteronuclear single quantum coherence (HSQC) spectra were used to assist in characterizing the metabolites. The authors reported that by using this methodology it was possible to identify a larger number of metabolites from liver tissue than previous NMR-based investigations and described the use of more concentrated extracts in a 1.7 mm cryogenic probe as the main advantage, when compared to a 5 mm probe (40 μL rather than 300–500 μL). Nagato *et al.*[82] also reported the use of a 1.7 mm microprobe instead of a 5 mm probe as a way to decrease the amount of sample needed for NMR analysis. Whereas the 5 mm probe requires up to 25 mg of sample, the 1.7 mm microprobe only requires

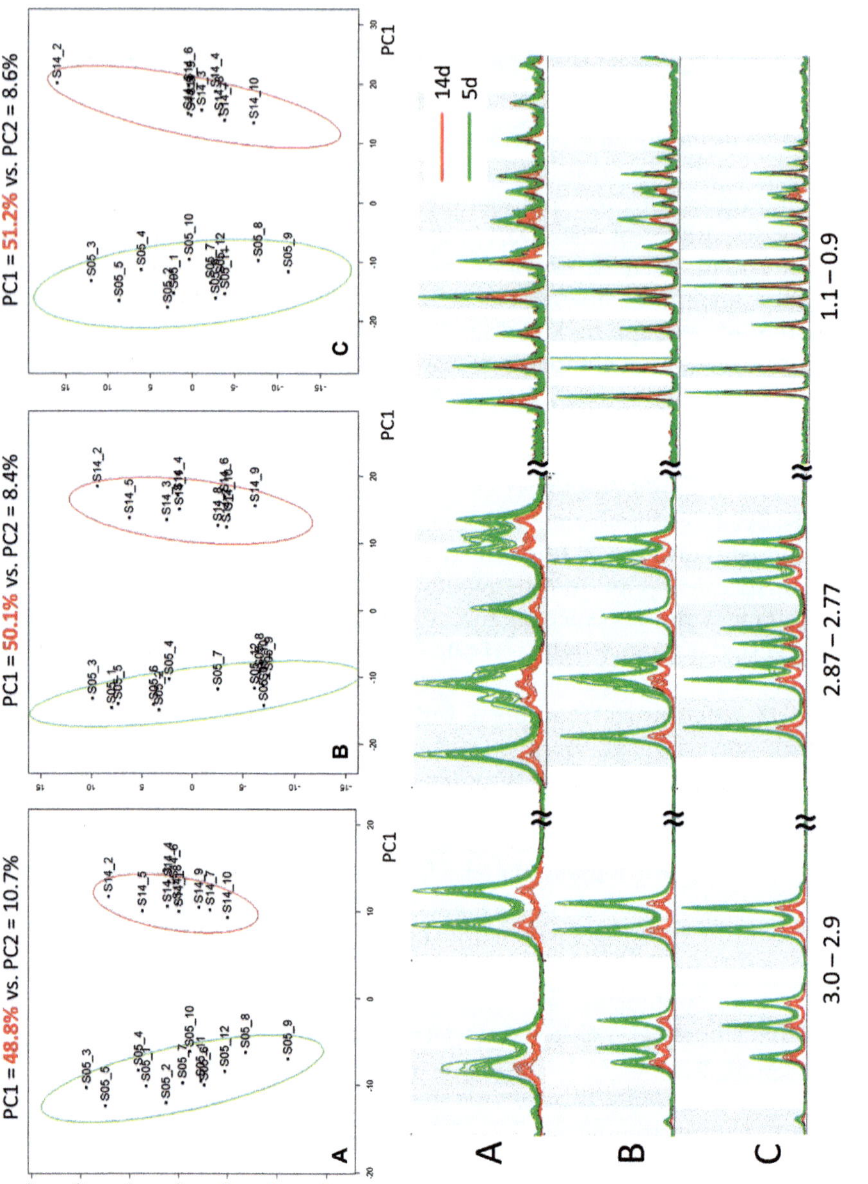
PC1 = 48.8% vs. PC2 = 10.7%
PC1 = 50.1% vs. PC2 = 8.4%
PC1 = 51.2% vs. PC2 = 8.6%
PC1
A
B
C
14d
5d
3.0 – 2.9
2.87 – 2.77
1.1 – 0.9

1.0–1.5 mg of sample. Therefore, the use of a 1.7 mm microprobe allows a more precise and detailed characterization of the polar metabolome in systems where sample size is a limitation.

8.3.4 *In vivo* NMR Spectroscopy

In vivo analysis of the polar metabolome of intact tissues can be performed by Magnetic Resonance Spectroscopy (MRS). This is a non-invasive technique that provides the most comprehensive, physiologically relevant data on metabolic activity in the tissues of any MR method. However, significant line broadening of resonances is observed due to the reduced molecular motions, local B_0 variations driven by magnetic susceptibility, differences between tissue types, and local paramagnetic broadening effects of metals and oxygen.[83] ^{1}H and ^{31}P are the most common nuclei observed in *in vivo* MRS, but less abundant spin-active nuclei, such as ^{13}C, are observed by the introduction of the isotopically enriched metabolic substrates. Additionally, dissolution dynamic nuclear polarization (dDNP), a hyperpolarization technique, increases carbon polarization by more than four orders of magnitude compared to thermal equilibrium for *in vivo* ^{13}C MRS by transferring polarization from high γ spins (unpaired electrons) to low γ nuclei (^{13}C) in a metabolic substrate at low temperature ($\sim$1–2 K) *via* microwave irradiation.[79]

Excised tissues present a heterogeneous nature and are ideal for studying *via* High Resolution Magic Angle Spinning (HR-MAS), which represents a hybrid between traditional solid-state MAS NMR applications and solution-state NMR methods. The principle of this technique is to reduce line widths and to increase the quality of the spectrum by the rapid rotation in the magic angle (54.7° relative to the main magnetic field B_0), which significantly minimize the effects of dipole coupling and anisotropic chemical shift, as well as the difference in magnetic susceptibility due to the heterogeneity of the sample.[84] Augustijn *et al.*[85] used one- and two-dimensional HR-MAS NMR to obtain the polar metabolome profile from intact leaves of *Arabidopsis thaliana* Col-0 (wild-type Columbia) and of the two mutants (VP16-02-003 and VP16-05-014) with enhanced growth characteristics and putatively higher sensitivity to biotic stress. Wang *et al.*[84] showed that HR-MAS NMR-based

Figure 8.4 PCA for each ^{1}H NMR dataset of 22 wheat samples acquired on a given NMR instrument and zoom in on a spectral region. Top: Scores plot of PCA for semi-polar methanolic extracts of spikelets at 5 DAF (green ellipse) or 14 DAF (red ellipse), for 400 MHz JEOL instrument (A, 224 buckets), 500 MHz Bruker instrument (B, 397 buckets), 600 MHz Bruker instrument (C, 465 buckets). Ellipses represent 95% of confidence level. Bottom: Comparison of several spectral regions (3–2.9, 2.87–2.77, and 1.1–0.9 ppm) for 5 (green) and 14 (red) DAF for 400, 500, and 600 MHz spectrometers with bucket positions for the entire dataset.
Reproduced from ref. 79, https://doi.org/10.1007/s11306-019-1488-3, under the terms of the CC BY 4.0 license, http://creativecommons.org/licenses/by/4.0/.

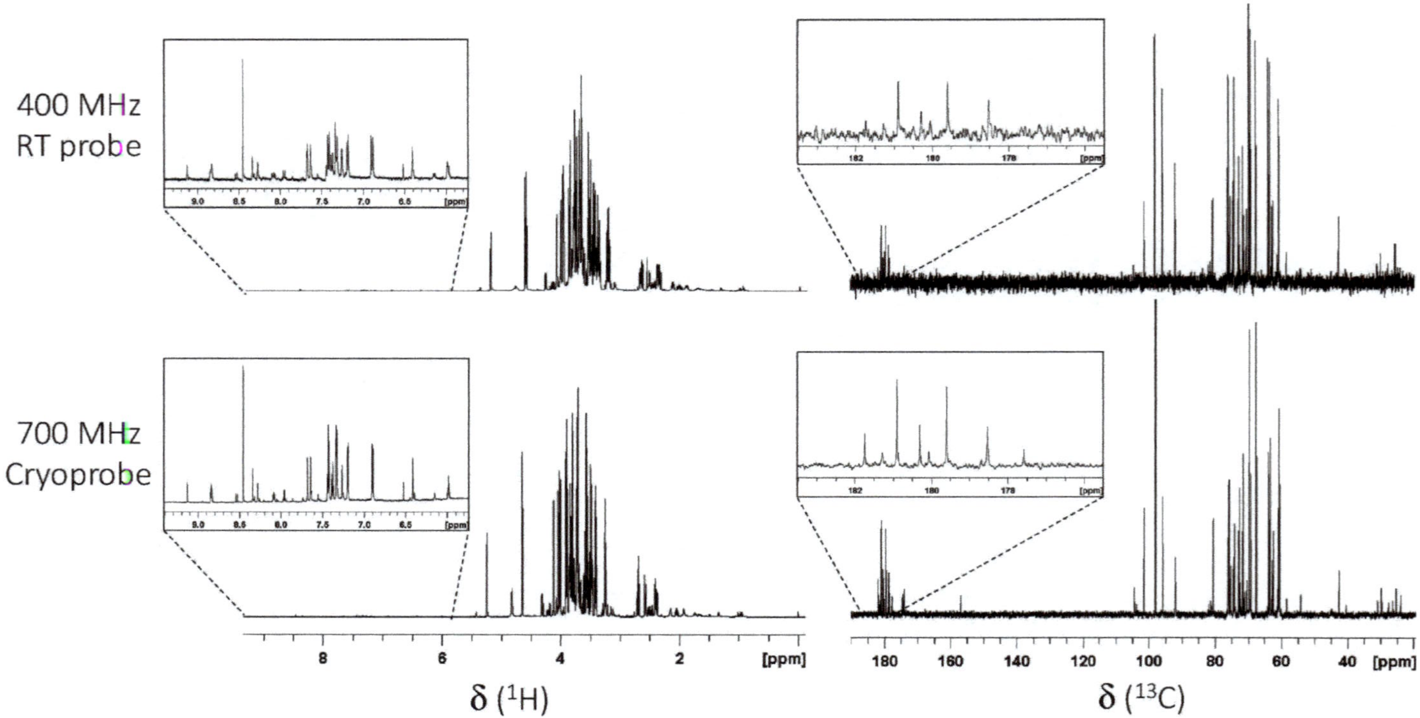

Figure 8.5 ^1H (left) and ^{13}C (right) NMR spectra of a cucumber (*Cucumis sativus*) polar extract, recorded at 400 MHz with a room temperature probe (top) and at 700 MHz with a cryogenic probe (bottom). The two ^1H spectra were recorded under identical conditions (128 scans, 5 s recovery delay and 2.5 s acquisition time, 12 ppm spectral width, processing with 0.3 Hz exponential multiplication and zero filling to 64 K). The two ^{13}C spectra were recorded with identical conditions (Waltz-16 inverse-gated ^1H decoupling, 512 scans, 20 s recovery delay and 0.8 s acquisition time, 236 ppm spectral width, processing with 3.0 Hz exponential multiplication and zero filling to 256 K).
Reproduced from ref. 80 with permission from Elsevier, Copyright 2018.

metabolomics provided evidence of distinct metabolic characteristics and metabolic pathway disturbances of pancreatic ductal adenocarcinoma (PDAC) patients and Sragué–Dawley (SD) rat models, whereas Flores *et al.*[86] evaluated factors affecting the quality of HR-MAS NMR analysis of plant tissues, such as particle size, sample and solvent amounts, solvent polarity, swelling time, rotor manipulation, and pulse sequence setting.

8.4 Strategies for the Identification of Polar Metabolites

8.4.1 Available Literature for Metabolite Identification in NMR-based Metabolomics

Metabolite identification is perhaps one of the most crucial steps in any metabolomics approach. It should provide a comprehensive depiction of the metabolome by allowing the simultaneous identification of a broad panel of quantifiable metabolites. Hopefully, these metabolites will provide insights into the understanding of the biological question under investigation. Nevertheless, NMR-based metabolomics face an intrinsically difficult challenge due to the considerably high degree of signal overlap in the NMR spectra. Despite its complexity, the identification of a great number of polar metabolites can be achieved by carefully analyzing the NMR spectroscopic data and the structural information available in the literature and NMR databases.

Available NMR data related to assignments of metabolites covers the most diverse and heterogeneous panel of samples, including human body fluid samples such as plasma,[21,87] serum,[88–90] saliva[91–93] and urine,[13,94] human hippocampus,[26] human pleural effusions,[95] ovarian tumor cyst,[96] reconstructed human epidermis,[97] endometrial tissue,[98] colorectal tumor tissue fecal samples,[7,49] mammalian mitochondria,[99] murine brain,[22,23] murine liver,[29,44] murine urine,[100,101] murine hippocampus,[102] dairy cow serum,[103] chicken (*Pectoralis major*), beef (*Semitendinosus*), and pork (*Longissimus thoracis*) meat,[104] plant leaves,[105] seeds,[12,28] seed embryo,[30] pomegranate fruits,[106] sprouting tubers,[42] bacteria,[107] fungi,[45] and yeast.[108,109]

Forty metabolites were identified in a plasma sample from an endometriosis patient (Figure 8.6),[87] whereas forty-six metabolites were identified in chickpea root samples inoculated with *Fusarium oxysporum* (Figure 8.7 and Table 8.1).[110] These two studies share in common a great number of polar metabolites, which include several amino acids, carbohydrates, and organic acids. Nevertheless, the presence of some metabolites seems to be dependent on the nature of the samples. Therefore, it is advisable that authors seek the available NMR data from studies with a similar nature for more accurate and efficient metabolite identification.

A typical NMR table reported in the literature should contain the chemical shift, coupling constant, and multiplicity, along with further NMR

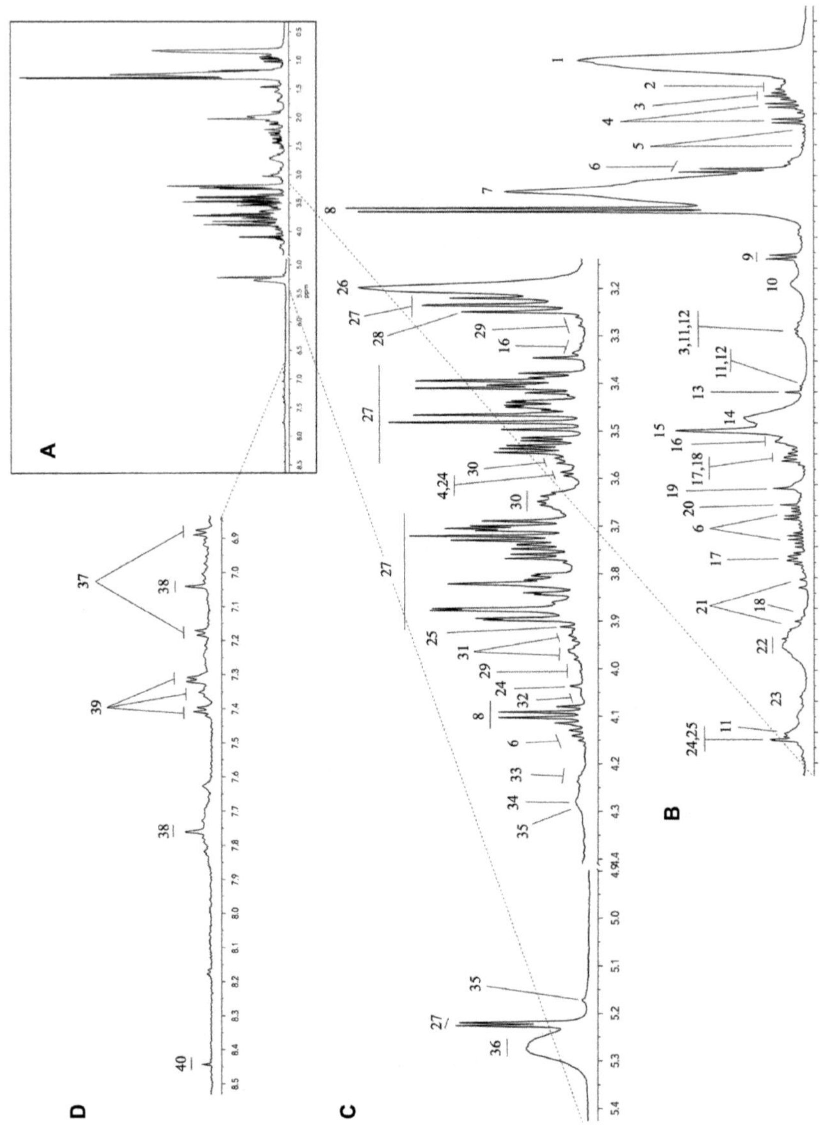
A
B
C
D

experiments used to support the metabolite identification, thus providing helpful tools during the assignment of metabolite identity in related studies (Table 8.1).

8.4.2 Available Databases for Metabolite Identification in NMR-based Metabolomics

Assessing the immense number of manuscripts describing NMR spectral data for metabolite identification purposes can be quite exhausting and challenging. Taking that into account, it is advisable to include in any NMR-based metabolomics approach the use of NMR spectral databases to aid metabolite identification and quantification. NMR spectral databases provide access to valuable information for a few thousand compounds, providing crucial information vital for NMR-based metabolomics studies. Offline NMR database: Commercial databases for standard spectra and chemical shift search include ACD/Labs, Chenomx, and Sigma Aldrich FTNMR library. Nevertheless, they will not be discussed in detail.

8.4.2.1 Biological Magnetic Resonance Bank (BMRB)

The Biological Magnetic Resonance Bank (BMRB: www.bmrb.wisc.edu) provides a collection of experimental and derived NMR spectral data of biological molecules. Amongst other things, the BMRB collection provides free access to quantitative NMR spectral information and derived data for over 250 metabolites.[111] BMRB has provided support for over 900 NMR-based metabolomics studies, including wastewater,[112] human gut microbiome,[113] zebrafish,[114] *Zea mays* embryo,[115] pig seminal plasma,[116] and urine[117] samples.

Figure 8.6 Representative 600-MHz ^{1}H-nuclear magnetic resonance Carr–Purcell–Meiboom–Gill spin-echo pulse spectrum of a plasma sample from an endometriosis patient. (A) Full spectrum (δ 8.5–0.5 ppm); (B–D) Magnifications of spectra regions from 0.7–3.10 ppm, 3.20 5.40 ppm, and 6.90–8.50 ppm regions, respectively. Assigned metabolites: 1, lipoproteins: high-density lipoprotein (HDL), low-density lipoprotein (LDL), very-low-density lipoprotein (VLDL) ($-CH_3$); 2, isoleucine; 3, leucine; 4, valine; 5, fucose; 6, 3-hydroxybutyrate; 7, lipoproteins: HDL, LDL, VLDL ($-CH_2$)*n*; 8, lactate; 9, alanine; 10, fatty acids ($-CH_2-CH_2CO$); 11, lysine; 12, arginine; 13, acetate; 14, unsaturated fatty acids ($-CH_2-CH_2-CH{=}CH-$); 15, *N*-acetyl groups; 16, proline; 17, glutamine; 18, methionine; 19, fatty acids ($-CH_2-CH_2CO$); 20, acetoacetate; 21, citrate; 22, polyunsaturated fatty acids ($-CH{=}CH-CH_2-CH{=}CH-$); 23, asparagine; 24, creatinine; 25, creatine; 26, choline-containing metabolites (choline, phosphocholine, glycerophosphocholine); 27, glucose; 28, trimethylamine-*N*-oxide; 29, myo-inositol; 30, glycerol; 31, serine; 32, choline; 33, threonine, 34, glycerophosphocholine; 35, glycerol of lipids; 36, fatty acids ($-CH{=}CH-$); 37, tyrosine; 38, histidine; 39, phenylalanine; 40, formate. Reproduced from ref. 87 with permission from Elsevier, Copyright 2016.

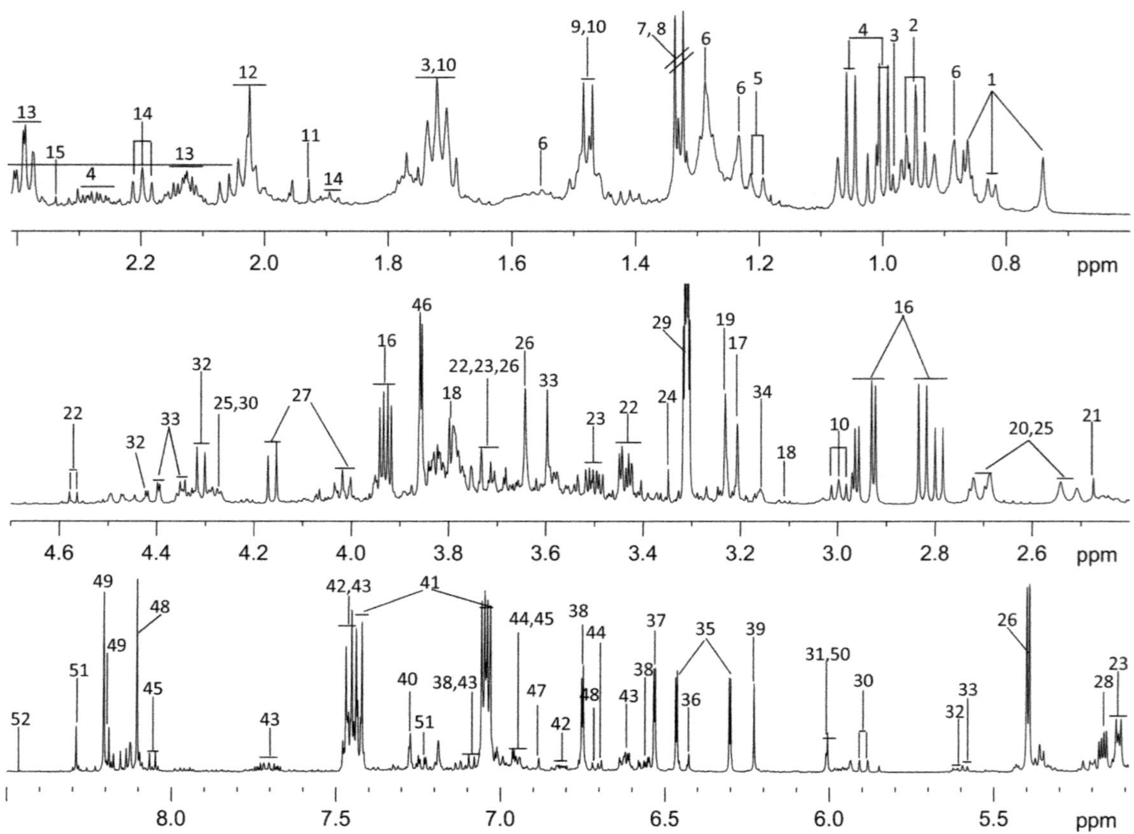

Figure 8.7 Typical NMR spectrum of a chickpea root extract. Assignments of metabolites from ^{1}H NMR analysis are shown in Table 8.1. Reproduced from ref. 110, https://doi.org/10.1111/pbi.12522, under the terms of the CC BY 4.0 license, https://creativecommons.org/licenses/by/4.0/.

8.4.2.2 The Human Metabolome Database (HMDB)

The Human Metabolome Database (HMDB: www.hmdb.ca) is perhaps the most important NMR spectral database for human related studies. It provides free access to an electronic database containing detailed information about small molecules found in the human body covering over 114 000 entries.[118–121] Some of these entries are hyperlinked to additional databases (KEGG, PubChem, MetaCyc, ChEBI, PDB, UniProt, and GenBank) providing a comprehensive systems biology depiction. The Human Metabolome DataBase allows the user to download ^{1}H, ^{13}C, COSY, TOCSY, HSQC, DEPT, and HMBC data of standard metabolites. The Human Metabolome Database also encompasses four additional databases: DrugBank (~2280 entries), T3DB (~3670 entries), SMPDB (~25 000 entries), and FooDB (~28 000 entries). HMDB has provided support for over 5000 studies; NMR-based metabolomics studies, including human blood,[122] human urine,[123] pluripotent stem cells,[124] colorectal cancer cells,[125] murine blood,[126] murine brain,[127] frog (*Rana amurensis*),[128] *Macrotyloma uniflorum* (Lam.) Verdc,[129] and tomato samples.[130]

8.4.2.3 Birmingham Metabolite Library (BML)

The Birmingham Metabolite Library (BML: www.bml-nmr.org/) provides free access to a collection of experimental NMR spectra of 208 metabolites.[131] NMR spectra of each metabolite were recorded using a Bruker 500 MHz spectrometer. For that, solutions of the standard compounds were prepared at different pH values (6.6, 7.0 and 7.4), and the acquisition included 1D and 2D *J*-resolved spectra, applying varying water suppression pulse sequences. BML has provided support for nearly 100 NMR-based metabolomics studies, mainly for plants,[132–135] but also for urine[136] and blood[137,138] samples.

8.4.2.4 Madison–Qingdao Metabolomics Consortium Database (MMCD)

The Madison–Qingdao Metabolomics Consortium Database (MMCD: http://mmcd.nmrfam.wisc.edu/) provides free access to a collection of information on small molecules of biological interest assembled from public databases and scientific literature.[139] It includes the chemical formula, names and synonyms, structure, physical and chemical properties, NMR data of standard compounds, as well as NMR chemical shifts determined by empirical and/or theoretical approaches. Some of these entries are hyperlinked to additional databases (KEGG and PubChem) providing a comprehensive systems biology depiction. MMCD has provided support for over 400 NMR-based metabolomics studies, including plant,[140,141] blood,[142] fungi,[143] hepatocellular carcinoma cells,[144] murine liver,[145] intramuscular postmortem fluid,[146] and food-based samples.[147]

Table 8.1 Assignments of metabolites from ^{1}H-NMR analysis. Reproduced from ref. 110, https://doi.org/10.1111/pbi.12522, under the terms of the CC BY 4.0 license, https://creativecommons.org/licenses/by/4.0/.

Key	Metabolites	Assignment	δH^{1} (multiplicity)[a]	δC^{13}	Assigned with
1	Phytosterol	CH_3	0.74 (s), 0.78 (s), 0.83 (s), 0.86 (s)	23.1, 21.9, 17.7, 15.8	HSQC, HMBC
2	Isoleucine	αCH	3.68 (d)	62.6	JRE, COSY, TOCSY, HSQC, HMBC
		βCH	1.98 (m)	39.4	
		γCH_2	1.24 (m), 1.45 (m)	27.4	JRE, COSY, TOCSY, HMBC
		γ′CH_3	1.02 (d, 7.0 Hz)	17.6	JRE, COSY, TOCSY, HSQC, HMBC
		δCH_3	0.94 (t, 7.3 Hz)	13.9	
3	Leucine	βCH_2	1.73 (m)	42.4	JRE, COSY, TOCSY, HMBC, HSQC
		γCH	1.69 (m)	27.4	
		δCH_3	0.98 (d, 6.5 Hz)	24.5	JRE, COSY, TOCSY, HSQC
		δ′CH_3	0.96 (d, 5.9 Hz)	24.5	
4	Valine	αCH	3.62 (d, 4.3 Hz)	63.4	JRE, COSY, TOCSY, HSQC, HMBC
		βCH	2.28 (m)	32.1	JRE, COSY, TOCSY, HSQC, HMBC
		γCH_3	1.01 (d, 6.9 Hz)	19.6	JRE, COSY, TOCSY, HSQC, HMBC
		γ′CH_3	1.06 (d, 6.9 Hz)	20.9	
		COOH		177.3	HMBC
5	D-3-hydroxybutyrate	CH	4.16 (dt)	68.4	HMBC
		CH_2	2.41 (dd)	46.5	COSY, TOCSY
		CH_2	2.31 (dd)	46.5	
		γCH_3	1.20 (d, 10.2 Hz)	24.6	JRE, COSY, TOCSY, HSQC
6	Lipid	CH_3	0.88 (s)	19.3	TOCSY, HSQC, HMBC
		$(CH_2)_n$	1.29 (m), 1.59 (m)	20.1	
7	Lactate	αCH	4.13 (q)	[c]	JRE, COSY, TOCSY
		βCH_3	1.34 (d, 6.56 Hz)	22.9	JRE, COSY, TOCSY, HSQC
		COOH		183.1	HMBC
8	Threonine	α-CH	3.53 (d, 2.91 Hz)	63.5	JRE, COSY,
		β-CH2	4.23 (m)	69.1	TOCSY, HSQC, HMBC
		γ-CH3	1.34 (d. 6.5 Hz)	22.9	
		COOH		[c]	
9	Alanine	αCH	3.79 (q)	53.4	JRE, COSY, TOCSY, HSQC, HMBC
		βCH_3	1.49 (d, 7.3 Hz)	19.1	
		COOH		[c]	HMBC

10	Lysine	αCH	3.73 (t)	57.4	JRE, COSY, TOCSY, HSQC, HMBC
		βCH$_2$	1.91 (m)	[c]	
		γCH$_2$	1.49 (m)	25.8	
		εCH$_2$	1.73 (m)	29.0	
		εCH$_2$	3.01 (t, 7.5 Hz)	42.2	
		COOH		[c]	HMBC
11	Acetate	CH$_3$	1.94 (s)	25.7	JRE, HSQC, HMBC
		COOH		[c]	
12	Acetamide	CH$_3$	2.03 (s)	17.6	JRE, HSQC
		CO		[c]	HMBC
12	Glutamate	αCH	3.77 (m)	57.6	JRE, COSY, TOCSY, HSQC, HMBC
		βCH$_2$	2.06 (m), 2.1 (m)	29.8	
		γCH$_2$	2.34 (dt, 3.8, 11.6 Hz)	36.4	
		δCO		184.3	HMBC
		COOH		177.2	
13	Glutamine	αCH	2.15 (m)	29.6	
		βCH$_2$	2.40 (m)	36.9	
		γCH$_2$	3.77 (m)	57.1	
14	γ-Aminobutyrate	αCH$_2$	2.19 (t, 7.6 Hz)	32.2	JRE, COSY, TOCSY, HSQC
		βCH$_2$	1.9 (m)	26.0	JRE, COSY, TOCSY, HSQC, HMBC
		γCH$_2$	3.12 (t)	44.2	
		COOH		186.1	HMBC
15	Pyruvate	CH$_3$	2.35 (s)	28.3	HSQC
		CO		[c]	
		COOH		[c]	
16	Asparagine	αCH	3.94 (dd)	54.7	JRE, COSY, TOCSY, HSQC, HMBC
		βCH$_2$	2.84 (dd, 8.3 Hz) 2.94 (dd, 4.01 Hz)	37.5	
		CO		176.5	HMBC
		COOH		176.5	
17	Choline	N(CH$_3$)$_3$	3.21 (s)	56.5	JRE, COSY, TOCSY, HSQC, HMBC
		OCH$_2$	[c]	[c]	
		NCH$_2$	3.51 (m)	70.4	HMBC
18	Ethanolamine	αCH$_2$	3.12 (t)	44.3	JRE, COSY, TOCSY, HSQC, HMBC
		βCH$_2$	3.81 (t)	63.6	

Table 8.1 (*Continued*)

Key	Metabolites	Assignment	δH^1 (multiplicity)[a]	δC^{13}	Assigned with
19	Phosphocholine	$N(CH_3)_3$	3.23 (s)	56.5	TOCSY, HSQC
		OCH_2	*c*	*c*	
		NCH_2	3.61 (m)	*c*	
20	Citrate	CH_2	2.54 (d,16.5 Hz)	47.1	TOCSY, HSQC, HMBC
		CH_2	2.68 (d, 16.5 Hz)	78.2	
21	Succinate	CH_2	2.49 (s)	35.2	JRE, HSQC
		COOH		182.2	HMBC
22	β-Glucose	C_1H	4.59 (d, 7.8 Hz)	99.2	JRE, COSY, TOCSY, HSQC, HMBC
		C_2H	3.25 (t, 9.3 Hz)	77.6	
		C_3H	3.44 (dd)	79.0	
		C_4H	3.38 (dd)	56.1	
		C_5H	3.41 (t)	72.8	
		C_6H	3.73 (dd), 3.83 (dd)	63.1	
23	α-Glucose	C_1H	5.12 (d, 3.9 Hz)	95.4	JRE, COSY, TOCSY, HSQC
		C_2H	3.54 (*c*)	72.2	COSY, TOCSY, HSQC, HMBC
		C_3H	3.71 (*c*)	76.0	
		C_4H	3.43 (*c*)	72.8	
		C_5H	3.84 (*c*)	74.5	
		C_6H	3.73 (*c*)	64.2	
24	Methanol	CH_3	3.34 (s)	52.1	JRE, HSQC
25	Malate	CH_2	2.50 (dd)	44.4	JRE, TOCSY, HSQC, HMBC
		CH_2	2.73 (dd, 9.2 Hz, 9.2 Hz)	44.4	
		CH	4.29 (dd)	72.5	
26	Sucrose	C_1H	3.66 (s)	64.9	JRE, TOCSY, HSQC, HMBC
		C_2H	3.80 (m)	63.6	
		C_3H	3.81 (m)	84.9	
		C_4H	*c*	79.6	
		C_5H	5.41 (d)	95.4	
			c	106.7	HMBC
27	Fructose	C_1H	3.80 (m)	63.6	JRE, TOCSY, HSQC, HMBC
		C_2H	3.84 (m)	84.6	
		C_3H	4.03 (dd)	77.2	
		C_4H	4.17 (d)	79.8	

28	Trehalose	C_1H	3.48 (d)	[c]	TOCSY, HSQC, HMBC
		C_2H	3.69 (d)	[c]	
		C_3H	3.80 (m)	75.4	
		C_4H	5.16 (d)	95.4	
29	D-Methanol	CH_3	3.32 (m)	51.4	HSQC
30	Uridine	C_1		169.4	HMBC
		C_2		155.0	
		C_3	7.88 (d, 8.1 Hz)	145.1	JRE, COSY, TOCSY, HSQC, HMBC
		C_4	5.91 (d, 3.9 Hz)	104.7	
		C_5	5.93 (d)	104.7	JRE, HMBC
31	5′-CMP	C_1H	7.85 (d, 7.5 Hz)	144.9	JRE, COSY, TOCSY, HSQC, HMBC
		C_2H	6.06 (d, 6.9 Hz)	99.5	
		C_3		169.1	HMBC
		C_4		160.5	
32	U		5.64, 3.70, 4.34	81.6, 66.7	
33	U		5.61, 3.66, 4.32	81.8	
34	U		3.16	45.4, 178.6	
35	U		6.32, 6.48	102.3, 97.5	HSQC, HMBC
36	Genistein	CH	6.51 (d, 2.1 Hz)	96.2	HSQC
		CH	6.34 (d, 2.1 Hz)		Spiking
		CH	8.126 (s)		
		CH	7.39 (s)		
		CH	5.41 (d, 3.8 Hz)		
37	Fumarate	CH	6.53 (s)	138.7	JRE, HSQC
		COOH		177.5	HMBC
38	Luteolin	C_4H	6.56 (d, 2.19 Hz)	108.6	JRE, TOCSY, HSQC
		C_3H	6.78 (d, 2.46 Hz)	[c]	Spiking
		C_2H	7.09 (d, 8.7 Hz)	[c]	
		C_1H	7.45 (d, 8.7 Hz)	[c]	
39	Orotate	CH_3, CH	1.18 (t), 3.62 (m), 3.71 (s),	101.7	HSQC
		CH, CH	4.82 (s), 6.1 (s), 6.24 (s),		Spiking
40	Salicin[b]	CH	7.28 (d)	128.3	JRE, HSQC
		CH	4.68 (q)	103.7	HSQC

Table 8.1 (*Continued*)

Key	Metabolites	Assignment	δH[1] (multiplicity)[a]	δC[13]	Assigned with
41	Clotrimazole	CH	7.02 (d, 8.7 Hz)	117.1	JRE, TOCSY, HSQC
		CH	7.44 (d, 8.7 Hz)	133.5	Spiking
		CH	7.16 (dd)		
		CH	3.60 (s), 3.644 (s)		
			3.72 (t)		
			8.08 (d, 9.0 Hz)		
42	U		6.82 (m)	113.9	
			7.46	[c]	
43	Tryptophan	C4H, ring	7.74 (d, 8.28 Hz)	[c]	JRE, COSY, TOCSY
		C5H, ring	[c]	[c]	
		C6H, ring	7.09 (t, 7.37)	[c]	
		C7H, ring	7.46 (d)	[c]	
44	U				
45	Nicotinate[b]	4CH	8.06 (d)	[c]	
46	Guanidoacetate[b]	\	3.86 (s)	160.2	JRE, HSQC
47	Quinone[b]	CH	6.88 (s)	[c]	JRE, TOCSY
48	4-Nitrophenol	CH	8.116 (s)	126.0	JRE, TOCSY, HMBC
		CH	6.86 (s)	[c]	
49	Hypoxanthine	C_2H	8.19 (s)	[c]	JRE, TOCSY, HSQC, HMBC
		C_7H	8.21 (s)	139.7	
50	Adenosine	C_1H	8.34 (s)	143.4	HSQC, HMBC
		C_4H	8.25 (s)	[c]	HSQC
		C_6H	6.07 (d, 5.9 Hz)	91.4	JRE, COSY, TOCSY, HSQC, HMBC
51	Inosine	CH, CH, CH	8.28 (s), 3.91 (d, 3.06 Hz),	[c]	JRE
			6.05 (d, 6.0 Hz)		Spiking
52	Formate	CH	8.46 (s)	[c]	JRE

[a]Multiplicity: s-singlet; d-doublet; dd-doublet of doublets; t-triplet; q-quartet; U-unidentified signal.
[b]Tentatively assigned.
[c]Signals or multiplicities were not determined.

8.4.2.5 Yeast Metabolome Database (YMDB)

The Yeast Metabolome Database (YMDB: www.ymdb.ca) provides free access to a collection of small molecule metabolites found in *Saccharomyces cerevisiae* metabolism under optimal conditions or produced by *S. cerevisiae* in the production of wines, beers, and spirits covering over 16 000 small molecules, 900 enzymes, and 149 associated transporters.[148,149]

8.4.2.6 The Pseudomonas aeruginosa Metabolome Database (PAMDB)

The *Pseudomonas aeruginosa* Metabolome Database (PAMDB, www.pseudomonas.umaryland.edu) provides free access to a collection of small molecule metabolites found in *P. aeruginosa* covering over 4300 metabolites and over 900 pathways. PAMDB provides a detailed description of the metabolites, including structural and physiochemical information.[150]

8.4.2.7 The Escherichia coli Metabolome Database (ECMDB)

The *Escherichia coli* Metabolome Database (ECMDB, www.ecmdb.ca/about) provides free access to a collection of small molecule metabolites found in *E. coli* covering over 3700 metabolites and over 1400 associated enzymes. PAMDB provides extensive metabolomic data and metabolic pathway diagrams, including physical properties, reactions, pathways, references, external links, and associated proteins or enzymes.[151,152]

8.4.2.8 NMRShiftDB

NMRShiftDB provides free access to ^{1}H and ^{13}C NMR data for over 40 000 metabolites. Additionally, the user can perform spectrum predictions (^{13}C, ^{1}H, and other nuclei) as well as search for spectra, structures, and other properties.[153,154]

References

1. A. H. Emwas, *Methods Mol. Biol.*, 2015, **1277**, 161–193.
2. R. A. Silva, T. C. S. Pereira, A. R. Souza and P. R. Ribeiro, *Clin. Chim. Acta*, 2020, **502**, 269–279.
3. P. R. Ribeiro, R. R. Barbosa and C. P. De Almeida, *Microbial Interventions in Agriculture and Environment: Volume 1: Research Trends, Priorities and Prospects*, 2019, pp. 29–50.
4. F. V. Carvalho and P. R. Ribeiro, *Food Res. Int.*, 2019, **125**, 108615.
5. H. Tran, M. McConville and P. Loukopoulos, *J. Vet. Diagn. Invest.*, 2020, 635–647.
6. S. Muroya, S. Ueda, T. Komatsu, T. Miyakawa and P. Ertbjerg, *Metabolites*, 2020, **10**, 188.

7. Y. Lin, C. Ma, C. Liu, Z. Wang, J. Yang, X. Liu, Z. Shen and R. Wu, *Oncotarget*, 2016, **7**, 29454–29464.
8. Y. Li, Q. He, Z. Geng, S. Du, Z. Deng and E. Hasi, *R. Soc. Open Sci.*, 2018, **5**, 171722.
9. E. Moser, E. Laistler, F. Schmitt and G. Kontaxis, *Front. Phys.*, 2017, **5**, 33.
10. A. B. Barnes, M. L. Mak-Jurkauskas, Y. Matsuki, V. S. Bajaj, P. C. A. van der Wel, R. DeRocher, J. Bryant, J. R. Sirigiri, R. J. Temkin, J. Lugtenburg, J. Herzfeld and R. G. Griffin, *J. Magn. Reson.*, 2009, **198**, 261–270.
11. L. C. Robosky, M. D. Reily and D. Avizonis, *Anal. Bioanal. Chem.*, 2007, **387**, 529–532.
12. N. Ellis, C. Hattori, J. Cheema, J. Donarski, A. Charlton, M. Dickinson, G. Venditti, P. Kaló, Z. Szabó, G. B. Kiss and C. Domoney, *Front. Plant Sci.*, 2018, **9**, 1022.
13. A. H. Emwas, R. Roy, R. T. McKay, D. Ryan, L. Brennan, L. Tenori, C. Luchinat, X. Gao, A. C. Zeri, G. A. N. Gowda, D. Raftery, C. Steinbeck, R. M. Salek and D. S. Wishart, *J. Proteome Res.*, 2016, **15**, 360–373.
14. C. Lucas-torres, T. Bernard, G. Huber, P. Berthault, Y. Nishiyama, P. S. Kandiyal, B. Elena-herrmann, L. Molin, F. Solari, A. K. Bouzier-sore and A. Wong, *Metabolites*, 2020, **10**, 54.
15. K. E. Haslauer, D. Hemmler, P. Schmitt-Kopplin and S. S. Heinzmann, *Anal. Chem.*, 2019, **91**, 11063–11069.
16. S. Kostidis, R. D. Addie, H. Morreau, O. A. Mayboroda and M. Giera, *Anal. Chim. Acta*, 2017, **980**, 1–24.
17. Y. B. Monakhova and B. W. K. Diehl, *Magn. Reson. Chem.*, 2017, **55**, 996–1005.
18. S. F. Graham, P. K. Kumar, T. Bjorndahl, B. Han, A. Yilmaz, E. Sherman, R. O. Bahado-Singh, D. Wishart, D. Mann and B. D. Green, *Biochim. Biophys. Acta, Mol. Basis Dis.*, 2016, **1862**, 1675–1684.
19. W. D. Koning and K. V. Dam, *Anal. Biochem.*, 1992, **204**, 118–123.
20. M. Rizzi, M. Baltes, U. Theobald and M. Reuss, *Biotechnol. Bioeng.*, 1997, **55**, 592–608.
21. Y. S. Guo and J. Z. Tao, *J. Anim. Sci.*, 2018, **96**, 1769–1777.
22. T. Cruz, M. Gleizes, S. Balayssac, E. Mornet, G. Marsal, J. L. Millán, M. Malet-Martino, L. G. Nowak, V. Gilard and C. Fonta, *J. Neurochem.*, 2017, **140**, 919–940.
23. S. F. Graham, N. L. Rey, A. Yilmaz, P. Kumar, Z. Madaj, M. Maddens, R. O. Bahado-Singh, K. Becker, E. Schulz, L. K. Meyerdirk, J. A. Steiner, J. Ma and P. Brundin, *J. Proteome Res.*, 2018, **17**, 2460–2469.
24. K. D. Courtney, D. Bezwada, T. Mashimo, K. Pichumani, V. Vemireddy, A. M. Funk, J. Wimberly, S. S. McNeil, P. Kapur, Y. Lotan, V. Margulis, J. A. Cadeddu, I. Pedrosa, R. J. DeBerardinis, C. R. Malloy, R. M. Bachoo and E. A. Maher, *Cell Metab.*, 2018, **28**, 793–800.e792.
25. Z. Schofield, M. A. C. Reed, P. N. Newsome, D. H. Adams, U. L. Günther and P. F. Lalor, *World J. Gastroenterol.*, 2017, **23**, 2685–2695.

26. J. Detour, C. Bund, C. Behr, H. Cebula, E. A. Cicek, M. P. Valenti-Hirsch, B. Lannes, B. Lhermitte, A. Nehlig, P. Kehrli, F. Proust, E. Hirsch and I. J. Namer, *Epilepsia*, 2018, **59**, 607–616.
27. I. Lauri, F. Savorani, N. Iaccarino, P. Zizza, L. M. Pavone, E. Novellino, S. B. Engelsen and A. Randazzo, *Metabolites*, 2016, **6**, 4.
28. V. C. Brito, C. P. de Almeida, R. R. Barbosa, M. G. A. Carosio, A. G. Ferreira, L. G. Fernandez, R. D. de Castro, H. Hilhorst, W. Ligterink and P. R. Ribeiro, *Ind. Crops Prod.*, 2020, **145**, 112110.
29. N. Happyana and O. Kayser, *Planta Med.*, 2016, **82**, 1217–1223.
30. H. Rolletschek, J. Schwender, C. König, K. D. Chapman, T. Romsdahl, C. Lorenz, H. P. Braun, P. Denolf, K. Van Audenhove, E. Munz, N. Heinzel, S. Ortleb, T. Rutten, S. McCorkle, T. Borysyuk, A. Guendel, H. Shi, M. Vander Auwermeulen, S. Bourot and L. Borisjuk, *Plant Cell*, 2020, **32**, 2383–2401.
31. G. Lazzarino, M. Nuutinen, B. Tavazzi, D. Di Pierro and B. Giardina, *Anal. Biochem.*, 1989, **181**, 239–241.
32. I. Maksimovic, S. Zhang, I. Vuckovic and M. V. Irazabal, *Methods Cell Biol.*, 2019, **154**, 15–29.
33. M. M. Chen, A. L. Li, M. C. Sun, Z. Feng, X. C. Meng and Y. Wang, *J. Zhejiang Univ. Sci. B*, 2014, **15**, 333–342.
34. L. P. de Jonge, R. D. Douma, J. J. Heijnen and W. M. van Gulik, *Metabolomics*, 2012, **8**, 727–735.
35. R. V. Kapoore and S. Vaidyanathan, *Metabolites*, 2018, **8**, 72.
36. H. Bi, K. W. Krausz, S. K. Manna, F. Li, C. H. Johnson and F. J. Gonzalez, *Anal. Bioanal. Chem.*, 2013, **405**, 5279–5289.
37. M. Carnicer, G. Vieira, T. Brautaset, J. C. Portais and S. Heux, *Microb. Cell Fact.*, 2016, **15**, 92.
38. J. Kronthaler, G. Gstraunthaler and C. Heel, *OMICS: J. Integr. Biol.*, 2012, **16**, 90–97.
39. C. E. McHugh, T. L. Flott, C. R. Schooff, Z. Smiley, M. A. Puskarich, D. D. Myers, J. G. Younger, A. E. Jones and K. A. Stringer, *Metabolites*, 2018, **8**, 93.
40. G. A. Nagana Gowda and D. Raftery, *Anal. Chem.*, 2014, **86**, 5433–5440.
41. H. Wu, A. D. Southam, A. Hines and M. R. Viant, *Anal. Biochem.*, 2008, **372**, 204–212.
42. H. X. Zou, D. Zhao, H. Wen, N. Li, W. Qian and X. Yan, *PLoS One*, 2020, **15**, e0235415.
43. F. Angilè, L. Del Coco, C. R. Girelli, L. Basso, L. Rizzo, S. Piraino, L. Stabili and F. P. Fanizzi, *Molecules*, 2020, 25, 806.
44. A. Amiel, M. Tremblay-Franco, R. Gautier, S. Ducheix, A. Montagner, A. Polizzi, L. Debrauwer, H. Guillou, J. Bertrand-Michel and C. Canlet, *Metabolites*, 2019, **10**, 9.
45. J. Oh, D. H. Yoon, J. G. Han, H. K. Choi and G. H. Sung, *Saudi J. Biol. Sci.*, 2018, **25**, 1128–1134.

46. M. H. Li, H. D. Xu, Y. Liu, T. Chen, L. Jiang, Y. H. Fu and J. S. Wang, *Toxicol. Res.*, 2016, **5**, 1039–1052.
47. N. J. Kruger, M. A. Troncoso-Ponce and R. G. Ratcliffe, *Nat. Protoc.*, 2008, **3**, 1001–1012.
48. H. Zhang, L. Qiao, X. Li, Y. Wan, L. Yang and H. Wang, *Oncol. Rep.*, 2016, **36**, 3436–3448.
49. Y. Lin, C. Ma, T. Bezabeh, Z. Wang, J. Liang, Y. Huang, J. Zhao, X. Liu, W. Ye, W. Tang, T. Ouyang and R. Wu, *Int. J. Cancer*, 2019, **145**, 1679–1689.
50. J. Jáčová, M. Jořenek, K. Pospíšková, L. Najdekr, L. Zajoncová, D. Friedecký and T. Adam, *J. Chromatogr. A*, 2019, **1605**, 360355.
51. B.-J. Webb-Robertson, Y.-M. Kim, E. M. Zink, K. A. Hallaian, Q. Zhang, R. Madupu, K. M. Waters and T. O. Metz, *Metabolomics: Off. J. Metabolomic Soc.*, 2014, **10**, 897–908.
52. E. C. Y. Chan, K. K. Pasikanti and J. K. Nicholson, *Nat. Protoc.*, 2011, **6**, 1483–1499.
53. S. Saito, Y. Hirao, A. F. Quadery, B. Xu, A. Elguoshy, H. Fujinaka, S. Koma, K. Yamamoto and T. Yamamoto, *Methods Protoc*, 2019, **2**, 46.
54. J. Moreda-Piñeiro and A. Moreda-Piñeiro, *TrAC, Trends Anal. Chem.*, 2015, **71**, 265–274.
55. D. Vuckovic, I. de Lannoy, B. Gien, R. E. Shirey, L. M. Sidisky, S. Dutta and J. Pawliszyn, *Angew. Chem., Int. Ed.*, 2011, **50**, 5344–5348.
56. M. Karimpour, I. Surowiec, J. Wu, S. Gouveia-Figueira, R. Pinto, J. Trygg, A. M. Zivkovic and M. L. Nording, *Anal. Chim. Acta*, 2016, **908**, 121–131.
57. N. Gholizadeh, J. Pundavela, R. Nagarajan, A. Dona, S. Quadrelli, T. Biswas, P. B. Greer and S. Ramadan, *Urol. Oncol.*, 2020, **38**, 150–173.
58. L. Goldoni, T. Beringhelli, W. Rocchia, N. Realini and D. Piomelli, *Anal. Biochem.*, 2016, **501**, 26–34.
59. A. Le Guennec, F. Tayyari and A. S. Edison, *Anal. Chem.*, 2017, **89**, 8582–8588.
60. E. R. Andersson, R. D. Day, J. M. Loewenstein, C. M. Woodley and T. B. Schock, *Metabolites*, 2019, **9**, 32.
61. C. G. Faulkes, T. R. Eykyn and D. Aksentijevic, *Biol. Lett.*, 2019, **15**, 20190710.
62. N. Villa-Ruano, R. Velásquez-Valle, L. G. Zepeda-Vallejo, N. Pérez-Hernández, M. Velázquez-Ponce, V. M. Arcos-Adame and E. Becerra-Martínez, *Food Res. Int.*, 2018, **106**, 870–877.
63. S. Roques, C. Deborde, N. Richard, L. Sergent, F. Kurz, S. Skiba-Cassy, B. Fauconneau and A. Moing, *Metabolomics*, 2018, **14**, 155.
64. H. Xu, L.-Y. Ruan, C. Chen, J.-T. Fan, J.-F. Chen, W.-L. Zhao, H.-H. Meng, W.-Y. Liu, W.-L. Zhao, Q. Zheng, Z.-C. Liu and J.-S. Wang, *J. Ethnopharmacol.*, 2020, **254**, 112403.
65. G. A. Nagana Gowda, Y. N. Gowda and D. Raftery, *Anal. Chem.*, 2015, **87**, 706–715.

66. Y. Kodani, T. Miyakawa, T. Komatsu and M. Tanokura, *Sci. Rep.*, 2017, **7**, 1297.
67. M. Lemaire-Chamley, F. Mounet, C. Deborde, M. Maucourt, D. Jacob and A. Moing, *Metabolites*, 2019, **9**, 93.
68. M. Maulidiani, B. Y. Sheikh, A. Mediani, L. S. Wei, I. S. Ismail, F. Abas and N. H. Lajis, *Phytochem. Lett.*, 2015, **13**, 308–318.
69. P. Pramai, N. A. Abdul Hamid, A. Mediani, M. Maulidiani, F. Abas and S. Jiamyangyuen, *J. Food Drug Anal.*, 2018, **26**, 47–57.
70. A. Tomassini, G. Curone, M. Solè, G. Capuani, F. Sciubba, G. Conta, A. Miccheli and D. Vigo, *Nat. Prod. Res.*, 2019, **33**, 1085–1091.
71. K. Ito, Y. Tsuboi and J. Kikuchi, *Commun. Chem.*, 2020, **3**, 80.
72. N. Miyaguti, D. Stanisic, S. C. P. Oliveira, G. S. Dos Santos, B. S. Manhe, L. Tasic and M. C. C. Gomes-Marcondes, *Nutrients*, 2020, **12**, DOI: 10.3390/nu12072106.
73. I. C. A. R. de Melo, I. L. L. de Souza, L. H. C. Vasconcelos, M. T. Scotti, B. A. da Silva, J. Schripsema, A. H. Fonseca and E. J. Oliveira, *J. Ethnopharmacol.*, 2020, **253**, 112678.
74. M. Maulidiani, A. Mediani, F. Abas, Y. S. Park, Y.-K. Park, Y. M. Kim and S. Gorinstein, *Talanta*, 2018, **184**, 277–286.
75. A.-H. Emwas, R. Roy, R. T. McKay, L. Tenori, E. Saccenti, G. A. N. Gowda, D. Raftery, F. Alahmari, L. Jaremko, M. Jaremko and D. S. Wishart, *Metabolites*, 2019, **9**, 123.
76. F. Bhinderwala, P. Evans, K. Jones, B. R. Laws, T. G. Smith, M. Morton and R. Powers, *Anal. Chem.*, 2020, **92**, 9536–9545.
77. H. D. dos Santos, Y. A. Alvarenga and E. F. Boffo, *Microchem. J.*, 2020, **152**, 104293.
78. N. MacKinnon, W. Ge, P. Han, J. Siddiqui, J. T. Wei, T. Raghunathan, A. M. Chinnaiyan, T. M. Rajendiran and A. Ramamoorthy, *NMR-based metabolomic profiling of urine: Evaluation for application in prostate cancer detection.*
79. C. Deborde, J.-X. Fontaine, D. Jacob, A. Botana, V. Nicaise, F. Richard-Forget, S. Lecomte, C. Decourtil, K. Hamade, F. Mesnard, A. Moing and R. Molinié, *Metabolomics*, 2019, **15**, 28.
80. C. Deborde, A. Moing, L. Roch, D. Jacob, D. Rolin and P. Giraudeau, *Prog. Nucl. Magn. Reson. Spectrosc.*, 2017, **102–103**, 61–97.
81. F. Fathi, A. Brun, K. H. Rott, P. Falco Cobra, M. Tonelli, H. R. Eghbalnia, E. Caviedes-Vidal, W. H. Karasov and J. L. Markley, *Metabolites*, 2017, **7**, 61.
82. E. G. Nagato, B. P. Lankadurai, R. Soong, A. J. Simpson and M. J. Simpson, *Magn. Reson. Chem.*, 2015, **53**, 745–753.
83. R. Canese, G. Palombelli, M. Chirico, P. Sestili, M. Bagnoli, S. Canevari, D. Mezzanzanica, F. Podo and E. Iorio, *NMR Biomed.*, 2019, **32**, e4016.
84. S. Wang, S. Wen, P. Guo, H. Liu, J. Feng and H. Huang, *J. Pharm. Biomed. Anal.*, 2020, **190**, 113546.

85. D. Augustijn, N. V. Tol, B. J. van der Zaal, H. J. M. de Groot and A. Alia, *PLoS One*, 2019, **13**, e0209695.
86. I. S. Flores, B. C. B. Martinelli, V. S. Pinto, L. H. K. Queiroz, Jr and L. M. Lião, *Phytochem. Anal.*, 2019, **30**, 5–13.
87. S. Vicente-Munoz, I. Morcillo, L. Puchades-Carrasco, V. Paya, A. Pellicer and A. Pineda-Lucena, *Fertil. Steril.*, 2016, **106**, 1733–1741, .e1731.
88. K. Ameta, A. Gupta, D. Ameta, R. Sethi, D. Kumar, I. Ahmad and A. A. Mahdi, *Clin. Chim. Acta*, 2016, **456**, 56–62.
89. A. Gupta, N. Bansal, N. Mitash, D. Kumar, M. Kumar, S. N. Sankhwar, A. Mandhani and U. P. Singh, *J. Pharm. Biomed. Anal.*, 2020, **183**, 113134.
90. N. A. S. Miyaguti, D. Stanisic, S. C. P. de Oliveira, G. S. Dos Santos, B. S. Manhe, L. Tasic and M. C. C. Gomes-Marcondes, *Nutrients*, 2020, **12**, 1–15.
91. M. Bazzano, L. Laghi, C. Zhu, E. Lotito, S. Sgariglia, B. Tesei and F. Laus, *Metabolites*, 2020, **10**, 1–14.
92. M. Grimaldi, A. Palisi, G. Rossi, I. Stillitano, F. Faiella, P. Montoro, M. Rodriquez, R. Palladino, A. M. D'Ursi and R. Romano, *J. Pharm. Biomed. Anal.*, 2018, **160**, 436–442.
93. J. L. Pereira, D. Duarte, T. J. Carneiro, S. Ferreira, B. Cunha, D. Soares, A. L. Costa and A. M. Gil, *Oral Diseases*, 2019, **25**, 1545–1554.
94. L. H. Münger, A. Trimigno, G. Picone, C. Freiburghaus, G. Pimentel, K. J. Burton, F. P. Pralong, N. Vionnet, F. Capozzi, R. Badertscher and G. Vergères, *J. Proteome Res.*, 2017, **16**, 3321–3335.
95. L. Zennaro, P. Vanzani, L. Nicolè, R. Cappellesso and A. Fassina, *Cancer Cytopathol.*, 2017, **125**, 341–348.
96. M. Kyriakides, N. Rama, J. Sidhu, H. Gabra, H. C. Keun and M. El-Bahrawy, *Oncotarget*, 2016, **7**, 7216–7226.
97. F. M. Moussallieh, E. Moss, K. Elbayed, G. Lereaux, F. Tourneix and J. P. Lepoittevin, *Contact Dermatitis*, 2020, **82**, 137–146.
98. E. Subramani, M. Jothiramajayam, M. Dutta, D. Chakravorty, M. Joshi, S. Srivastava, A. Mukherjee, C. Datta Ray, B. N. Chakravarty and K. Chaudhury, *Hum. Reprod.*, 2016, **31**, 854–865.
99. W. J. Xu, H. Wen, H. S. Kim, Y. J. Ko, S. M. Dong, I. S. Park, J. I. Yook and S. Park, *Proc. Natl. Acad. Sci. U. S. A.*, 2018, **115**, 4152–4157.
100. L. An, Q. Lang, W. Shen, Q. Shi and F. Feng, *Mol. Med. Rep.*, 2016, **14**, 2093–2100.
101. J. Ji, L. Zhang, H. Zhang, C. Sun, J. Sun, H. Jiang, M. H. Abdalhai, Y. Zhang and X. Sun, *Toxicol. Res.*, 2016, **5**, 689–696.
102. B. Toczylowska, E. Zieminska, P. Senator and J. W. Lazarewicz, *Mol. Neurobiol.*, 2020, **57**, 3089–3105.
103. J. Zheng, L. Sun, S. Shu, K. Zhu, C. Xu, J. Wang and H. Wang, *J. Vet. Med. Sci.*, 2016, **78**, 1421–1428.
104. H. C. Kim, Y. J. Ko, M. Kim, J. Choe, H. I. Yong and C. Jo, *Food Sci. Anim. Resour.*, 2019, **39**, 1–12.

105. D. Augustijn, U. Roy, R. Van Schadewijk, H. J. M. De Groot and A. Alia, *PLoS One*, 2016, **11**, e0163258.
106. M. Hasanpour, S. Saberi and M. Iranshahi, *Planta Med.*, 2020, **86**, 212–219.
107. J. Zhou, C. Zhang, J. Han, C. Lu, Y. Li, T. Ming and X. Su, *Arch. Microbiol.*, 2020, **202**, 437–445.
108. F. Puig-Castellví, Y. Pérez, B. Piña, R. Tauler and I. Alfonso, *Anal. Chem.*, 2018, **90**, 12422–12430.
109. C. Peng, T. Viana, M. A. Petersen, F. H. Larsen and N. Arneborg, *Metabolomics*, 2018, **14**, 1–12.
110. Y. Kumar, L. Zhang, P. Panigrahi, B. B. Dholakia, V. Dewangan, S. G. Chavan, S. M. Kunjir, X. Wu, N. Li, P. R. Rajmohanan, N. Y. Kadoo, A. P. Giri, H. Tang and V. S. Gupta, *Plant Biotechnol. J.*, 2016, **14**, 1589–1603.
111. E. L. Ulrich, H. Akutsu, J. F. Doreleijers, Y. Harano, Y. E. Ioannidis, J. Lin, M. Livny, S. Mading, D. Maziuk, Z. Miller, E. Nakatani, C. F. Schulte, D. E. Tolmie, R. Kent Wenger, H. Yao and J. L. Markley, *Nucleic Acids Res.*, 2007, **36**, D402–D408.
112. E. Shumilina, T. Johansen, C. Glasner, J. Robert, T. Rustad and A. Dikiy, *Waste Biomass Valorization*, 2020, **11**, 1349–1357.
113. R. M. Shah, E. J. McKenzie, M. T. Rosin, S. R. Jadhav, S. V. Gondalia, D. Rosendale and D. J. Beale, *Metabolites*, 2020, **10**, 94.
114. Y. Pérez, M. Casado, D. Raldúa, E. Prats, B. Piña, R. Tauler, I. Alfonso and F. Puig-Castellví, *Anal. Bioanal. Chem.*, 2020, **412**, 5695–5706.
115. L. B. Pena, C. L. Matayoshi, A. A. E. Méndez, M. Arán, C. J. Moratto, J. M. Vázquez-Ramos and S. M. Gallego, *Plant Physiol. Biochem.*, 2020, **155**, 560–569.
116. Y. Mateo-Otero, P. Fernández-López, S. Gil-Caballero, B. Fernandez-Fuertes, S. Bonet, I. Barranco and M. Yeste, *Biomolecules*, 2020, **10**, 1–16.
117. A. R. Lima, J. Pinto, D. Barros-Silva, C. Jerónimo, R. Henrique, M. L. Bastos, M. Carvalho and P. Guedes Pinho, *Metabolomics*, 2020, **16**, 1–9.
118. D. S. Wishart, Y. D. Feunang, A. Marcu, A. C. Guo, K. Liang, R. Vázquez-Fresno, T. Sajed, D. Johnson, C. Li, N. Karu, Z. Sayeeda, E. Lo, N. Assempour, M. Berjanskii, S. Singhal, D. Arndt, Y. Liang, H. Badran, J. Grant, A. Serra-Cayuela, Y. Liu, R. Mandal, V. Neveu, A. Pon, C. Knox, M. Wilson, C. Manach and A. Scalbert, *Nucleic Acids Res.*, 2018, **46**, D608–D617.
119. D. S. Wishart, T. Jewison, A. C. Guo, M. Wilson, C. Knox, Y. Liu, Y. Djoumbou, R. Mandal, F. Aziat, E. Dong, S. Bouatra, I. Sinelnikov, D. Arndt, J. Xia, P. Liu, F. Yallou, T. Bjorndahl, R. Perez-Pineiro, R. Eisner, F. Allen, V. Neveu, R. Greiner and A. Scalbert, *Nucleic Acids Res.*, 2013, **41**, D801–D807.
120. D. S. Wishart, C. Knox, A. C. Guo, R. Eisner, N. Young, B. Gautam, D. D. Hau, N. Psychogios, E. Dong, S. Bouatra, R. Mandal, I. Sinelnikov, J. Xia, L. Jia, J. A. Cruz, E. Lim, C. A. Sobsey, S. Shrivastava, P. Huang,

P. Liu, L. Fang, J. Peng, R. Fradette, D. Cheng, D. Tzur, M. Clements, A. Lewis, A. de souza, A. Zuniga, M. Dawe, Y. Xiong, D. Clive, R. Greiner, A. Nazyrova, R. Shaykhutdinov, L. Li, H. J. Vogel and I. Forsythei, *Nucleic Acids Res.*, 2009, **37**, D603–D610.

121. D. S. Wishart, D. Tzur, C. Knox, R. Eisner, A. C. Guo, N. Young, D. Cheng, K. Jewell, D. Arndt, S. Sawhney, C. Fung, L. Nikolai, M. Lewis, M. A. Coutouly, I. Forsythe, P. Tang, S. Shrivastava, K. Jeroncic, P. Stothard, G. Amegbey, D. Block, D. D. Hau, J. Wagner, J. Miniaci, M. Clements, M. Gebremedhin, N. Guo, Y. Zhang, G. E. Duggan, G. D. MacInnis, A. M. Weljie, R. Dowlatabadi, F. Bamforth, D. Clive, R. Greiner, L. Li, T. Marrie, B. D. Sykes, H. J. Vogel and L. Querengesser, *Nucleic Acids Res.*, 2007, **35**, D521–D526.
122. T. Yeo, M. Sealey, Y. Zhou, L. Saldana, S. Loveless, T. D. W. Claridge, N. Robertson, G. DeLuca, J. Palace, D. C. Anthony and F. Probert, *Sci. Rep.*, 2020, **10**, 1–12.
123. J. Tian, G. Fu, Z. Xu, X. Chen, J. Sun and B. Jin, *Med. Hypotheses*, 2020, **143**, 110100.
124. B. Elena-Herrmann, E. Montellier, A. Fages, R. Bruck-Haimson and A. Moussaieff, *Sci. Rep.*, 2020, **10**, 1–11.
125. S. Ge, Q. Zhang, Y. Tian, L. Hao, J. Duan and B. Zhang, *Clin. Chim. Acta*, 2020, **510**, 291–297.
126. A. Tzimou, D. Benaki, S. Nikolaidis, E. Mikros, I. Taitzoglou and V. Mougios, *Biogerontology*, 2020, **21**, 577–591.
127. T. Zhang, H. Zheng, K. Fan, N. Xia, J. Li, C. Yang, H. Gao and Y. Yang, *Metab. Brain Dis.*, 2020, **35**, 1165–1173.
128. S. V. Shekhovtsov, N. A. Bulakhova, Y. P. Tsentalovich, E. A. Zelentsova, L. V. Yanshole, E. N. Meshcheryakova and D. I. Berman, *Sci. Rep.*, 2020, **10**, 1–11.
129. M. Gautam, S. Katoch and R. K. Chahota, *Food Res. Int.*, 2020, **137**, 109600.
130. A. C. Abreu and I. Fernández, *Molecules*, 2020, **25**, 3738.
131. C. Ludwig, J. M. Easton, A. Lodi, S. Tiziani, S. E. Manzoor, A. D. Southam, J. J. Byrne, L. M. Bishop, S. He, T. N. Arvanitis, U. L. Günther and M. R. Viant, *Metabolomics*, 2012, **8**, 8–18.
132. A. Verma, R. Parihar and B. Baishya, Identification of metabolites in coriander seeds (Coriandrum Sativum L.) aided by ultrahigh resolution total correlation NMR spectroscopy, *Magn. Reson. Chem.*, 2019, **57**, 304–316.
133. M. A. Salem, L. Perez de Souza, A. Serag, A. R. Fernie, M. A. Farag, S. M. Ezzat and S. Alseekh, Metabolomics in the Context of Plant Natural Products Research: From Sample Preparation to Metabolite Analysis, *Metabolites*, 2020, **10**(1), 37.
134. N. L. Ma, W. A. Che Lah, N. Abd. Kadir, M. Mustaqim and Z. Rahmat, *et al.*, Susceptibility and tolerance of rice crop to salt threat: Physiological and metabolic inspections, *PLoS One*, 2018, **13**(2), e0192732.
135. F. Casu, A. M. Watson, J. Yost, J. W. Leffler, T. G. Gaylord, F. T. Barrows, P. A. Sandifer, M. R. Denson and D. W. Bearden, Investigation of graded-level soybean meal diets in red drum (Sciaenops ocellatus) using

NMR-based metabolomics analysis, *Comp. Biochem. Physiol., Part D: Genomics Proteomics*, 2019, **29**, 173–184.
136. W. H. Hsu, S. J. Wang, Y. M. Chao, C. J. Chen, Y. F. Wang, J. L. Fuh, S. P. Chen and Y. L. Lin, Urine metabolomics signatures in reversible cerebral vasoconstriction syndrome, *Cephalalgia*, 2020, **40**(7), 735–747.
137. A. Rodriguez-Martinez, J. M. Posma, R. Ayala, N. Harvey, B. Jimenez, A. L. Neves, J. C. Lindon, K. Sonomura, T. A. Sato, F. Matsuda, P. Zalloua, D. Gauguier, J. K. Nicholson and M. E. Dumas, J-Resolved 1H NMR 1D-Projections for Large-Scale Metabolic Phenotyping Studies: Application to Blood Plasma Analysis, *Anal. Chem.*, 2017, **89**(21), 11405–11412.
138. R. Narasimhan, R. Coras and S. B. Rosenthal, *et al.*, Serum metabolomic profiling predicts synovial gene expression in rheumatoid arthritis, *Arthritis Res. Ther.*, 2018, **20**, 164.
139. Q. Cui, I. A. Lewis, A. D. Hegeman, M. E. Anderson, J. Li, C. F. Schulte, W. M. Westler, H. R. Eghbalnia, M. R. Sussman and J. L. Markley, *Nat. Biotechnol.*, 2008, **26**, 162–164.
140. L. Xia, M. Xiaodong, C. Yunhe, L. Junxiang, Z. Junzhu, Z. Feifei, S. Zhenyuan and H. Lei, Transcriptomic and metabolomic insights into the adaptive response of Salix viminalis to phenanthrene, *Chemosphere*, 2021, **262**, 127573.
141. S. Ryu, T. Muramatsu and K. Furihata, *et al.*, NMR-based metabolic profiling and comparison of Japanese persimmon cultivars, *Sci. Rep.*, 2019, **9**, 15011.
142. S. Wen, B. Zhan, J. Feng, W. Hu, X. Lin, J. Bai and H. Huang, Non-invasively predicting differentiation of pancreatic cancer through comparative serum metabonomic profiling, *BMC Cancer*, 2017, **17**(1), 708.
143. C. Liu, F. Chen, J. Zhang, L. Liu, H. Lei, H. Li, Y. Wang, Y. C. Liao and H. Tang, *J. Proteome Res.*, 2019, **18**(9), 3317–3327.
144. C. Lin, J. Dong, Z. Wei, K. K. Cheng, J. Li, S. You, Y. Liu, X. Wang and Z. Chen, 1H NMR-Based Metabolic Profiles Delineate the Anticancer Effect of Vitamin C and Oxaliplatin on Hepatocellular Carcinoma Cells, *J. Proteome Res.*, 2020, **19**(2), 781–793.
145. M. H. Li, X. Feng, D. J. Deng Ba, C. Chen, L. Y. Ruan, Y. X. Xing, L. Y. Chen, G. J. Zhong and J. S. Wang, Hepatoprotection of Herpetospermum caudigerum Wall. against CCl_4-induced liver fibrosis on rats, *J. Ethnopharmacol.*, 2019, **229**, 1–14.
146. J. Heimer, D. Gascho, C. Tappero, M. J. Thali and N. Zoelch, Non-invasive analysis and identification of an intramuscular fluid collection by postmortem 1H-MRS in a case of a fatal motor vehicle accident, *Int. J. Leg. Med.*, 2020, **134**(3), 1167–1174.
147. A. Cukkemane, P. Kumar and B. Sathyamoorthy, A metabolomics footprint approach to understanding the benefits of synbiotics in functional foods and dietary therapeutics for health, communicable and non-communicable diseases, *Food Res. Int.*, 2020, **128**, 108679.
148. M. Ramirez-Gaona, A. Marcu, A. Pon, A. C. Guo, T. Sajed, N. A. Wishart, N. Karu, Y. Djoumbou Feunang, D. Arndt and D. S. Wishart, YMDB 2.0:

a significantly expanded version of the yeast metabolome database, *Nucleic Acids Res.*, 2017, **45**(D1), D440–D445.

149. T. Jewison, C. Knox, V. Neveu, Y. Djoumbou, A. C. Guo, J. Lee, P. Liu, R. Mandal, R. Krishnamurthy, I. Sinelnikov, M. Wilson and D. S. Wishart, *YMDB: The yeast metabolome database.*
150. W. Huang, L. K. Brewer, J. W. Jones, A. T. Nguyen, A. Marcu, D. S. Wishart, A. G. Oglesby-Sherrouse, M. A. Kane and A. Wilks, *Nucleic Acids Res.*, 2018, **46**, gkx1061.
151. T. Sajed, A. Marcu, M. Ramirez, A. Pon, A. C. Guo, C. Knox, M. Wilson, J. R. Grant, Y. Djoumbou and D. S. Wishart, ECMDB 2.0: A richer resource for understanding the biochemistry of E. coli, *Nucleic Acids Res.*, 2016, **44**(D1), D495–D501.
152. A. C. Guo, T. Jewison, M. Wilson, Y. Liu, C. Knox, Y. Djoumbou, P. Lo, R. Mandal, R. Krishnamurthy and D. S. Wishart, ECMDB: the E. coli Metabolome Database, *Nucleic Acids Res.*, 2013, **41**(Database issue), D625–D630.
153. C. Steinbeck and S. Kuhn, NMRShiftDB – compound identification and structure elucidation support through a free community-built web database, *Phytochemistry*, 2004, **65**(19), 2711–2717.
154. C. Steinbeck, S. Krause and S. Kuhn, NMRShiftDB-constructing a free chemical information system with open-source components, *J. Chem. Inf. Comput. Sci.*, 2003, **43**(6), 1733–1739.

CHAPTER 9

State-of-the-art Mass Spectrometry-based Approaches to Explore the Polar Metabolome: Alzheimer's Disease as a Case Study

RAÚL GONZÁLEZ-DOMÍNGUEZ,*[a,b]
ÁLVARO GONZÁLEZ-DOMÍNGUEZ,[c,d] ANA SAYAGO[a,b] AND
ÁNGELES FERNÁNDEZ-RECAMALES[a,b]

[a] Department of Chemistry, Faculty of Experimental Sciences, University of Huelva, 21007, Spain; [b] International Campus of Excellence CeiA3, University of Huelva, 21007, Spain; [c] Department of Pediatrics, Hospital Universitario Puerta del Mar, Cádiz 11009, Spain; [d] Institute of Research and Innovation in Biomedical Sciences of the Province of Cádiz (INiBICA), Cádiz, Spain
*Email: raul.gonzalez@dqcm.uhu.es

9.1 The Complexity of the Human Metabolome

The human metabolome encompasses thousands of metabolites with huge chemical and functional heterogeneity, comprising more than 114 000 entities according to the last actualization of the Human Metabolome Database in 2018.[1] On the one hand, the endogenous metabolome represents the

New Developments in Mass Spectrometry No. 10
Advanced Mass Spectrometry-based Analytical Separation Techniques for Probing the Polar Metabolome
Edited by Rawi Ramautar

Published by the Royal Society of Chemistry, www.rsc.org

complete set of small molecules directly involved in catabolic and/or anabolic reactions within organism homeostasis, consisting of multiple chemical classes such as amino acids and derivatives (*e.g.* neurotransmitters, peptides), carbohydrates, biogenic amines, organic acids, nucleic acid derivatives, vitamins, lipids (*e.g.* fatty acids, steroids, phospholipids, glycerolipids) and many others. Accordingly, metabolites may be considered as suitable read-outs of the functional and phenotypic status of biological systems due to their direct participation in primary metabolic pathways, and because they are, in turn, affected by the proper functioning of higher biochemical organization levels, including the genome, the transcriptome and the proteome. However, emerging evidence has demonstrated over recent years that external factors also have a great impact on human health by complementing intrinsic processes within the framework of the exposome concept, which could be defined as the totality of external exposures and associated biological responses.[2,3] These exogenous compounds are in turn normally subjected to various metabolization and biotransformation reactions in the liver, kidney and intestine after their ingestion/absorption (*e.g.* phase I and II reactions, microbiota metabolism), thus leading to the generation of a myriad of different metabolites in biological fluids and tissues. Diet is one of the modifiable factors with major impact on the human metabolome and, consequently, on health status. In this vein, the food metabolome could be defined as the entire set of metabolites coming from the ingestion, absorption and biotransformation of food components, including nutrients (*e.g.* fiber, vitamins), bioactive compounds (*e.g.* polyphenols, carotenoids), additives (*e.g.* sweeteners) and food-processing by-products (*e.g.* Maillard reaction products).[4,5] Other lifestyle habits contributing to the human exposome include medication, smoking and consumption of alcohol or illicit drugs. Furthermore, the organism is continuously exposed to multiple xenobiotics with potential toxic effects (*e.g.* endocrine disruption, carcinogenesis), including household chemicals (*e.g.* personal care products, food packaging) and environmental pollutants (*e.g.* pesticides, volatile organic compounds).[6] Within this tangled interplay between endogenous and exogenous factors, the microbiota also plays a major role in metabolic regulation and host health, being involved in biotransformation processes for detoxification or absorption of xenobiotics, and the biosynthesis of some essential molecules (*e.g.* vitamins, neurotransmitters, secondary bile acids).[7,8]

This compositional diversity consequently results in the vast complexity of the human metabolome in terms of molecular weight (*ca.* in the range 50–1500 Da), polarity (*e.g.* from highly polar low molecular weight metabolites to hydrophobic lipids) and other physicochemical properties.[9] Furthermore, metabolite levels in biological fluids and tissues may span a wide range of concentrations, as environmental pollutants are normally found at levels three orders of magnitude lower than endogenous and food-derived metabolites.[10] In this vein, it should also be noted the high intra- and inter-individual variability of the human metabolome because of genetic (*e.g.* gender), temporal (*e.g.* circadian rhythm), environmental (*e.g.* dietary habits, domestic and workplace exposures) and microbial (*e.g.* eubiosis/dysbiosis, antibiotics consumption) factors, which introduce a higher level of complexity in metabolomics research toward the concept of metabotypes, *i.e.* the grouping of individuals based on similarities in their metabolic phenotype.[11]

Taking all this into account, one of the main challenges in metabolomics relies on the selection of the proper analytical methodology to target the metabolome's complexity. Nuclear magnetic resonance (NMR) spectroscopy and mass spectrometry (MS) are the most commonly employed techniques in metabolomics research, the latter normally coupled with liquid/gas chromatography or capillary electrophoresis. However, each one of these methods has inherent analytical limitations that impede the determination of the entire human metabolome in a single analysis, so more and more authors have emphasized in recent years the need of combining complementary approaches to achieve comprehensive metabolomics coverage.[12]

9.2 Analytical Approaches to Explore the Polar Metabolome

The complexity and dynamic range of the human metabolome make the application of powerful and versatile analytical platforms mandatory, with high sensitivity and selectivity, with the aim of detecting and (semi)-quantifying the maximum number of metabolites as possible for performing a comprehensive characterization of biological systems under study and for biomarker discovery purposes. Reproducibility is also essential to minimize analytical variability for ensuring that the data generated is a direct and meaningful expression of the biochemical status, thus enabling for cross-cohort and inter-laboratory comparisons, as well as for conducting meta-analyses in epidemiological studies. To this end, the development of targeted quantitative approaches is of special interest for measuring absolute metabolite concentrations, which inherently provide enhanced analytical reproducibility than non-targeted metabolomics. However, analytical throughput is sometimes overlooked in metabolomics, although it is a critical factor to consider especially in large-scale studies. High-throughput approaches aim to provide comprehensive metabolomics results by using simple, rapid and small-sample consuming methods. This allows minimizing the total analysis time when dealing with large sample sizes, thus considerably reducing intra- and inter-batch variability, as well as the consumption of valuable biological samples and potentially contaminating chemicals, and consequently costs.

Nuclear magnetic resonance (NMR) and reversed-phase liquid chromatography coupled to mass spectrometry (RPLC–MS) are nowadays the main workhorses in metabolomics research, as evidenced by the large number of published papers on this regard. NMR is widely employed for high-throughput metabolomics because of its high reproducibility, reduced analysis time and simplicity (*e.g.* minimal requirements for sample preparation, analytical robustness), thus offering direct identification and quantification of a range of metabolites.[13] However, the interpretation of NMR spectra is usually a challenging and time-consuming task. Furthermore, the low sensitivity and spectral resolution of this technique considerably hinder the detection of minor metabolites in complex samples, despite the numerous methodological advances made with the aim of increasing metabolomics coverage (*e.g.* bi-dimensional spectroscopy: COSY, TOCSY) and sensitivity (*e.g.* use of cryogenically cooled probes, increased magnetic field strength). On the other hand, MS-based approaches provide significantly enhanced sensitivity and selectivity, which facilitate the detection of numerous metabolite classes at physiological levels. Among them, RPLC–MS has become the gold standard metabolomics platform for profiling low and medium polarity metabolites, accounting for up to 70% of the entire human metabolome.[14] This coupling is usually accomplished by using the electrospray ionization (ESI) source, which is applicable for the sensitive detection of compounds with very different polarities and molecular weights, although other ionization sources have also been proposed as efficient alternatives for analyzing non-polar compounds with lower ion suppression, such as atmospheric pressure chemical ionization (APCI) and atmospheric pressure photoionization (APPI).[15,16] However, RPLC faces important limitations in the analysis of polar metabolites (*e.g.* amino acids and derivatives, carbohydrates, amines, organic acids), which are implicated in a multitude of primary biochemical pathways. To overcome this hurdle, alternative analytical tools are available to approach the polar metabolome, namely gas chromatography (GC), hydrophilic interaction liquid chromatography (HILIC), capillary electrophoresis (CE) and direct mass spectrometry (DMS) based approaches.

9.2.1 Gas Chromatography–Mass Spectrometry

The coupling of gas chromatography with mass spectrometry (GC–MS) is a robust and versatile technique widely employed in metabolomics due to its reproducibility, sensitivity, chromatographic resolution and selectivity.[17] To date, most metabolomics methods based on GC–MS use electron impact (EI) as the ionization source, which generates reproducible and well characterized mass spectra. The standardization of EI-based methods, always performed at 70 eV, and the reproducibility across instruments from different manufacturers and with different mass analyzers (*e.g.* ion trap, quadrupole, time of flight) has enabled the construction of mass spectral libraries, such as the National Institute of Standards and Technology (NIST) Mass Spectral Database or the GOLM Metabolome Database, among others.[18] This methodological standardization and the availability of libraries considerably simplify the analytical workflow and facilitate metabolite identifications, which makes GC–MS the method of choice for many analysts. Furthermore, this technique also provides highly reproducible retention times and retention indices, the latter of which are determined by analyzing a reference standard mix (*e.g.* alkanes, fatty acid methyl esters),[19] which can be used as an orthogonal tool to increase the confidence in metabolite identifications. In recent years, the use of multidimensional gas chromatography has exponentially emerged in metabolomics research due to its excellent separation capacity, which allows significantly expanding the metabolome coverage, but also increasing the method sensitivity and dynamic range.[20,21] For this purpose, two columns with orthogonal retention mechanisms are coupled (*e.g.* polar and non-polar), so that the eluate from the first column is transferred to the second one using a modulator, where metabolites undergo a further separation step before reaching the detector (Figure 9.1).

Despite all these analytical strengths, GC–MS metabolomics platforms can only be employed for analyzing volatile metabolites with low molecular weight (*ca.* 50–600 Da). For this purpose, the application of a chemical derivatization process is usually required prior to the analysis with the aim of reducing the polarity of metabolites and increasing their volatility and thermal stability.[22] Furthermore, derivatization can also improve the method sensitivity and selectivity and facilitate the differentiation of

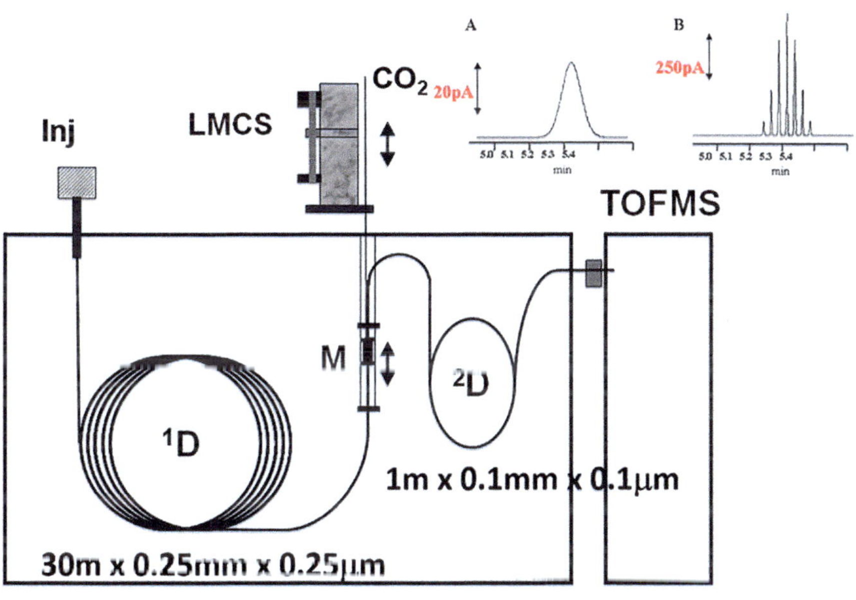

Figure 9.1 Scheme of a GC×GC–MS instrument (Inj: injection port; LMCS: cryogenic system; M: modulator; 1D: first dimension column; 2D: second dimension column; TOFMS: Time-of-flight Mass Spectrometer detector). Chromatograms A and B represent the chromatographic output on the first and second dimension, respectively.
Reproduced from ref. 21 with permission from Springer Nature, Copyright 2018.

analytes with similar fragmentation patterns. Numerous derivatization reagents have been proposed in metabolomics, but the most commonly employed approach is based on the application of a two-step derivatization protocol involving the protection of carbonyl moieties of keto acids and sugars by oximation and subsequent trimethylsilylation of active hydrogens (*e.g.* carboxylic acids, alcohols, amines).[23] The main advantages of this procedure include the comprehensive coverage of multiple chemical classes and its relative simplicity. Additionally, the application of the oximation step prior to the silylation reaction allows minimizing the generation of numerous by-products formed by ring condensation of reducing sugars, consequently improving mass fragmentation and chromatography separation. Nevertheless, the derivatization reaction must be tightly controlled since it represents an important source of variability and bias due to differences in derivatization efficiencies and loss of compounds and might also increase data complexity because of the formation of various derivatized peaks from a single metabolite.[24]

In summary, although the need for a time-consuming derivatization step and long analysis times (usually higher than in LC-based approaches) limit considerably the analytical throughput of GC–MS, this approach is still widely employed for metabolomics profiling of low molecular weight metabolites, such as amino acids and derivatives, sugars, organic and fatty acids, amines and many other primary metabolites. Furthermore, numerous efforts have been performed for simplifying the derivatization protocol by using automated methods, which consequently increase the integrity and reproducibility of data and enable more high-throughput analyses.[25,26]

9.2.2 Hydrophilic Interaction Liquid Chromatography–Mass Spectrometry

As an alternative to GC–MS and complementary to RPLC–MS, hydrophilic interaction liquid chromatography (HILIC) is increasingly being used in metabolomics to explore the polar primary metabolome. This chromatographic approach employs hydrophilic stationary phases common to normal phase chromatography, but sharing water-containing organic mobile phases similar to RPLC. The separation mechanisms in HILIC are complex and not completely understood to date, but the primary retention mechanism is believed to be the partitioning of analytes between the mobile phase and an aqueous layer semi-immobilized on the surface of the stationary phase, as proposed in the pioneering study by Alpert.[27] Besides this partitioning, weak electrostatic interactions and dipole–dipole interactions may also play a role in HILIC. However, the grade of contribution of each interaction mechanism in the final chromatographic separation process greatly depends on the mobile phase composition, the type of stationary phase and the physicochemical properties of the metabolites under study.[28] Mobile phases in HILIC are typically composed of a polar organic solvent and a low proportion of water or an aqueous buffer, thus providing enhanced method sensitivity due to higher ionization and nebulization efficiency. The use of aprotic organic solvents, mostly acetonitrile, is usually preferred, since protic solvents (*e.g.* alcohols) can compete in the solvation process of the column stationary phase and produce hydrogen bonding interactions. On the other hand, the aqueous content normally ranges between 5–30%, typically requiring a minimum of 2% water to enable the formation of the aqueous layer on the surface of the polar stationary phase. Additives and buffers are also commonly employed to control the mobile phase pH and ion strength, which elicit a great impact on chromatographic resolution and peak shape.[29] With regard to the stationary phase, bare silica is nowadays considered the gold-standard technique due to its versatility in terms of retention, peak shape and selectivity, but different chromatographic selectivity can be obtained by using other functionalized columns (*e.g.* amide, cyano, amino, diol, sulfoalkylbetaine).[29] For more details about the fundaments of HILIC, the reader is referred to Chapter 4.

Currently, most of these HILIC stationary phases are commercially available in ultra-high performance liquid chromatography (UHPLC) format. The use of sub-2 μm particles in UHPLC allows speeding up of the chromatographic process and increasing the separation efficiency compared with conventional high performance liquid chromatography (HPLC). This consequently results in faster analyses, higher sensitivity and reproducibility and improved chromatographic resolution, which enables detection of more metabolites with enhanced peak shape and reduced ion suppression.[30] Accordingly, UHPLC–MS is currently one of the most powerful analytical tools for high-throughput metabolomics with comprehensive metabolomics coverage.

Nonetheless, HILIC presents serious limitations that hinder its application in large-scale metabolomics studies. Poor chromatographic reproducibility is the main drawback of this separation technique, resulting in significant analytical drifts in terms of retention times and peak shapes along sequence runs.[29–31] To reach acceptable repeatability, shallow chromatographic gradients and long post-analysis equilibration times are required to maintain the equilibrium of the aqueous layer on the stationary phase, which consequently reduce analytical throughput considerably. Chromatographic resolution in HILIC is also highly influenced by the sample-diluent solvent, which should be as similar as possible to the initial mobile-phase composition to minimize deterioration of peak shapes, but also taking into consideration other important factors such as sample solubility and stability.[31] Furthermore, the lower loading capacity of HILIC compared with other methods, such as RPLC, makes the use of small injection volumes mandatory, thereby resulting in reduced method sensitivity. Altogether, and considering the complexity of separation mechanisms, significant analytical improvements are still required to facilitate the implementation of HILIC-based approaches in high-throughput metabolomics.

9.2.3 Capillary Electrophoresis–Mass Spectrometry

Capillary electrophoresis (CE) separates analytes on the basis of their intrinsic electrophoretic mobility, which is dependent on the mass and charge of the compounds. Therefore, CE stands out as a technique with complementary and orthogonal performance to chromatographic approaches because of the application of a different separation mechanism, showing great potential for the analysis of highly polar and ionic metabolites which are usually not resolved in GC and LC.[32] The fundamentals of CE–MS, including the various electrophoretic separation modes, have been discussed in Chapter 6. Until now, three complementary CE-based methods have predominantly been employed in metabolomics for the separation of cationic, anionic and neutral metabolites.[33] Capillary zone electrophoresis (CZE) is the most common approach for the analysis of cationic and anionic molecules by using acidic and basic BGEs, respectively. On the other hand, neutral metabolites are preferably profiled by micellar electrokinetic chromatography (MEKC) by using surfactants (*e.g.* sodium dodecyl sulfate, SDS). Although CE separations have traditionally been performed in bare fused-silica capillaries, the use of coating polymers allows minimizing the adsorption of matrix components into the capillary surface, which causes irreproducibility in migration times, and increases the separation selectivity. For instance, capillaries coated with positively charged polymers change the direction of the EOF and, hence, facilitate the analysis of anionic metabolites.[33]

The main strengths of CE–MS include: (i) small-sample volume requirements; (ii) simple sample treatment, only requiring the elimination of interferences that could induce capillary clogging; and (iii) high separation resolution.[32] However, this technique has not yet been successfully implemented in high-throughput metabolomics research, mainly due to technical problems related with the coupling of CE to MS and associated analytical drawbacks such as reduced robustness and sensitivity. The interfacing of CE with ESI--MS is very challenging due to the difficulty of combining both electrical circuits, and because low CE liquid flows hinder the formation of stable electrosprays.[34] To overcome this hurdle, the sheath–liquid interface is the most common CE–MS configuration in metabolomics, based on connecting the outlet of the CE capillary with an ESI needle and adding a co-axial sheath-liquid to increase the nebulization and ionization efficacy. Although stainless steel ESI needles are most often employed, these can oxidize and corrode when using reverse polarity and negative MS ionization for analyzing anionic metabolites, so that the use of platinum needles is preferred in that case.[35] The use of the sheath-liquid inherently results in the dilution of the CE effluent and, consequently, in reduced sensitivity. Furthermore, the nebulizing gas employed in the sheath-liquid interface can cause suction effects and therefore peak broadening. To solve these limitations, numerous efforts have been made during recent years to generate miniaturized sheath-liquid interfaces with the aim of reducing dilution effects,[36] as well as the elimination of the nebulizing gas in these interfaces for improved separation efficiency.[37]

Nonetheless, despite all these methodological improvements, the robustness, reproducibility and sensitivity of this technique are still significantly lower than that provided by GC–MS and LC–MS based approaches, so that the application of CE–MS in metabolomics is nowadays considerably relegated to the background.

9.2.4 Direct Mass Spectrometry

Complementing conventional hyphenated approaches based on the coupling of mass spectrometry with chromatographic and electrophoretic techniques, direct mass spectrometry (DMS) analysis has also demonstrated great utility in high-throughput metabolomics.[38] This instrumental configuration is based on the direct introduction of the sample extracts into the mass spectrometer, either by direct infusion (DI) using a syringe pump or by flow injection (FI) into a carrier solvent. Consequently, the absence of a previous time-consuming separation step has multiple advantages: 1) reduced total analysis time; 2) higher instrumental simplicity, because the use of direct injection avoids common troubles associated with chromatography and electrophoresis (*e.g.* column/capillary clogging and deterioration, retention/migration time drifts); and 3) increased metabolomic coverage, since there are no biased or selective retention mechanisms prior to MS detection (Figure 9.2).[39] All this therefore facilitates the implementation of this high-throughput metabolomics approach in epidemiological studies dealing with large sample sizes.[40]

However, the major limitations of DMS-based metabolomics platforms include reduced sensitivity due to ion suppression effects driven by the direct introduction of whole samples into the MS system and the impossibility of resolving chemical isomers. To resolve metabolites with very similar monoisotopic masses, the use of ultra-high-resolution instruments is recommended (*e.g.* FT-ICR-MS, Orbitrap). As an alternative, ion mobility MS allows performing multidimensional separations of isobaric compounds in the mass domain,[41] but its use is still not widespread in common metabolomics practice. Accordingly, direct MS-fingerprinting has been proposed as a suitable "first pass" screening tool for rapid and comprehensive metabolomics analysis,[42] complementing hyphenated methods that could provide higher sensitivity and selectivity for the profiling of specific metabolome compartments.

9.3 The Polar Metabolome in Alzheimer's Disease

Alzheimer's disease (AD) is the most prevalent neurodegenerative disorder worldwide, usually preceded by a pre-clinical phase known as mild cognitive impairment (MCI).[43,44] The exact etiology of AD is still poorly understood, in spite of the numerous efforts made by the scientific community over recent

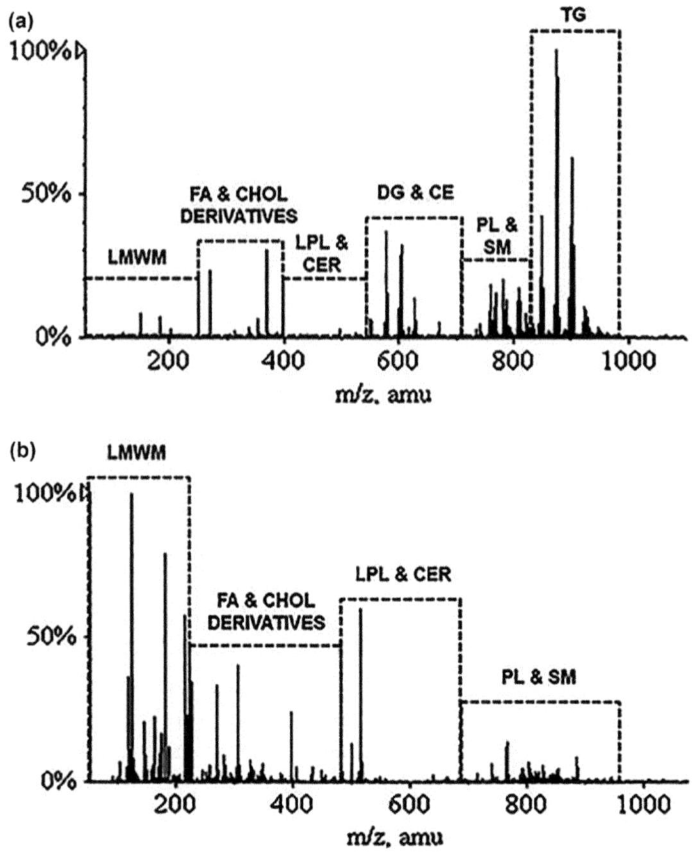

Figure 9.2 Typical metabolomics fingerprints obtained by DMS analysis of serum samples under positive (A) and negative (B) electrospray ionization. Reproduced from ref. 39 with permission from Springer Nature, Copyright 2017.

decades, but there is nowadays growing evidence about the involvement of multiple players during the onset and development of this neurodegenerative disorder, including genetic, environmental and ageing-related factors. Furthermore, it is also noteworthy that pathological mechanisms associated to AD are not exclusive of the central nervous system, but are also manifested at the systemic level.[45] Indeed, AD has been described as a multifactorial disease primarily characterized by the deposition of amyloid plaques and hyper-phosphorylated tau protein tangles, as well as by other cellular perturbations including oxidative stress, inflammatory processes, abnormal metal homeostasis and many others.[46–48] In this vein, metabolomics has demonstrated great potential over recent years for holistically deciphering the molecular changes behind pathological mechanisms occurring in the onset and progression of this neurodegenerative disorder.[49,50] However, most of these previously published works in AD metabolomics rely on the use of NMR or RPLC–MS to investigate metabolic alterations in brain, blood, cerebrospinal fluid (CSF) and other biological samples, whereas only a few authors focused on the polar metabolome, as summarized in the next sections of this chapter.

9.3.1 Application of GC–MS Metabolomics in Alzheimer's Disease Research

GC–MS based metabolomics has successfully been employed by numerous researchers to investigate the implication of a range of primary metabolite classes in AD development by analyzing different biological samples. Xu *et al.* reported widespread metabolic perturbations in seven brain regions from AD patients (hippocampus, entorhinal cortex, middle-temporal gyrus, cingulate gyrus, sensory cortex, motor cortex and cerebellum), comprising significant alterations in levels of 55 metabolites related to glucose homeostasis, the urea cycle and amino acid metabolism.[51] Interestingly, some of these alterations were also detected in some regions traditionally considered to be not affected by this neurodegenerative disorder (*e.g.* cerebellum), thereby suggesting that pathological mechanisms associated to AD could be part of a global brain disease rather than be localized in specific brain areas. The analysis of serum samples has evidenced that many of these AD-related metabolic disturbances can also be detected at the systemic level and therefore be considered as potential biomarkers for diagnosis.[52] In that study, the authors identified 23 discriminant metabolites indicative of impairments in neurotransmission systems, energy-related pathways (*e.g.* tricarboxylic acid cycle, TCA), the urea cycle and some others. The potential of exhaled breath as a biological matrix for investigating the involvement of volatile compounds in neurodegenerative diseases has also been proposed in another work, taking into consideration that blood changes can be transmitted to the breath through lung exchange.[53] The analysis of breath samples by GC–MS from AD, MCI and Parkinson's disease (PD) patients showed increased phenol content in PD samples, but no significant alterations in AD and MCI samples. Of note, similar results have been obtained by studying different animal and cellular models of AD. Alterations in 17 metabolites were detected in the hippocampus of the senescence-accelerated mouse prone 8 (SAMP8), including energy-related metabolites, amino acids, lipids and some others.[54] These metabolic changes were investigated along a 10-month follow-up, from young mice (2 months old) to mature (7 months old) and, finally, aged mice (12 months old). More recently, Han *et al.* demonstrated that chronic unpredictable mild stress may aggravate cognitive impairment in APP/PS1 mice and sharpen hippocampal metabolic perturbations.[55] The molecular mechanisms behind this stress-related cognitive impairment in the APP/PS1 mouse model could be related to various pathways, including sphingolipid metabolism, homeostasis of ketone bodies and amino acid metabolism. Comparative

GC–MS analysis of brain and plasma samples from TASTPM mice revealed similar disturbances in both biological matrices with regard to energy metabolism and steroid biosynthesis, whereas other metabolites were only perturbed in the plasma (*e.g.* fatty acids) or brain (*e.g.* amino acids).[56] In another study, the urinary volatile metabolome was explored to decipher the impact of mutating the amyloid precursor protein (APP) gene in three transgenic mice strains.[57] Chemometric analysis provided accurate genotype classification (>84%) based on the urinary volatile profile, thus enabling the identification of several odorants differentially expressed in these AD animal models. GC–MS has also been employed to study the metabolic profile of AβPP-transfected Chinese hamster ovary (CHO) cells, which was characterized by significant mitochondrial dysfunction preceding extracellular Aβ accumulation.[58] Secondly, the authors investigated the therapeutic potential of two peroxisome proliferator-activated receptor gamma (PPARγ) agonists, namely rosiglitazone and pioglitazone, to attenuate these alterations.

To complement the performance of this metabolomics technique, the application of multiplatform approaches based on the combination of GC–MS and RPLC–MS has emerged during recent years as one of the most powerful strategies to achieve comprehensive metabolome characterizations. Serum samples from AD, MCI (stable and AD-converters) and control subjects were collected at baseline in a prospective study and were subjected to metabolomics analysis by RP–UHPLC–MS and GC×GC–MS, this latter equipped with two columns with complementary polarities and a cryogenic modulator.[59] Decreased levels of ether phospholipids, phosphatidylcholines, sphingomyelins and sterols were observed in AD patients at the baseline. On the other hand, 2,4-dihydroxybutanoic acid was identified as the most important predictor of disease progression, which was upregulated in AD converters indicating the potential involvement of hypoxia in early pathogenesis. Similarly, the use of this metabolomics multiplatform has also been described for discovering two panels of potential diagnostic plasma metabolite markers to differentiate AD and amnesic MCI patients from control subjects.[60] These panels comprised metabolites involved in the same biochemical pathways, namely fatty acid metabolism, one-carbon metabolism, amino acid metabolism and nucleic acid metabolism, thus suggesting the occurrence of common pathogenic mechanisms. In a more recent study, the same approach was employed to look for potential AD biomarkers in a Down syndrome (DS) population, who are well-known to almost universally develop AD during their lifespan.[61] Results suggested that AD in DS is characterized by a shift in energy metabolism toward the upregulation of anaerobic respiration, which was reflected in increased plasma levels of lactate, pyruvate and methyladipate and decreased uridine content. The analysis of CSF samples revealed increased cortisol levels in severe AD patients, as well as increased cysteine and decreased uridine in light AD.[62] The combination of these three metabolites with other amino acids provided predictive models with sensitivity and specificity above 80%. Motsinger-Reif *et al.* also analyzed CSF samples to evaluate the ability of metabolite markers to discriminate between AD and control subjects and to compare their discrimination performance with that provided by traditional Aβ and tau biomarkers.[63] Authors detected two differentiating metabolic features with potential diagnostic utility, which require further confirmatory and validation studies. With regard to the combination of GC–MS and RPLC–MS to investigate AD-related metabolic perturbations in transgenic mice, González-Domínguez *et al.* conducted a comprehensive metabolomics characterization of the APP/PS1 mouse model by studying multiple biological samples, including serum,[64] different brain regions (*i.e.* hippocampus, cortex, striatum, cerebellum, olfactory bulbs),[65] metabolically active organs (*i.e.* liver, kidney)[66] and organs involved in the immune function (*i.e.* spleen, thymus),[67] with the aim of getting deeper insights into the AD pathogenic mechanisms, both at the central and systemic levels. Metabolic perturbations identified in the APP/PS1 model were similar to that previously reported in human AD studies and widespread to all the biological compartments investigated, comprising altered lipid homeostasis (*e.g.* phospholipids, sphingolipids, cholesterol, fatty acids and acyl-carnitines), disturbances in energy metabolism, amino acid and nucleic acid metabolism and many others (Figure 9.3). This is in line with hippocampal disturbances previously reported in three transgenic models of AD, indicative of mitochondrial stress and altered energy metabolism.[68] These metabolomics results were supported by changes in mitochondrial morphology, loss of the integrity of synaptic mitochondria and energy production, as well as inhibition of mitochondrial trafficking contributing to increased neuronal susceptibility to excitotoxic cell death.

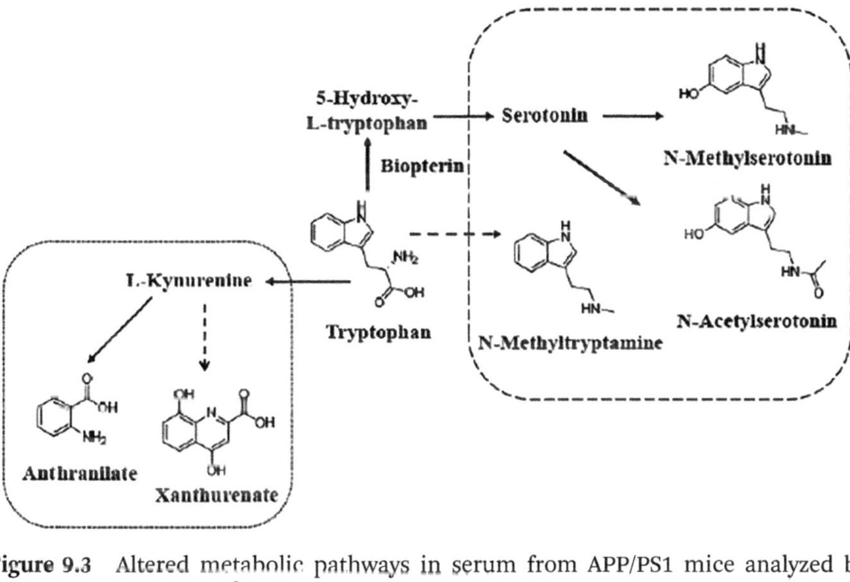

Figure 9.3 Altered metabolic pathways in serum from APP/PS1 mice analyzed by GC–MS and RPLC–MS.
Adapted from ref. 64 with permission from Elsevier, Copyright 2015.

9.3.2 Application of HILIC–MS Metabolomics in Alzheimer's Disease Research

The potential of HILIC–MS to explore AD-related metabolic perturbations in the polar metabolome was first demonstrated in two preliminary studies conducted in brain[69] and plasma[70] samples. The authors reported that amide-based columns provided higher accuracy and reproducibility than other HILIC sorbents. The subsequent application of this methodology evidenced that HILIC enables comprehensive metabolomics coverage, with 1264 and 1457 molecular features detected in brains under positive and negative ionization, respectively; whereas 1182 to 2284 features were detected in plasma. The authors reported that amide-based columns provided higher accuracy and reproducibility than other HILIC sorbents. Furthermore, statistical modeling enabled subject classification with satisfactory accuracy and prediction power, but significant metabolites driving this discrimination were not identified. In another work, Graham *et al.* compared plasma samples from MCI patients, MCI patients who developed AD upon subsequent follow-up and matched healthy controls.[71] Pathway analysis revealed disturbances in 22 metabolic pathways, some of them traditionally associated with the onset of AD, such as cholesterol, glucose or amino acid metabolism, but also providing new insights in two interlinked biochemical spaces: polyamine metabolism and L-arginine metabolism. These metabolomics alterations were capable of distinguishing MCI subjects converting to AD up to two years before formal clinical diagnosis. More recently, HILIC–MS urinary metabolomics highlighted the central role of aromatic amino acid metabolism in the AD transgenic CRND8 mouse model.[72] A total of 73 differential metabolites were identified, covering perturbations in tryptophan metabolism (upregulation of the serotonin pathway, downregulation of the kynurenine pathway), aromatic L-amino acid decarboxylase deficiency (accumulation of *N*-acetylvanilalanine and 3-methoxytyrosine) and changes in microbiota and glycine conjugation metabolization processes.

As an alternative to the GC + RPLC–MS multiplatform described in the previous section, the combination of orthogonal HILIC and RPLC has also been employed in AD research to maximize metabolomics coverage. Comprehensive metabolomics analysis of frontal cortex brain samples from AD and control subjects demonstrated perturbations in the metabolism of glycerophospholipids and in six metabolic pathways of the central metabolism.[73] Notably, the authors uncovered a significant impairment in mitochondrial aspartate metabolism, which was correlated with dementia and AD pathology. Kim *et al.* have recently described that brain amino acid profiles represent a characteristic signature of AD pathogenesis, so that the modulation of amino acid metabolism might be a possible therapeutic approach to treat this neurodegenerative disease.[74] In summary, they detected increased content of hypotaurine, myo-inositol and oxo-proline in AD brains, while glutamate and *N*-acetyl-aspartate were decreased. The analysis of CSF samples using this multi-metabolomics approach indicated significant alterations in levels of 17 metabolites, similar to those described in the brain (*e.g.* amino acids, taurine), thus demonstrating the potential of this biological fluid to reflect the neurochemistry of the brain.[75] These discriminant metabolites were statistically modeled to predict the progression of AD along four diagnostic groups: healthy control subjects, stable MCI patients, MCI subjects who developed AD after a 2 year follow-up and AD patients. Combined analysis of CSF and plasma samples provided a more comprehensive characterization of the molecular mechanisms behind AD and MCI pathogenesis.[76] A total of 342 and 351 discriminant metabolic features were detected in plasma and CSF, respectively, which were correlated with multiple altered canonical pathways: energy metabolism, mitochondrial function, neurotransmission, amino acid and lipid metabolism and many others. In another study, Armirotti *et al.* were surprisingly not able to differentiate AD patients and healthy controls on the basis of their plasma metabolomics profiles obtained by RPLC + HILIC–MS, but a clear discrimination was observed between MCI and controls.[77]

9.3.3 Application of CE–MS Metabolomics in Alzheimer's Disease Research

Although its use is not so widespread in metabolomics, some authors have described the application of CE–MS for characterizing perturbations in the polar metabolome associated with AD pathogenesis in various biological samples. Ibáñez *et al.* employed this metabolomics platform to examine metabolic differences in CSF samples from subjects with different cognitive status, including patients with subjective cognitive impairment (SCI), MCI patients that remained stable within 2 years, MCI patients that progressed to AD after this follow-up period and AD patients.[78] Multivariate statistical analysis enabled correct classification of the samples in the diagnostic groups (97–100%), with 83% of diagnostic accuracy. The panel of discriminant metabolites comprised choline, valine, a tripeptide, dimethylarginine, creatine, serine, arginine, suberylglycine, carnitine and histidine, being considered as possible disease progression biomarkers. More recently, CE-based metabolomics analysis of CSF was employed to discriminate between AD and idiopathic normal pressure hydrocephalus (iNPH) patients, another neurodegenerative disorder with similar symptoms.[79] The authors found four metabolites significantly altered in the CSF of patients with AD compared to iNPH ones, which could be of great utility to classify potential patients with AD earlier and with greater precision. Similarly, combined CE-fingerprinting of serum and saliva samples was also proposed in another study to identify metabolite biomarkers for diagnosing different neurodegenerative dementias, including AD, frontotemporal lobe dementia (FTLD) and Lewy body disease (LBD).[80] Compared with healthy controls, six metabolites in serum and two metabolites in saliva were significantly altered in dementia patients, whereas 45 molecular features detected in serum could discriminate at least one pair of the diagnostic groups. González-Domínguez *et al.* analyzed serum samples from AD, MCI and control subjects to elucidate the pathological mechanisms behind the progression of AD.[81] Besides some alterations related to oxidative stress and deficiencies in energy metabolism, the most interesting and novel results were related to the association of several vascular risk factors with AD. Furthermore, a negative association was also detected between AD and serum levels of proline betaine, a marker of citrus consumption, which has been validated in a recently published metabolomics study focused on investigating diet-related metabolites associated with cognitive decline.[82] In line with these previous results, CE–MS metabolomics has also demonstrated that AD pathogenesis can be associated with dysregulated brain transmethylation and polyamine metabolism, including abnormalities in neurotransmitter signaling, urea cycle, aspartate–glutamate metabolism and glutathione synthesis.[83] These metabolomics perturbations were confirmed by transcriptomics analysis and were significantly correlated with disease severity in terms of neuritic plaque burden and neurofibrillary pathology.

9.3.4 Application of Other Metabolomics Platforms to Explore the Polar Metabolome in Alzheimer's Disease

Besides GC–MS, HILIC–MS and CE–MS, some authors have proposed the application of other complementary MS-based analytical tools to explore the AD-related polar metabolome, as briefly discussed in this section. In this context, different strategies have been described with the aim of enhancing the resolving power of RPLC for the analysis of polar metabolites. For instance, various manufacturers have developed novel RPLC columns with improved retention of polar compounds, which have successfully been employed in AD research for analyzing cationic metabolites in CSF[84] and purine metabolites in brain samples.[85] Kaddurah-Daouk *et al.* have also published various metabolomics studies based on ion pairing RPLC and subsequent electrochemical array detection for investigating AD perturbations in the hydrophilic metabolome.[86,87] This methodology was applied to determine redox-active metabolites in CSF samples from AD and MCI patients, which evidenced significant alterations in the metabolism of aromatic amino acids and the purine pathway. As an alternative, the application of derivatization reactions for increasing the hydrophobicity of polar metabolites has also emerged as an interesting approach to resolve the polar metabolome by means of RPLC-based metabolomics. Dansylation is nowadays one of the most commonly employed derivatization procedures for this purpose, which has been satisfactorily applied for salivary and urinary metabolomics in AD research.[88,89] Alternatively, Takayama *et al.* developed a novel approach for chiral metabolomics fingerprinting based on the labeling of enantiomers with optically-active derivatization reagents and subsequent precursor ion scan chromatography of the derivatives (Figure 9.4).[90] This method was used

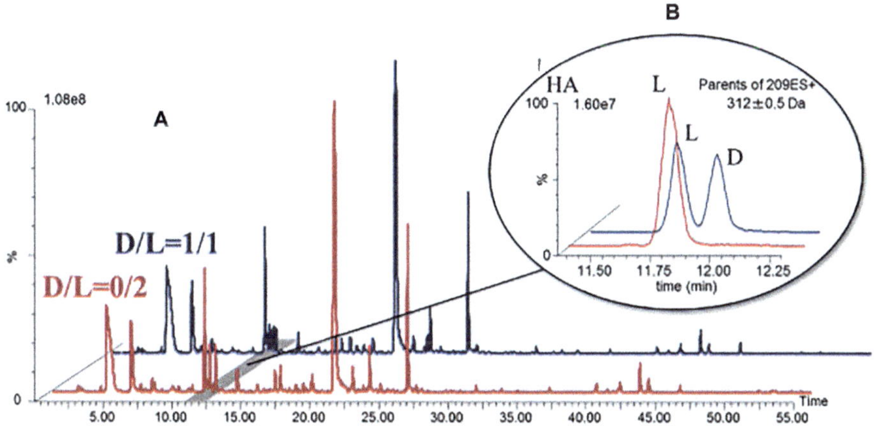

Figure 9.4 Chiral metabolomic chromatograms showing the separation of enantiomers for various amines and carboxyls.
Reproduced from ref. 90 with permission from Elsevier, Copyright 2015.

to determine chiral amines and carboxyls in brain samples as biomarker candidates for AD diagnosis. The authors found decreased content of L-phenylalanine and increased L-lactic acid in AD brains, thus demonstrating the potential of chiral separations in metabolomics.

On the other hand, direct mass spectrometry (DMS) analysis has also been widely employed in biomedical research for simultaneously determining a broad range of metabolites with different polarities and other physicochemical properties.[38] This instrumental configuration is based on direct introduction of sample extracts into the mass spectrometer without previous chromatographic or electrophoretic separation, which significantly reduces the total analysis time and minimizes the analytical bias inherent to separation techniques. Therefore, DMS stands out as a powerful screening tool for high-throughput metabolomics analysis, particularly in AD research as recently reviewed.[39,50] In this vein, González-Domínguez *et al.* have developed during recent years various complementary DMS-based approaches for serum metabolite fingerprinting, including direct infusion ESI–MS metabolomics,[91,92] direct infusion ESI–MS lipidomics[93] and flow injection APPI–MS metabolomics.[94] The application of these methodologies to serum samples from AD patients revealed significant perturbations in numerous low molecular weight polar metabolites involved in energy metabolism, neurotransmission and antioxidant systems. Furthermore, aberrant phospholipid homeostasis was also evident, with reduced circulating levels of phospholipid species containing polyunsaturated fatty acids, increased content of saturated ones and accumulation of the corresponding hydrolysis by-products (*e.g.* diglycerides).[95] These high-throughput platforms based on DMS analysis were subsequently employed to perform a multi-compartmental metabolomics investigation of APP/PS1 transgenic mice, by analyzing serum,[96,97] urine,[98] different brain regions (*i.e.* hippocampus, cortex, cerebellum, olfactory bulbs)[99] and other peripheral organs (*i.e.* liver, kidney, spleen, thymus),[100] with results in agreement with earlier studies performed in human AD cohorts[91–95] and complementary works carried out in the same animal model by using orthogonal GC–MS and RPLC–MS.[64–67] Similarly, Lin *et al.* also employed direct infusion MS with the aim of characterizing metabolic perturbations in the hippocampus[101] and cerebellum[102] of the CRND8 transgenic model, which were mainly related to altered metabolism of arachidonic acid and eicosanoids, amino acids, nucleotides and other metabolite classes.

9.4 Concluding Remarks

The couplings of mass spectrometry with gas chromatography (GC–MS), hydrophilic interaction liquid chromatography (HILIC–MS) and capillary electrophoresis (CE–MS) are nowadays the main workhorses for approaching the polar metabolome, which is usually underestimated when using RPLC–MS. Traditionally, GC–MS has been considered the gold standard technique for profiling primary metabolite classes because of its reproducibility, sensitivity and chromatographic resolution. On the other hand, the use of HILIC and CE is currently also widespread in metabolomics due to their usually higher analytical throughput (*e.g.* chemical derivatization to increase metabolite volatility is not needed). However, each one of these metabolomics approaches has their inherent analytical strengths and limitations, which consequently hinder comprehensive metabolome characterizations. For this reason, numerous authors have emphasized in recent years the need for combining complementary and orthogonal methods to maximize metabolomics coverage. Furthermore, significant technical developments are still needed for achieving more high-throughput metabolomics analysis, such as for instance: 1) improved and faster chromatographic and electrophoretic systems, especially for HILIC and CE based approaches that currently require long separation runs to maintain enough reproducibility; 2) the use of multi-dimensional separation techniques for increasing the metabolomics coverage in a single analysis; 3) the implementation of automated protocols for sample preparation and, if needed, derivatization with the aim of simplifying the analytical workflow; 4) the use of ultra-high-resolution and ion mobility MS for enhanced resolution of metabolites with similar masses.

References

1. D. S. Wishart, Y. D. Feunang, A. Marcu, A. C. Guo, K. Liang, R. Vázquez-Fresno, T. Sajed, D. Johnson, C. Li, N. Karu, Z. Sayeeda, E. Lo, N. Assempour, M. Berjanskii, S. Singhal, D. Arndt, Y. Liang, H. Badran, J. Grant, A. Serra-Cayuela, Y. Liu, R. Mandal, V. Neveu, A. Pon, C. Knox, M. Wilson, C. Manach and A. Scalbert, *Nucleic Acids Res.*, 2018, **46**, D608.
2. S. M. Rappaport and M. T. Smith, *Science*, 2010, **330**, 460.

3. G. W. Miller and D. P. Jones, *Toxicol. Sci*, 2014, **137**, 1.
4. M. M. Ulaszewska, C. H. Weinert, A. Trimigno, R. Portmann, C. Andres Lacueva, R. Badertscher, L. Brennan, C. Brunius, A. Bub, F. Capozzi, M. Cialiè Rosso, C. E. Cordero, H. Daniel, S. Durand, B. Egert, P. G. Ferrario, E. J. M. Feskens, P. Franceschi, M. Garcia-Aloy, F. Giacomoni, P. Giesbertz, R. González-Domínguez, K. Hanhineva, L. Y. Hemeryck, J. Kopka, S. E. Kulling, R. Llorach, C. Manach, F. Mattivi, C. Migné, L. H. Münger, B. Ott, G. Picone, G. Pimentel, E. Pujos-Guillot, S. Riccadonna, M. J. Rist, C. Rombouts, J. Rubert, T. Skurk, P. S. C. Sri Harsha, L. Van Meulebroek, L. Vanhaecke, R. Vázquez-Fresno, D. S. Wishart and G. Vergères, *Mol. Nutr. Food Res.*, 2019, **63**, e1800384.
5. R. González-Domínguez, M. Urpi-Sarda, O. Jáuregui, P. W. Needs, P. A. Kroon and C. Andrés-Lacueva, *J. Agric. Food Chem.*, 2020, **68**, 1851.
6. M. Bouhifd, T. Hartung, H. T. Hogberg, A. Kleensang and L. Zhao, *J. Appl. Toxicol.*, 2013, **33**, 1365.
7. G. Peron, N. Hidalgo-Liberona, R. González-Domínguez, M. Garcia-Aloy, S. Guglielmetti, S. Bernardi, B. Kirkup, P. A. Kroon, A. Cherubini, P. Riso and C. Andrés-Lacueva, *J. Agric. Food Chem.*, 2020, **68**, 1780.
8. J. K. Nicholson, E. Holmes, J. Kinross, R. Burcelin, G. Gibson, W. Jia and S. Pettersson, *Science*, 2012, **336**, 1262.
9. C. W. W. Beecher, *Metabolic Profiling: Its Role in Biomarker Discovery and Gene Function Analysis*, Springer, New York, 2003, 311.
10. S. M. Rappaport, D. K. Barupal, D. S. Wishart, P. Vineis and A. Scalbert, *Environ. Health Perspect.*, 2014, **122**, 769.
11. M. Palmnäs, C. Brunius, L. Shi, A. Rostgaard-Hansen, N. E. Torres, R. González-Domínguez, R. Zamora-Ros, Y. L. Ye, J. Halkjær, A. Tjønneland, G. Riccardi, R. Giacco, G. Costabile, C. Vetrani, J. Nielsen, C. Andres-Lacueva and R. Landberg, *Adv. Nutr.*, 2020, **11**, 524.
12. A. Gonzalez-Dominguez, E. Duran-Guerrero, A. Fernandez-Recamales, A. M. Lechuga-Sancho, A. Sayago, M. Schwarz, C. Segundo and R. Gonzalez-Dominguez, *Curr. Top. Med. Chem.*, 2017, **17**, 3289.
13. A. H. M. Emwas, R. M. Salek, J. L. Griffin and J. Merzaban, *Metabolomics*, 2013, **9**, 1048.
14. H. G. Gika, G. A. Theodoridis, R. S. Plumb and I. D. Wilson, *J. Pharm. Biomed. Anal.*, 2014, **87**, 12.
15. H. Tian, J. Bai, Z. An, Y. Chen, R. Zhang, J. He, X. Bi, Y. Song and Z. Abliz, *Rapid Commun. Mass Spectrom.*, 2013, **27**, 2071.
16. S.-S. Cai and J. A. Syage, *Anal. Chem.*, 2006, **78**, 1191.
17. D. J. Beale, F. R. Pinu, K. A. Kouremenos, M. M. Poojary, V. K. Narayana, B. A. Boughton, K. Kanojia, S. Dayalan, O. A. H. Jones and D. A. Dias, *Metabolomics*, 2018, **14**, 152.
18. N. Schauer, D. Steinhauser, S. Strelkov, D. Schomburg, G. Allison, T. Moritz, K. Lundgren, U. Roessner-Tunali, M. G. Forbes, L. Willmitzer, A. R. Fernie and J. Kopka, *FEBS Lett.*, 2005, **579**, 1332.

19. N. Strehmel, J. Hummel, A. Erban, K. Strassburg and J. Kopka, *J. Chromatogr. B*, 2008, **871**, 182.
20. E. A. H. Keppler, C. L. Jenkins, T. J. Davis and H. D. Bean, *TrAC, Trends Anal. Chem.*, 2018, **109**, 275.
21. D. J. Beale, F. R. Pinu, K. A. Kouremenos, M. M. Poojary, V. K. Narayana, B. A. Boughton, K. Kanojia, S. Dayalan, O. A. H. Jones and D. A. Dias, *Metabolomics*, 2018, **14**, 152.
22. K. Dettmer, P. A. Aronov and B. D. Hammock, *Mass Spectrom. Rev.*, 2007, **26**, 51.
23. R. González-Domínguez, Á. González-Domínguez, A. Sayago and Á. Fernández-Recamales, *Methods Mol. Biol.*, 2018, **1750**, 125.
24. F. Xu, L. Zou, C. N. Ong, L. Zou, C. N. Ong and C. N. Ong, *TrAC, Trends Anal. Chem.*, 2010, **29**, 269.
25. Q. Gu, F. David, F. Lynen, K. Rumpel, J. Dugardeyn, D. Van Der Straeten, G. Xu and P. Sandra, *J. Chromatogr. A*, 2011, **1218**, 3247.
26. H. Abbiss, C. Rawlinson, G. L. Maker and R. Trengove, *Metabolomics*, 2015, **11**, 1908.
27. A. J. Alpert, *J. Chromatogr.*, 1990, **499**, 177.
28. M. R. Gama, R. G. da Costa Silva, C. H. Collins and C. B. Bottoli, *TrAC, Trends Anal. Chem.*, 2012, **37**, 48.
29. B. Buszewski and S. Noga, *Anal. Bioanal. Chem.*, 2012, **402**, 231.
30. P. D. Rainville, G. Theodoridis, R. S. Plumb and I. D. Wilson, *TrAC, Trends Anal. Chem.*, 2014, **61**, 181.
31. J. Ruta, S. Rudaz, D. V. McCalley, J. L. Veutheyand and D. Guillarme, *J. Chromatogr. A*, 2010, **1217**, 8230.
32. C. Barbas, E. P. Moraes and A. Villaseñor, *J. Pharm. Biomed. Anal.*, 2011, **55**, 823.
33. M. G. M. Kok, G. W. Somsen and G. J. de Jong, *TrAC, Trends Anal. Chem.*, 2014, **61**, 223.
34. J. S. Toraño, R. Ramautar and G. de Jong, *J. Chromatogr. B*, 2019, **1118–1119**, 116.
35. M. van Mever, T. Hankemeier and R. Ramautar, *Electrophoresis*, 2019, **40**, 2349.
36. V. Gonzalez-Ruiz, S. Codesido, S. Rudaz and J. Schappler, *Electrophoresis*, 2018, **39**, 853.
37. N. Drouin, J. Pezzatti, Y. Gagnebin, V. Gonzáalez-Ruiz, J. Schappler and S. Rudaz, *Anal. Chim. Acta*, 2018, **1032**, 178.
38. R. González-Domínguez, A. Sayago and M. A. Fernández-Recamales, *Bioanalysis*, 2017, **9**, 131.
39. R. González-Domínguez, *J. Anal. Test.*, 2017, **1**, 213.
40. E. Chekmeneva, G. dos Santos Correia, M. Gómez-Romero, J. Stamler, Q. Chan, P. Elliott, J. K. Nicholson and E. Holmes, *J. Proteome Res.*, 2018, **17**, 3492.
41. P. Dwivedi, P. Wu, S. J. Klopsch, G. J. Puzon, L. Xun and H. H. Hill Jr, *Metabolomics*, 2008, **4**, 63.

42. J. Draper, A. J. Lloyd, R. Goodacre and M. Beckmann, *Metabolomics*, 2013, **9**, 4.
43. C. Reitz, C. Brayne and R. Mayeux, *Nat. Rev. Neurol.*, 2011, **7**, 137.
44. R. C. Petersen, *J. Intern. Med.*, 2004, **256**, 183.
45. J. K. Morris, R. A. Honea, E. D. Vidoni, R. H. Swerdlow and J. M. Burns, *Biochim. Biophys. Acta*, 2014, **1842**, 1340.
46. K. Blennow, M. J. de Leon and H. Zetterberg, *Lancet*, 2006, **368**, 387.
47. R. B. Maccioni, J. P. Munoz and L. Barbeito, *Arch. Med. Res.*, 2001, **32**, 367.
48. R. González-Domínguez, T. García-Barrera and J. L. Gómez-Ariza, *Metallomics*, 2014, **6**, 292.
49. R. González-Domínguez, A. Sayago and Á. Fernández-Recamales, *J. Chromatogr. B: Anal. Technol. Biomed. Life Sci.*, 2017, **1071**, 75.
50. R. González-Domínguez, A. Sayago and Á. Fernández-Recamales, *Metabolites*, 2018, **8**, 52.
51. J. Xu, P. Begley, S. J. Church, S. Patassini, K. A. Hollywood, M. Jüllig, M. A. Curtis, H. J. Waldvogel, R. L. Faull, R. D. Unwin and G. J. Cooper, *Biochim. Biophys. Acta*, 2016, **1862**, 1084.
52. R. González-Domínguez, T. García-Barrera and J. L. Gómez-Ariza, *J. Pharm. Biomed. Anal.*, 2015, **107**, 75.
53. P. W. Ko, K. Kang, J. B. Yu, J. S. Huh, H. W. Lee and J. O. Lim, *Sens. Lett.*, 2014, **12**, 1198.
54. H. Wang, K. Lian, B. Han, Y. Wang, S. H. Kuo, Y. Geng, J. Qiang, M. Sun and M. Wang, *J. Alzheimer's Dis.*, 2014, **39**, 841.
55. B. Han, J. H. Wang, Y. Geng, L. Shen, H. L. Wang, Y. Y. Wang and M. W. Wang, *Cell. Physiol. Biochem.*, 2017, **41**, 1766.
56. Z. P. Hu, E. R. Browne, T. Liu, T. E. Angel, P. C. Ho and E. C. Y. Chan, *J. Proteome Res.*, 2012, **11**, 5903.
57. B. A. Kimball, D. A. Wilson and D. W. Wesson, *Sci. Rep.*, 2016, **6**, 19495.
58. K. L. Chang, H. N. Pee, W. P. Tan, G. S. Dawe, E. Holmes, J. K. Nicholson, E. C. Chan and P. C. Ho, *J. Alzheimer's Dis.*, 2015, **44**, 215.
59. M. Orešič, T. Hyötyläinen, S. K. Herukka, M. Sysi-Aho, I. Mattila, T. Seppänan-Laakso, V. Julkunen, P. V. Gopalacharyulu, M. Hallikainen, J. Koikkalainen, M. Kivipelto, S. Helisalmi, J. Lötjönen and H. Soininen, *Transl. Psychiatry*, 2011, **1**, e57.
60. G. Wang, Y. Zhou, F. J. Huang, H. D. Tang, X. H. Xu, J. J. Liu, Y. Wang, Y. L. Deng, R. J. Ren, W. Xu, J. F. Ma, Y. N. Zhang, A. H. Zhao, S. D. Chen and W. Jia, *J. Proteome Res.*, 2014, **13**, 2649.
61. T. J. Gross, E. Doran, A. K. Cheema, E. Head, I. T. Lott and M. Mapstone, *Dev. Neurobiol.*, 2019, **79**, 622.
62. C. Czech, P. Berndt, K. Busch, O. Schmitz, J. Wiemer, V. Most, H. Hampel, J. Kastler and H. Senn, *PLoS One*, 2012, **7**, e31501.
63. A. A. Motsinger-Reif, H. Zhu, M. A. Kling, W. Matson, S. Sharma, O. Fiehn, D. M. Reif, D. H. Appleby, P. M. Doraiswamy, J. Q. Trojanowski, R. Kaddurah-Daouk and S. E. Arnold, *Acta Neuropathol. Commun.*, 2013, **1**, 28.

64. R. González-Domínguez, T. García-Barrera, J. Vitorica and J. L. Gómez-Ariza, *Biochimie*, 2015, **110**, 119.
65. R. González-Domínguez, T. García-Barrera, J. Vitorica and J. L. Gómez-Ariza, *Biochim. Biophys. Acta*, 2014, **1842**, 2395.
66. R. González-Domínguez, T. García-Barrera, J. Vitorica and J. L. Gómez-Ariza, *Mol. BioSyst.*, 2015, **11**, 2429.
67. R. González-Domínguez, T. García-Barrera, J. Vitorica and J. L. Gómez-Ariza, *Electrophoresis*, 2015, **36**, 577.
68. E. Trushina, E. Nemutlu, S. Zhang, T. Christensen, J. Camp, J. Mesa, A. Siddiqui, Y. Tamura, H. Sesaki, T. M. Wengenack, P. P. Dzeja and J. F. Poduslo, *PLoS One*, 2012, **7**, e32737.
69. S. F. Graham, O. P. Chevallier, D. Roberts, C. Holscher, C. T. Elliot and B. D. Green, *Anal. Chem.*, 2013, **85**, 1803.
70. K. Inoue, H. Tsuchiya, T. Takayama, H. Akatsu, Y. Hashizume, T. Yamamoto, N. Matsukawa and T. Toyo'oka, *J. Chromatogr. B: Anal. Technol. Biomed. Life Sci.*, 2015, **974**, 24.
71. S. F. Graham, O. P. Chevallier, C. T. Elliot, C. Hölscher, J. Johnston, B. McGuinness, P. G. Kehoe, A. P. Passmore and B. D. Green, *PLoS One*, 2015, **10**, e0119452.
72. Z. Tang, L. Liu, Y. Li, J. Dong, M. Li, J. Huang, S. Lin and Z. Cai, *Curr. Alzheimer Res.*, 2016, **13**, 764.
73. G. Paglia, M. Stocchero, S. Cacciatore, S. Lai, P. Angel, M. T. Alam, M. Keller, M. Ralser and G. Astarita, *J. Proteome Res.*, 2016, **15**, 608.
74. Y. H. Kim, H. S. Shim, K. H. Kim, J. Lee, B. C. Chung, N. W. Kowall, H. Ryu and J. Lee, *Exp. Neurobiol.*, 2019, **28**, 376.
75. C. Ibañez, C. Simo, D. K. Barupal, O. Fiehn, M. Kivipelto, A. Cedazo-Mínguez and A. Cifuentes, *J. Chromatogr. A*, 2013, **1302**, 65.
76. E. Trushina, T. Dutta, X. M. Persson, M. M. Mielke and R. C. Petersen, *PLoS One*, 2013, **8**, e63644.
77. A. Armirotti, A. Basit, N. Realini, C. Caltagirone, P. Bossù, G. Spalletta and D. Piomelli, *Anal. Biochem.*, 2014, **455**, 48.
78. C. Ibáñez, C. Simó, P. J. Martín-Álvarez, M. Kivipelto, B. Winblad, A. Cedazo-Mínguez and A. Cifuentes, *Anal. Chem.*, 2012, **84**, 8532.
79. Y. Nagata, A. Hirayama, S. Ikeda, A. Shirahata, F. Shoji, M. Maruyama, M. Kayano, M. Bundo, K. Hattori, S. Yoshida, Y. I. Goto, K. Urakami, T. Soga, K. Ozaki and S. Niida, *Biomark. Res.*, 2018, **6**, 5.
80. M. Tsuruoka, J. Hara, A. Hirayama, M. Sugimoto, T. Soga, W. R. Shankle and M. Tomita, *Electrophoresis*, 2013, **34**, 2865.
81. R. González-Domínguez, A. García, T. García-Barrera, C. Barbas and J. L. Gómez-Ariza, *Electrophoresis*, 2014, **35**, 3321.
82. D. Y. Low, S. Lefèvre-Arbogast, R. González-Domínguez, M. Urpi-Sarda, P. Micheau, M. Petera, D. Centeno, S. Durand, E. Pujos-Guillot, A. Korosi, P. J. Lucassen, L. Aigner, C. Proust-Lima, B. P. Hejblum, C. Helmer, C. Andres-Lacueva, S. Thuret, C. Samieri and C. Manach, *Mol. Nutr. Food Res.*, 2019, **63**, e1900177.

83. U. V. Mahajan, V. R. Varma, M. E. Griswold, C. T. Blackshear, Y. An, A. M. Oommen, S. Varma, J. C. Troncoso, O. Pletnikova, R. O'Brien, T. J. Hohman, C. Legido-Quigley and M. Thambisetty, *PLoS Med.*, 2020, **17**, e1003012.
84. K. T. Myint, K. Aoshima, S. Tanaka, T. Nakamura and Y. Oda, *Anal. Chem.*, 2009, **81**, 1121.
85. B. Ansoleaga, M. Jové, A. Schlüter, P. Garcia-Esparcia, J. Moreno, A. Pujol, R. Pamplona, M. Portero-Otín and I. Ferrer, *Neurobiol. Aging*, 2015, **36**, 68.
86. R. Kaddurah-Daouk, S. Rozen, W. Matson, X. Han, C. M. Hulette, J. R. Burke, P. M. Doraiswamy and K. A. Welsh-Bohmer, *Alzheimer's Dementia*, 2011, **7**, 309.
87. R. Kaddurah-Daouk, H. Zhu, S. Sharma, M. Bogdanov, S. G. Rozen, W. Matson, N. O. Oki, A. A. Motsinger-Reif, E. Churchill, Z. Lei, D. Appleby, M. A. Kling, J. Q. Trojanowski, P. M. Doraiswamy, S. E. Arnold and Pharmacometabolomics Research Network, *Transl. Psychiatry*, 2013, **3**, e244.
88. J. Zheng, R. A. Dixon and L. Li, *Anal. Chem.*, 2012, **84**, 10802.
89. J. Peng, K. Guo, J. Xia, J. Zhou, J. Yang, D. Westaway, D. S. Wishart and L. Li, *J. Proteome Res.*, 2014, **13**, 4457.
90. T. Takayama, T. Mochizuki, K. Todoroki, J. Z. Min, H. Mizuno, K. Inoue, H. Akatsu, I. Noge and T. Toyo'oka, *Anal. Chim. Acta*, 2015, **898**, 73.
91. R. González-Domínguez, T. García-Barrera and J. L. Gómez-Ariza, *Chem Papers*, 2012, **66**, 829.
92. R. González-Domínguez, T. García-Barrera and J. L. Gómez-Ariza, *Anal. Bioanal. Chem.*, 2014, **406**, 7137.
93. R. González-Domínguez, T. García-Barrera and J. L. Gómez-Ariza, *J. Pharm. Biomed. Anal.*, 2014, **98**, 321.
94. R. González-Domínguez, T. García-Barrera and J. L. Gómez-Ariza, *Talanta*, 2015, **131**, 480.
95. R. González-Domínguez, T. García-Barrera and J. L. Gómez-Ariza, *J. Proteomics*, 2014, **104**, 37.
96. R. González-Domínguez, T. García-Barrera, J. Vitorica and J. L. Gómez-Ariza, *J. Pharm. Biomed. Anal.*, 2015, **107**, 378.
97. R. González-Domínguez, T. García-Barrera, J. Vitorica and J. L. Gómez-Ariza, *Metabolomics*, 2015, **11**, 1175.
98. R. González-Domínguez, R. Castilla-Quintero, T. García-Barrera and J. L. Gómez-Ariza, *Anal. Biochem.*, 2014, **465**, 20.
99. R. González-Domínguez, T. García-Barrera, J. Vitorica and J. L. Gómez-Ariza, *J. Pharm. Biomed. Anal.*, 2015, **102**, 425.
100. R. González-Domínguez, T. García-Barrera, J. Vitorica and J. L. Gómez-Ariza, *Electrophoresis*, 2015, **36**, 2237.
101. S. Lin, H. Liu, B. Kanawati, L. Liu, J. Dong, M. Li, J. Huang, P. Schmitt-Kopplin and Z. Cai, *Anal. Bioanal. Chem.*, 2013, **405**, 5105.
102. S. Lin, B. Kanawati, L. Liu, M. Witting, M. Li, J. Huang, P. Schmitt-Kopplin and Z. Cai, *Talanta*, 2014, **118**, 45.

CHAPTER 10

Analytical Techniques for Material-limited Metabolomics: Recent Developments and Applications

ANDREW J. CHETWYND*[a] AND RAWI RAMAUTAR[b]

[a] School of Geography Earth and Environmental Sciences, University of Birmingham, Edgbaston, Birmingham, UK; [b] Biomedical Microscale Analytics, Leiden Academic Centre For Drug Research, Leiden University, Einsteinweg 55, 2333 CC, Leiden, The Netherlands
*Email: A.J.Chetwynd@bham.ac.uk

10.1 Introduction

During the past two decades, metabolomics has established itself as a key omics technique to help elucidate mechanisms of disease, treatment and toxic exposure. In metabolomics studies the sample, source, material and size/volume being analysed are important factors to consider as the sample preparation and analysis techniques are sample dependent, and sample characteristics can lead to different conclusions being made.[1,2] In the majority of clinical metabolomics studies relatively large volume biofluids are utilised such as urine, faeces and blood products as these keep costs low and are non-invasive (Figure 10.1). However, there are drawbacks to these sample matrices, for example, distinct tissue specific metabolome variations may become diluted in large biofluid volumes leading to less specific biomarker

New Developments in Mass Spectrometry No. 10
Advanced Mass Spectrometry-based Analytical Separation Techniques for Probing the Polar Metabolome
Edited by Rawi Ramautar

Published by the Royal Society of Chemistry, www.rsc.org

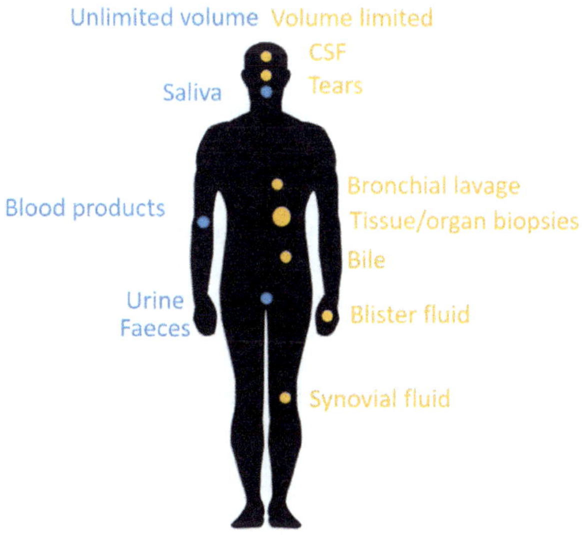

Figure 10.1 Selection of biofluids and tissue samples used in clinical metabolomics. Blue – large volume samples and gold – volume/biomass limited. Adapted from ref. 6, https://doi.org/10.3390/jcm9072161, under the terms of the CC BY 4.0 license, https://creativecommons.org/licenses/by/4.0/.

suites being detected. This leads to regulatory hurdles going from the lab to the clinic as highly specific biomarkers are required to inform clinical decisions. As such the metabolomics community is beginning to look for more subtle changes in the metabolome at the source of the disease such as tissue biopsies, limited cell numbers,[3] vesciles[4] or cerebrospinal fluid[5] to name just a few, where sample volumes are very limited (Figure 10.1).

A key bottleneck in the analyses of these sample types is the fact that they are volume or biomass limited given the more invasive nature of their collection and requirement for other clinical tests to be performed on them. This leaves metabolomics researchers with potentially less than a single microlitre for metabolomic analysis. This poses significant challenges for conventional liquid chromatography–mass spectrometric (LC–MS) and nuclear magnetic spectroscopy (NMR) based approaches. As such, the ability to acquire accurate and reliable metabolomic analysis of these volume/biomass limited samples becomes an important sample preparation and analytical problem to overcome. Over the past few years, advances have been made to enhance sensitivity of analysis for this type of metabolomic study either through improved sample preparation such as sample clean up or derivatisation or through improved separation and mass spectrometric methods. The latter of which can be seen in an array of advances in techniques such as capillary electrophoresis mass spectrometry (CE–MS)[7] and nanoflow liquid chromatography, nanospray electrospray ionisation–mass spectrometry (nLC–nESI–MS).[8,9] Both of these techniques are capable of not only sampling from small sample volumes but take advantage of the low chromatographic dilution and enhanced ionisation efficiency afforded by the nanoscale slow rate being employed.

These advances are making volume and biomass limited metabolomics studies more common place and have the potential to revolutionise and revitalise the translation of metabolomics from the lab to the bedside. Furthermore, these techniques also allow for fewer animals to be used in preclinical studies because the pooling of samples between animals will no longer be required, thus enabling metabolomic researchers to play a larger role in the three Rs; replacement, reduction and refinement. This chapter will discuss the current limitations in volume-limited metabolomics and discuss the state-of-the-art techniques that have been developed and applied to realise the potential of this sample type in metabolomics. This chapter will cover studies from 2017–2020 and focus on sample collection, preparation and nanoscale separations, nano liquid chromatography and capillary electrophoresis linked to mass spectrometry with the focus on separation techniques, which enable comprehensive separation of metabolites and thus increased metabolome coverage.

10.2 Sampling and Sample Preparation

10.2.1 Sample Collection

For the most part metabolomics sample collection has remained unchanged over the course of metabolomic history. While there has been debate over whether blood products should be collected as plasma or serum, and the choice of anticoagulant in the former has been caused for discussion, the collection of blood products has been relatively basic.[10] The same is true for urine samples where the main consideration is the type of urine being collected, 24 h samples or spot urine at a given time or first morning void, each of which has its own merit.[11] However, with the move to either limited blood sampling or collection of tissue metabolites, advances have been made in sample collection; here, two novel approaches for blood and tissue sample collection are described.

Dried blood spots (DBS) have become an increasingly popular method for collecting small volumes of blood from finger or tail pricks and can be used in remote areas of the world or to reduce the burden on experimental animals.[12] While sample collection using this approach has proven to be stable, the issues of varying haematocrit remain thus making reproducible sample volume collection difficult and require additional sample preparation or computational approaches for volume normalisation.[12,13] However, a recent development, volumetric absorptive micro sampling (VAM) alleviates this haematocrit problem and reliably and accurately collects blood samples ranging from 10–30 μL.[14] VAMs have been used for targeted metabolomics to analyse 36 metabolites to determine the method suitability. Once collected within the VAMs tip it requires drying for at least 2 hours. This however may potentially lead to sample degradation. As such, this targeted analysis found that for 36 metabolites in blood samples stored at ambient temperature over 15 days the majority remained stable. Metabolites are eluted using a 70% acetonitrile wash in water which can then be evaporated and reconstituted in any desired volume to maximise concentration, while maintaining excellent recovery profiles of >85% for most metaboltes.[13,14] A global metabolomics study using this VAMs technique was able to detect >5000 LC–MS features and tentatively annotated 133 metabolites from 10 μL of blood. However, this study found that over just 24 h at room temperature led to the degradation of 75% of the metabolites. This suggests that long term storage and shipment from remote areas should be done at lower temperatures to maintain sample integrity.[13]

The future gold standard for animal studies and potential future human studies would be to collect "live" metabolomics data. Currently there is a risk that in animal studies, where samples are collected post mortem or when biopsies are taken in a clinical setting, the metabolome may degrade between collection and sample preparation.[15] One possible solution to this problem is *in vivo* solid phase micro extraction (SPME) which is starting to see use in the metabolomics field.[15,16] Here a surgical blade is coated with a C18 SPME sorbent in unison with polyacrylonitrile to increase biocompatibility.[16] This blade is then inserted directly into the muscle tissue of a sedated organism to sample the metabolome in a minimally invasive manner, and the total sample "mass" is just the metabolites adsorbed to the C18 SPME material. Collecting samples in this manner is non-depleting and thus enables sample collection in real time during normal cellular metabolism while causing minimal metabolomic disruption.[15,17] *In vivo* SPME has proven to be effective for the analysis of a wide range of metabolites including low abundance steroids and lipids in addition to a wide array of the exposome such as phthalates and pesticides. These studies demonstrate particularly good recoveries for non-polar analytes as a result of the non-polar nature of the C18 phase being used as a sorbent.[16] In a global metabolomics study of fish exposure to benzo[a]pyrene, more than 300 features were detected, and the alterations to the fish muscle metabolome could be tracked using *in vivo* SPME in conjunction with LC–MS.[15] Going forward it is possible that these SPME phases could be miniaturised further to the nanoscale as nanoparticles have shown a propensity to selectively adsorb metabolites to their surface and may act as a more efficient sorbent than microscale particles due to their increased surface area to volume ratio for a given mass.[18,19]

10.2.2 Sample Preparation

In general sample preparation is overlooked in metabolomics with the majority of studies using "dilute and shoot" for biofluids or single/biphasic protein precipitations of blood products and tissue samples prior to LC–MS or NMR analysis.[1,20] While these approaches are relatively quick and often said to reduce some bias by minimising metabolite loss, they can be poorly suited for volume limited samples given the analytical techniques required for the analysis of these samples. Due to the narrow bore capillaries and columns used in CE–MS and nLC–MS it is critically important that the samples undergo a filtration and desalting step to remove any particulate matter or salts which may precipitate and block the capillaries and columns during separation.

In early studies utilising nLC–nESI–HRMS, mixed mode solid phase extraction (SPE) was used to remove particulate matter, desalt and acts as a pre-concentration step. These approaches have been used extensively with urine,[21,22] cells[23] and faecal[24,25] metabolomics. In plasma and tissue samples a lipid precipitation step has also been incorporated into the analysis to remove lipids which co-eluted and thus ion suppressed low abundance metabolites.[26,27] However these methods are particularly well suited to larger sample volumes due to the large sorbet beds of the SPE and lipid removal plates, however they demonstrate that pre-concentration and sample clean-up, to prevent source fouling, using more elaborate sample preparation techniques are not only reproducible but enable a greater coverage of the metabolome.[22,27]

Since this early introduction of SPE to sample clean-up of volume limited SPME[28,29] and µQuEChERs[30,31] have become more commonly utilised as a sample clean-up and concentration step. Here, smaller sorbent beds are used, and less material is used in general while maintaining highly reproducible results with excellent recoveries of metabolites.[30,31] The latter has found a particular niche in the environmental metabolomics lab where it has been successfully applied to prepare biomass limited samples from benthic invertebrates[31] and single amphipods.[30] In both cases the samples were pre-concentrated following µQuEChERs on a nanoLC trapping column which allows a large volume of diluted sample to be loaded prior to injection onto an analytical nLC column.[30,31] This however, as with SPE, has the disadvantage of losing core metabolites such as the highly polar ones in this case due to the C18 phase of the nLC trapping and analytical columns.

The second approach to improve sample preparation for small volumes is to chemically modify the metabolites *via* a derivatisation protocol. In recent years chemically modifying metabolites to improve sensitivity has grown in interest. As with SPE approaches, there has been some hesitation within the metabolomics community through fear of losing metabolites or reducing sample throughout, however, even traditional metabolite extraction techniques such as biphasic extractions add significant time to the overall process.[32] Due to the chemical diversity of the metabolome and the specific functional groups various derivatisation approaches target, it may sometimes be necessary to perform multiple derivatisation steps or decide *a priori* which metabolite classes to target.[32,33] These methods improve metabolite sensitivity *via* the addition of a hydrophobic moiety, which subsequently increases the retention time for the metabolite meaning they elute in a more organic mobile phase composition. Due to the higher percentage of organic solvent in the mobile phase at this stage the ionisation efficiency increases, and thus metabolites can be detected at lower concentrations.[33–35] There is also the added benefit that by increasing the interaction between metabolite and stationary phase and subsequently delaying elution that polar metabolites can easily be characterised using reversed phase column chemistry thus eliminating the need for the less robust HILIC.[36] In 2017 it was demonstrated that by using dansylyation derivatisation this preparation technique can be widely applied to volume limited samples following the detection of 80 metabolites from just 100 MCF-7 cells, when used in conjunction with nLC–nESI–HRMS.[33] Furthermore this study utilised differential chemical isotope labelling (CIL) with ^{12}C-/^{13}C-dansylation, which allows for more accurate and precise relative quantification of metabolites.[32,33] Recently, Willacey *et al.* demonstrated another form of derivatisation using dimethylaminophenacyl bromide (DmPABr).[35] Unlike several previous methods, this derivatisation approach is less selective and can derivatise carboxylic acids in addition to primary and secondary amines and thiols (Figure 10.2). This means that not only is a wider portion of the metabolome derivatised but it is also more stable than previous benzoyl chloride derivatisation methods.[34,35] These authors demonstrated that this new derivatisation process enabled the detection of 37 polar metabolites in as little as 250 to 1×10^4 HepG2 cells using reversed phase LC–MS and returned better peak shapes than HILIC. Furthermore, using this labelling approach most metabolites became ionisable in positive ESI mode, which in general is more

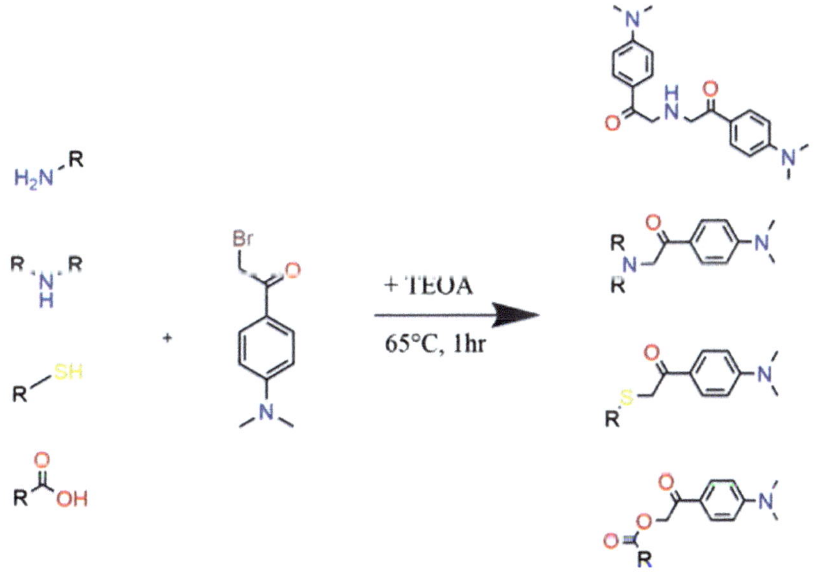

Figure 10.2 Examples of the derivatisation reaction of DmPABr with the primary amine, secondary amine, thiol and carboxylic acid, respectively. Reproduced from ref. 34 with permission from Elsevier, Copyright © 2020 The Authors.

stable and more sensitive, compared to negative ESI. If used in combination with micro, nanoLC or CE it is possible that more metabolites could be uncovered due to the low flow rates and subsequently increased sensitivity of these techniques.[34] This would have the added benefit of increasing sample throughput by eliminating the requirement to run in both positive and negative ionisation modes.

10.3 Micro- and Nanoscale Separation Techniques

The use of miniaturised flow systems such as nanoLC and CE has for several years been maligned due to poor retention/migration time reproducibility. However, the advent of direct nanoflow pumps over split flow to generate nanoflow rates in nanoLC and improved thermal regulation to reduce joule heating in CE have led to much more stable retention/migration characteristics in recent years.[8,37]

The sensitivity gains in nanoscale separations and ionisation make them ideal candidates to not only analyse low abundance metabolites in normal metabolomics studies but also to characterise the metabolome in volume/biomass restricted samples. This, in part, is due to their unique capabilities to sample from very small volumes, as both can take repeated injections from as little as 5 uL of sample and in the case of CE just a few 10s of nL are required for a single injection. The advantages of the platforms come from the separation itself and the ionisation process. In terms of the separation in both nLC and CE, small volumes are subject to significantly less chromatographic dilution, which enables more concentrated peaks to be eluted from the column/capillary and into the mass spectrometer. Furthermore, upon elution the droplets of the Taylor cone are orders of magnitude smaller when produced from nanoscale flow rates compared to analytical flow rates. This reduced volume requires less desolvation, Rayleigh limit and Coulomb explosion cycles for the charged analytes to enter the gas phase. Coupled to the fact that nESI emitters are positioned much closer to the MS aperture meaning more charged ions enter the mass analyser compared to in conventional ESI–MS.[8,38,39]

In early studies of volume-limited samples, direct infusion mass spectrometry (DIMS) was applied using nanoscale infusion rates (10–20 nL per min) into electrospray ionisation–high-resolution mass spectrometers (ESI–HRMS).[40] This approach enables the detection of metabolites due to the increased ionisation efficiency resulting from the low ESI flow rates. However, definitive identification of metabolites remains problematic due to a lack of separation meaning that isobaric isomers are difficult to distinguish. Using this approach coupled with SPME more than 100 metabolites have been detected in a single caviar egg, while also improving detection limits for drugs in a single microliter of blood.[41] Improvements to DIMS were made with the development of SIM-stitching, which is more sensitive and enables a more comprehensive analysis by using fragmentation.[42] Indeed, in a similar fashion a recent study incorporating SPME fibres into facemasks enabled the detection of a range of metabolites in exhaled breath using direct analysis real time mass spectrometry.[29] However, the definitive identification of metabolite remains very difficult particularly in the case of isobaric compounds. In addition, the infusion of all analytes into the MS at the same time results in the loss of low abundance metabolites due to ion suppression, detector saturation and in the case of trapping instruments the dominance of high abundance metabolites within traps relative to low abundance metabolites.

With the introduction of nLC–nESI–MS for metabolomics in the early 2010s many of these issues were alleviated at the expense of sample throughput, however unambiguous metabolite identification became possible, and this approach has been successfully used to increase sensitivity in metabolomics studies.[8] These nLC–nESI–MS methods have been applied to urine,[21] faeces,[24,25] a range of tissues,[26,30,31] blood products[26,27] and cells[23] using SPE or μQuEChERs and lipid extraction protocols as sample clean-up. In these early stages on the translation of nLC–nESI–MS to metabolomics, there were very few column chemistries commercially available and were typically all reversed phase due to nLC–MS being the dominant technique in proteomics. This meant that many polar metabolites were not retained and those that were suffered from poor peak shapes and peak capacities.[8,43] However, recent new developments have led to a wide array of column chemistries now being commercially available for nLC with an eye on the metabolomics community.[44–46] As with reversed phase nLC the first nano-HILIC work appears to have been undertaken in proteomics where a small molecule mix was used to characterise the peak capacities and separation characteristics prior to use on bottom-up proteomic samples.[47] The small molecule test solution returned peak capacities of 130 with excellent peak area reproducibility which would transfer well to the metabolomics community.[47] The first true metabolomics study we are aware of that used a nano HILIC column was investigating the small molecules present in dissolved organic matter within artic soils.[48] While this is not a volume-limited sample, given the difficulties in obtaining such samples they are precious. In this work, a nanoscale zwitterionic ZIC pHILIC column was used in combination with a nano-reversed phase and demonstrated the importance of having these polar phases for metabolomics at the nanoscale. In total nearly 13 000 features were detected with the majority being detected with the ZIC pHILIC while > 75% of peaks detected on the reversed phase column were retained for less than 2 minutes (Figure 10.3). This suggests that many features were lost in the solvent front with the reversed phase column whereas with the HILIC method the features were more distributed along the chromatogram (Figure 10.3).

In this work, Ladd *et al.* were able to detect a wide array of metabolites including amino acids, lipids, sugars, plant and microbial metabolites while also being able to distinguish subtle changes in the soil metabolome relative to the depth at which they were collected.[48] Going forward into the next few years it is likely that the advances in column chemistries and mobile phase

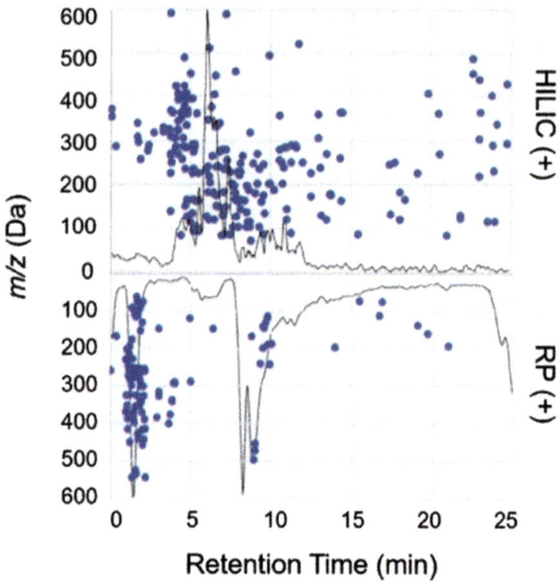

Figure 10.3 Retention characteristics of metabolites detected in artic soil extracts using ZIC pHILIC (top) and C18 reversed phase column (bottom); each blue dot represents a metabolite.
Reproduced from ref. 48, https://doi.org/10.1038/s41598-019-42118-9, under the terms of the CC BY 4.0 license, http://creativecommons.org/licenses/by/4.0/.

additives will also enable chiral separations of metabolites using nanoLC columns linked to MS in order to elucidate mechanisms of disease in the chiral metabolome.[49]

CE has undergone a revolution in recent years within the metabolomics community, the advent of sheath-less CE led to increased sensitivity *via* a significant reduction of the CE effluent with the elimination of sheath liquid.[50] CE–MS is ideally suited for volume limited analysis due to its very low flow rates, *i.e.* typically 50 nL per min. Furthermore, the minute injection volumes (<10 nL) make CE–MS an ideal platform for volume-limited metabolomics.[51,52] This has recently been demonstrated with the analysis of nucleotides from just an extract of 500 HEP-G2 cells, however, in this case the BGE was modified to a high pH separation which enabled the detection of attomolar concentrations of these metabolites in positive ESI mode as opposed to the more conventional negative ESI mode.[53]

While in the past the increased ionisation efficiency of CE–MS has dominated the mechanisms behind the increased sensitivity of CE–MS approaches, recently innovative injection techniques have been developed and applied for volume-limited metabolomics to increase sample loading and sensitivity. Recent examples of these techniques include transient isotachophoresis (t-ITP),[52] pH junctions[54] and field amplified sample injections.[55] Zhang *et al.* demonstrated the applicability of t-ITP to the analysis of limited cell numbers, with the volumes being sampled being equivalent to between 0.25 and 5 HepG2 cells. This was achieved by using a standard LLE extraction where the samples were reconstituted in 280 mM ammonium acetate thus increasing the ionic strength of the sample solution. This enables t-ITP by permitting the concentration/stacking of sample following a hydrodynamic injection. In this study, t-ITP increased the theoretical plate count for the majority of metabolites by an order of magnitude and improved the LODs of all metabolites with over a third having sub-nanomolar LODs compared to traditional hydrodynamic injection without t-ITP. Injecting the equivalent of 5 HEP-G2 cells allowed for the reliable and stable detection of 67 metabolites, and with the equivalent of 0.25 Hep-G2 cells 24 metabolites could be detected. This injection protocol thus offers significant help in the detection of volume-limited samples.[52] Van Mever *et al.* used a Box–Behnken design to fine-tune a pH junction style pre-concentration for zwitterionic metabolites in brain microdialysate.[54] Brain microdialysate offers a metabolomics window into the central nervous system, previously blood products have been the primary biofluid of interest for neurological issues.[56–58] However, with the ability to not only collect these volume limited samples but also accurately and reliably characterise them its possible improved diagnostic and treatment approaches may be developed for neurological diseases. Here, the Box–Behnken design approach was used to determine the volume of ammonium hydroxide prior to sample injection, sodium hydroxide concentration and sample volume.[54] Following only a 1 : 1 dilution of 10 μL brain microdialysate, 48 metabolites could be detected including the highly unstable GABA which had a LOD of 49 nM by CE–MS. This approach also enabled a sample stacking approach which works in a similar way to trapping columns with nLC–MS whereby multiple injections can be made to load a greater sample volume. In this case, 300 nL was injected thereby significantly improving sample loading but at the same time retaining high quality separations on the CE.[54] Field amplified sample injection (FASI) takes advantage of the mobility of metabolites within a solution to inject samples into the capillary. This approach concentrates metabolites at the inlet of the capillary as less sample buffer is injected diluting the sample. To achieve this, an electrokinetic injection is utilised whereby a voltage is applied for a set injection period during which the inlet is suspended in the sample solution. The approach has been demonstrated to increase sensitivity between 100 and 300-fold when compared to a standard hydrodynamic injection.[55] An important consideration for FASI is to reconstitute samples in a low ionic strength/poorly conductive solvent to enhance the electrokinetic concentration of metabolites. As such, recent work optimised this sample solvent to minimise formic acid concentrations while not reducing so much as hindering ionisation and limiting the aqueous content of the sample. This led to the sample buffer being settled on 0.01% formic acid, 4.99% water and 95% methanol.[55] Furthermore, the salt content of the sample will increase the ionic strength of the sample solution, thus an adequate washing procedure is required to precipitate salts but also maintain the solubility of the metabolites. This was achieved by Wei-Liao and colleagues by incorporating a 4:1 isopropanol:acetonitrile wash which enhanced sensitivity by 4.5–6-fold by eliminating excess salts from the sample solution.[55]

One major limitation to tissue analysis, other than the fact that they are biomass limited, is that samples are typically homogenised prior to analysis during metabolite extraction. As a result, spatial resolution is lost, and potentially important information is lost such as that defined along the edges of tumorous tissue.[59] The metabolomics community has begun to combat this with a combination of techniques such as MALDI,[60] ToF-SIMS,[61] REIMS,[62] DESI[63] and LESA,[64] all of which have been successfully applied to metabolomics. However, these techniques suffer from similar issues to direct infusion metabolomics in that there is no metabolite separation prior to mass spectrometric analysis. As a result, the analysis of isobaric metabolites becomes albeit impossible without advanced fragmentation techniques and even then, remains problematic.[65] To mitigate this, a recent demonstration of surface sampling CE–MS (SS-CE–MS) has been provided to characterise on-line 12 µm thick brain, spinal cord and kidney sections in addition to onion epidermis. Here, two capillaries are used, one to provide BGE to the surface of the tissue section and a second analytical column to inject from the BGE on the sample surface (Figure 10.4). When both capillaries are lowered to the surface of the tissue section an electrokinetic injection is applied, this approach enables a spatial resolution of between 360 and 600 µm. This is combined with the ability to separate isobaric metabolites and to spatially resolve metabolites. Within the brain cortex, 476 metabolites were putatively identified and included a wide range of metabolite classes such as amino acids, neurotransmitters and carnitines.[65]

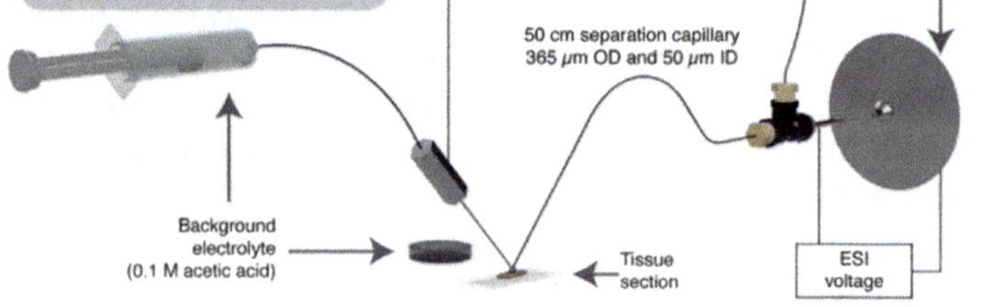

Figure 10.4 Schematic of SS-CE–MS instrumentation for tissue sampling. Reproduced with permission from Spatially Defined Surface Sampling Capillary Electrophoresis Mass Spectrometry Copyright 2019 American Chemical Society.[65]

Following this proof-of-concept there is significant scope to expand this work to obtain separation prior to MS analysis of tissue biopsies and histological sample to elucidate spatially disperse metabolomes.

Traditionally, CE has been restricted to the metabolome niche of the highly polar and charged metabolites thus to some extent restricting its appeal to the metabolomics community. However, methods are now being developed to significantly expand the chemical space open to CE–MS analysis *via* the application of non-aqueous capillary electrophoresis (NACE). In 2018 the first metabolomics NACE method was described by Azab *et al.*, here, using a bare fused silica capillary and a BGE comprising of 35 mM ammonium acetate in 70% acetonitrile, 15% methanol, 10% water and 5% isopropanol, high peak capacities and excellent peak shapes were possible for a wide range of fatty acids. Using a modified Bligh and Dyer method with methyl-*tert*-butyl-ether (MTBE) where the lipid portion of the extracted serum and plasma lies on the top of the biphase, exceptional recoveries of lipid standards were obtained ranging from 82–96%. The analytical repeatability was 7.3% and 12.1% RSD for peak areas for inter and intraday reproducibility across 18 fatty acid standards, with a relative migration time RSD of 0.2–2% across 78 injections over a 3 day period. This study was applied to 50 µL blood plasma and serum samples and enabled the reliable detection of 50 and 14 fatty acids from 50 extracted plasma and serum samples respectively.[66] This study also introduced a method for multi-segment injections (MSI) for metabolomics, which has previously been demonstrated to increase throughput in proteomics,[67] here it was used to enable 6 metabolomics samples and a QC to be injected in a single CE separation.[66] This allows much greater throughput of analysis with 7 samples per run giving an average of 4 minutes per sample (Figure 10.5),[66] this in turn improves capillary life time by reducing the time spent with high voltages applied and as a consequence reduces sample analysis costs. While a relatively large volume of blood plasma and serum was used in the development of this method, it will be equally applicable to volume limited samples (<3 µL) such as dried blood spots and infant sweat.[66]

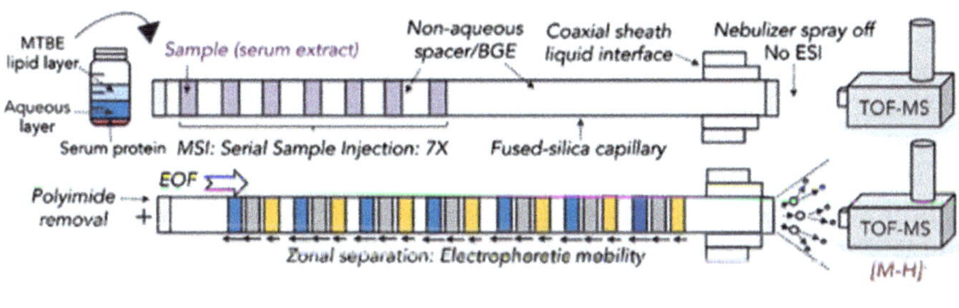

Figure 10.5 Schematic of biphasic extraction for NACE with multi-segment injections for increased sample throughput.
Reproduced from ref. 66 with permission from American Chemical Society, Copyright 2018.

10.4 Applications

In recent years, these methods have gone beyond method development and been applied to a wide range of metabolomics studies where samples have been either volume or biomass-limited; these are detailed in Table 10.1.

In the final section we discuss selected applications that have taken advantage of these advances in sample preparation and analysis to investigate single cell metabolomics and biomass limited tissue samples that represent the future of biopsy analysis.

10.4.1 Single-cell Metabolomics

A major goal in recent years has been the ability to perform metabolomics on single cells; this represents the pinnacle of sensitivity in metabolomics and offers the ability to investigate cellular heterogeneity. This will eliminate the assumptions currently being made in cellular metabolomics that all cells are homogeneous and may help to uncover metabolomics differences in cell cycle stages to cell senescence in cancer studies.[71] In these studies the extremes of volume-limited samples become realised with cell volumes ranging from femto to nanolitre in scale.[55]

Nakatini and colleagues applied an nLC–MS/MS approach utilising a PFP column, a newer column chemistry for nLC to the analysis of single HELA cells. These cells have a typical volume of just 4 picolitres, making them among the smallest cells used in cell culture labs for clinical purposes. In this work, intact cells were injected, in a 100 nL injection, without sample preparation, and cell lysis was achieved by cell membrane disruption in the organic mobile phase. In total, 22 individual HELA cells were analysed, and 16 amino acids and 2 nucleotides were detected. This study also compared nLC–MS to microLC–MS and demonstrated a 3–130-fold increase in sensitivity with nLC–MS which accounts for how these single cellular metabolites could be detected.[71] In a CE–MS based single cell study FASI–CE–MS was applied to *Aplysia californica* neuron cells, which on average are less than 50 μm in diameter. These cells were isolated and washed to desalt the cell and suspended in low ionic strength buffer, following a 30 s, 20 kV electrokinetic injection 21 metabolites could be detected and quantified including neurotransmitters.[55] The final single-cell applications take advantage of a novel sample introduction protocol, large volume dual preconcentration by isotachophoresis (LDIS). In order to sample a single cell a new thin-walled (10 μm) tapered emitter was developed, and the cells were lysed in methanol, dried and reconstituted in 5 μL HCl. This sample buffer volume significantly dilutes the cellular metabolites by going from pL to μL volumes. As such, a large volume injection of around 1200 nL is required so as not to lose sample during the injection process. This is preceded by a 3 M KOH leading electrolyte injection to enable isotachophoresis focusing and concentration.[73] When used on chemical standards, LDIS increased sensitivity by up to 600-fold *vs* standard sample introduction and returned LODs

Table 10.1 Volume and biomass-limited metabolomics studies using novel analytical approaches between 2017 and 2020.

Sample	Sample size	Sample preparation	Analytical technique	Notes	Ref.
Whole *Chironomus riparius*	12 mg	μQuEChERs	nanoLC–MS with column trapping	< 1150 features detected	31
Whole *Gammarus fossarum* organism	ND	μQuEChERs	NanoLC–MS	> 10 000 features detected	30
Blood	10 μL	VAMs	LC–MS/MS	36 metabolites detected	14
Blood	10 μL	VAMs	LC–MS	133 metabolites identified	13
Plasma	1 μL	LLE	nLC–IM–MS	PASEF fragmentation with trapping ion mobility. >1100 lipids identified	68
Plasma and serum	50 μL	LLE	MSI–NACE–MS	Multi segment non-aqueous CE–MS. 20 lipids detected	66
Plasma	5 μL	LLE	CE–MS	12 nucleotides in positive ESI mode	3
Brain microdialysate	10 μL	1 : 1 dilution with H_2O	CE–MS	Dynamic pH junction stacking. 48 metabolites detected	54
Muscle tissue	N/A	*In vivo* SPME	LC–MS	Broad range of (xeno)metabolites detected	16
Muscle tissue	N/A	*In vivo* SPME	LC–MS	> 300 metabolites	15
Kidney tissue	20 μm histological slice	LLE	CE–MS	Dopant enriched desolvation gas. >100 metabolites detected	69
Liver tissue	1 mg	MeOH extraction of lipids	nanoLC–IM–MS	PASEF fragmentation with trapping ion mobility. >975 lipids identified	68

Table 10.1 (*Continued*)

Sample	Sample size	Sample preparation	Analytical technique	Notes	Ref.
Placenta tissue	2 mg	LLE	MSI–CE–MS	NACE and polar CE–MS. 130 metabolites detected	70
Brain, spinal cord, kidney and onion epidermis	360–600 μL surface of 20 μm tissue slice	N/A	SS–CE–MS	Surface sampling CE–MS. 476 metabolites spatially resolved in brain slices	65
Single *Aplysia californica* neuron cell	Few fL	Desalting and cell isolation	FASD CE–MS	21 metabolites detected	55
Single HELA cell	4 pL	N/A	nanoLC–MS	18 metabolites detected	71
Single Hela	5 μL after extraction	Cell lysis in MeOH	LDIS–CE–MS	20 amino acids quantified, 450 features detected	72
100 cells	ND	CIL	nanoLC–MS with column trapping	80 metabolites detected	33
500 HepG2 cells	ND	LLE	CE–MS	+ ve ESI for all nucleotides and some energy metabolites (FAD, NADH)	3
5×10^5 HELA cells	ND	LLE	nLC–IM–MS	PASEF fragmentation with trapping ion mobility. >1351 lipids identified	68
HEP G2 cells	42 nL	LLE	CE–MS	tITP injections 67 metabolites in 5 cells 24 metabolites from 0.25 cells	52

on the fmol level. When used to analyse single HELA cells, 20 amino acids were quantified with a range of concentrations spanning 2.2–32.7 fmol. Interestingly evidence of cellular heterogeneity was observed with significant variation in concentration for several amino acids seen between different cells which may represent the cells being at different stages in the cell cycle.[72]

10.4.2 Biomass-limited Tissue Applications

As with the single cell work, biomass-limited tissue samples represent a potential large leap forward in metabolomic characterisation of disease. Applications here offer the ability to compliment traditional clinical assays on biopsies for cancer or other tissue-specific diseases.

Interestingly in these applications, the use of LLE has been retained as a basic sample preparation technique, however, in the first example MSI and NACE advances have been utilised for placental tissue analysis. While in the latter two applications mass spectrometry has been hyphenated with sheathless CE or nanoLC to enable in-depth analysis of the metabolome.

The growth in the metabolomic analysis of biomass-limited tissue biopsies is exemplified with the recent work by Saoi *et al.*[70] In this work MSI–NACE–MS was applied to around 2 mg dried placental samples to monitor fetal development and elucidate metabolomic characteristics of gender. The NACE method described previously to characterise the lipidome was combined with the more traditional CE–MS cationic and anionic metabolic profiling approaches to analyse the highly charged and polar metabolome, the first such study to combine both polar and non-polar metabolomics. Across the 3 separations a total of 130 metabolites were detected with 65 of these being definitively identified using commercial chemical standards. Alongside polar metabolites such as carnitines, amino acids and organic acids, a range of non-polar lipids and bile acids were also detected. The application of MSI allowed 6 sample and 1 QC injection (Figure 10.6) for each separation, which not only increases sample throughput but also allows robust quality control for each separation performed. The results of the study showed that 3 carnitines and uric acid proved to be highly specific biomarkers of gender in placental tissue.[70] In the future this approach could be used to characterise problems during pregnancy by sampling a small portion of the placenta during or after birth to provide additional information and in the latter minimise sampling on a newborn.

The liver is an organ of interest for a wide range of pathologies and as such is relatively frequently biopsied. These samples are often used for histological analysis or other clinical assay to provide a diagnosis, thus the tissue available for any metabolomics is only a small portion of an already small biopsy sample. Here a nLC–MS method is described for the analysis of liver sample (1 mg) in addition to plasma (1 μL) and HELA cells (500 000).[68] Using a standard LLE extraction lipids and polar metabolites were isolated prior to injection on an Easy nanoLC 12000 with a C18

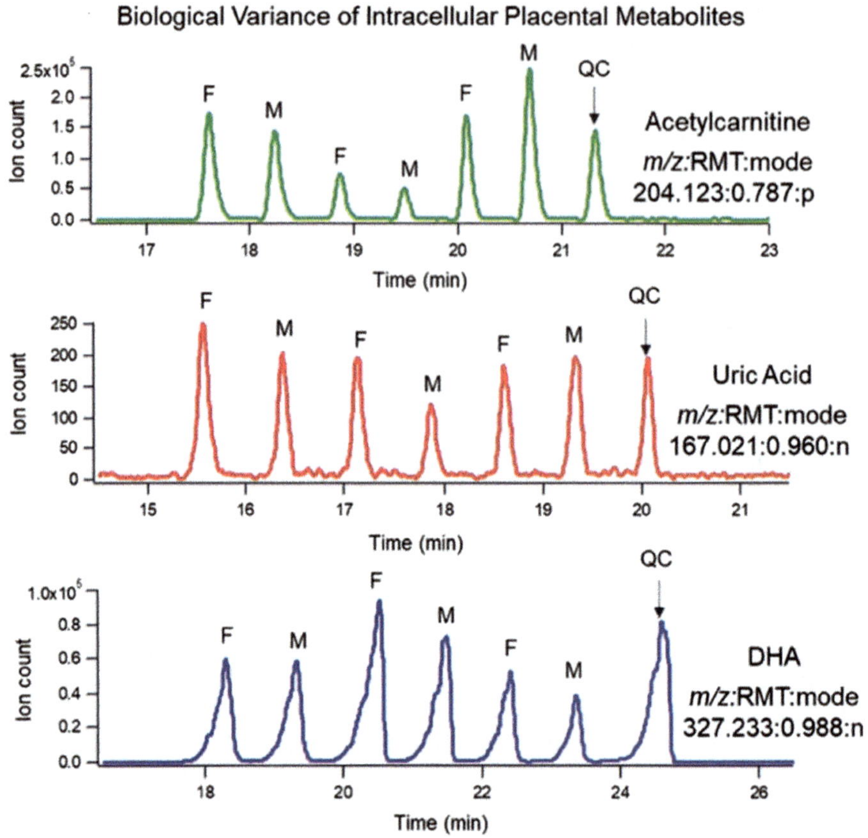

Figure 10.6 Extracted ion electropherograms of acetyl carnitine, uric acid and DHA from the 6 samples and QC in a single MSI separation; M and F refer to male or female sample injection.
Reproduced from ref. 70, https://doi.org/10.1038/s41598-020-66222-3, under the terms of the CC BY 4.0 license, http://creativecommons.org/licenses/by/4.0/.

nanoLC column with a 75 μm internal diameter. This nanoscale separation technique was hyphenated with a recent advance in mass spectrometry, *i.e.* trapped ion mobility (TIMS), which enables the novel fragmentation method parallel acquisition serial fragmentation (PASEF) to provide not only structural information but also collisional cross sectional areas from the ion mobility cell. Here, trapped ion mobility is used to accumulate 15 precursor ions prior to rapid fragmentation every 100 ms, allowing more ions to be fragmented than a typical DDA or DIA method.[68,74,75] This nLC–TIMS–PASEF–TOF–MS approach enabled the detection of >1850 unique lipids across all samples with 1108 seen in the plasma, 976 in liver and 1351 in the HELA cells.[68] This proof-of-principle for biomass-limited tissue is a vital stepping-stone to clinical metabolomics on tissue biopsies and could also be translated to nanoHILIC columns or CE to characterise the polar metabolome in these samples.

Alongside bulk tissue biopsies, histological slices are also an important clinical sample, which could potentially be repurposed following histological examination for metabolomics. In 2019, histological slices (20 μm thick) from a mouse model of polycystic kidney disease were analysed using sheath-less CE–MS.[69] Following metabolite extraction just 2.5 μL of sample was present in the injection vial making it ideal for CE–MS. However, to increase sensitivity and boost the number of detectable metabolites the desolvation gas was doped with acetonitrile in a procedure known as dopant enrich nitrogen (DeN). In the field of proteomics this approach has been demonstrated to increase sensitivity to glycopeptides by up to 25-fold.[76] In this work with sheath-less CE–MS, the use of DeN enabled the detection of about 100 metabolites in kidney histological slices. Using multivariate data analysis a number of key metabolites were highlighted which were able to discriminate between polycystic kidney model samples and controls. Thus demonstrating that it is possible to perform meaningful metabolomic analysis of histological sized samples. As such, small biopsy samples may be used in the future in human clinical metabolomic studies. Furthermore, no preconcentration techniques were employed in this study, which may have further increased the number of metabolites detected.[69]

10.5 Conclusions and Perspectives

Metabolomic studies are currently at a turning point, with focus moving away from large volume biofluids to volume and biomass-limited samples. While these samples may be more intrusive for patients, they offer renewed hope of finding more disease-specific metabolite panels to help elucidate mechanisms of disease and treatment. Recent advances in nLC–nESI–MS and CE–MS now enable the in-depth and reproducible analysis of volume-limited samples for both clinical and environmental metabolomics studies. These methods also have economic, ethical and environmental benefits to be realised. The low flow rates and reduction in the use of organic solvents reduce significantly the environmental burden of metabolomics, while multi-segment analysis in CE–MS allows more samples to be injected in a single, thus increasing the high-throughput capability of this approach for metabolomics studies. Finally, the use of large-scale animal studies can be reduced as pooling across individual animals will no longer be required to achieve a single analysable sample thus increasing the 3Rs credentials of metabolomics. The main limiting step in metabolomics still lies in the identification of metabolites with many features going unidentified in metabolomics studies. This is a major drawback now that analytical methods have advanced to allow more samples to be analysed, in order to truly advance the field, it is vital that new computational approaches are developed to improve peak picking of low abundance metabolites and to identify them.

References

1. A. J. Chetwynd, W. B. Dunn and G. Rodriguez-Blanco, in *Advances in Experimental Medicine and Biology*, 2017, vol. 965, pp. 19–44.
2. K. Segers, S. Declerck, D. Mangelings, Y. Vander Heyden and A. Van Eeckhaut, *Bioanalysis*, 2019, **11**, 2297–2318.
3. W. Zhang, F. Guled, T. Hankemeier and R. Ramautar, *Electrophoresis*, 2020, **41**, 360–369.
4. C. Williams, M. Palviainen, N. C. Reichardt, P. R. M. Siljander and J. M. Falcón-Pérez, *Metabolites*, 2019, **9**, 276.
5. R. Ramautar, R. Shyti, B. Schoenmaker, L. De Groote, R. J. E. Derks, M. D. Ferrari, A. M. J. M. Van Den Maagdenberg, A. M. Deelder and O. A. Mayboroda, *Anal. Bioanal. Chem.*, 2012, **404**, 2895–2900.
6. L. V Trevor, K. Riches-Suman, A. L. Mahajan and M. J. Thornton, *J. Clin. Med.*, 2020, **9**, 2161.
7. W. Zhang, T. Hankemeier and R. Ramautar, *Curr. Opin. Biotechnol.*, 2017, **43**, 1–7.
8. A. J. Chetwynd and A. David, *Talanta*, 2018, **182**, 380–390.
9. C. Aydoğan, F. Rigano, L. K. Krčmová, D. S. Chung, M. Macka and L. Mondello, *Anal. Chem.*, 2020, **92**(17), 11485–11497.
10. T. Barri and L. O. Dragsted, *Anal. Chim. Acta*, 2013, **768**, 118–128.
11. M. A. Fernández-Peralbo and M. D. Luque de Castro, *TrAC, Trends Anal. Chem.*, 2012, **41**, 75–85.
12. E. A. Palmer, H. J. Cooper and W. B. Dunn, *Anal. Chem.*, 2019, **91**(22), 14306–14313.
13. C. Volani, G. Caprioli, G. Calderisi, B. B. Sigurdsson, J. Rainer, I. Gentilini, A. A. Hicks, P. P. Pramstaller, G. Weiss, S. V. Smarason and G. Paglia, *Anal. Bioanal. Chem.*, 2017, **409**, 6263–6276.
14. M. G. M. Kok, C. Nix, G. Nys and M. Fillet, *Talanta*, 2019, **197**, 49–58.
15. A. Roszkowska, M. Yu, V. Bessonneau, L. Bragg, M. Servos and J. Pawliszyn, *Environ. Sci. Technol. Lett.*, 2018, **5**, 431–435.
16. A. Roszkowska, M. Yu, V. Bessonneau, J. Ings, M. McMaster, R. Smith, L. Bragg, M. Servos and J. Pawliszyn, *Environ. Pollut.*, 2019, **249**, 109–115.
17. H. L. Lord, X. Zhang, F. M. Musteata, D. Vuckovic and J. Pawliszyn, *Nat. Protoc.*, 2011, **6**, 896–924.
18. A. J. Chetwynd and I. Lynch, *Environ. Sci.: Nano*, 2020, **7**, 1041–1060.
19. A. J. Chetwynd, W. Zhang, J. A. Thorn, I. Lynch and R. Ramautar, *Small*, 2020, **16**(21), 2000295.
20. J. Sostare, R. Di Guida, J. Kirwan, K. Chalal, E. Palmer, W. B. Dunn and M. R. Viant, *Anal. Chim. Acta*, 2018, **1037**, 301–315.
21. A. J. Chetwynd, A. Samarawickrama, J. H. Vera, S. A. Bremner, A. Abdul-Sada, Y. Gilleece, S. G. Holt and E. M. Hill, *JAIDS, J. Acquired Immune Defic. Syndr.*, 2017, **74**, e45–e53.
22. A. J. Chetwynd, A. Abdul-Sada and E. M. Hill, *Anal. Chem.*, 2015, **87**, 1158–1165.

23. D. R. Jones, Z. Wu, D. Chauhan, K. C. Anderson and J. Peng, *Anal. Chem.*, 2014, **86**, 3667–3675.
24. A. J. Chetwynd, L. A. Ogilvie, J. Nzakizwanayo, F. Pazdirek, J. Hoch, C. Dedi, D. Gilbert, A. Abdul-Sada, B. V. Jones and E. M. Hill, *J. Chromatogr. A*, 2019, **1600**, 127–136.
25. M. L. Stoll, R. Kumar, E. J. Lefkowitz, R. Q. Cron, C. D. Morrow and S. Barnes, *Genes Immun.*, 2016, **17**, 400–405.
26. A. David, A. Lange, A. Abdul-Sada, C. R. Tyler and E. M. Hill, *Environ. Sci. Technol.*, 2017, **51**, 616–624.
27. A. David, A. Abdul-Sada, A. Lange, C. R. Tyler and E. M. Hill, *J. Chromatogr. A*, 2014, **1365**, 72–85.
28. J. Bongaerts, D. De Bundel, D. Mangelings, I. Smolders, Y. Vander Heyden and A. Van Eeckhaut, *J. Pharm. Biomed. Anal.*, 2018, **161**, 192–205.
29. Z.-C. Yuan, W. Li, L. Wu, D. Huang, M. Wu and B. Hu, *Anal. Chem.*, 2020, **92**(17), 11543–11547.
30. C. Bonnefoy, A. Fildier, A. Buleté, C. Bordes, J. Garric and E. Vulliet, *Talanta*, 2019, **202**, 221–229.
31. A. Berlioz-Barbier, A. Buleté, A. Fildier, J. Garric and E. Vulliet, *Chemosphere*, 2018, **196**, 347–353.
32. S. Zhao, H. Li, W. Han, W. Chan and L. Li, *Anal. Chem.*, 2019, **91**, 12108–12115.
33. X. Luo and L. Li, *Anal. Chem.*, 2017, **89**, 11664–11671.
34. C. C. W. Willacey, N. Karu, A. C. Harms and T. Hankemeier, *Microchem. J.*, 2020, 105445.
35. C. C. W. Willacey, M. Naaktgeboren, E. Lucumi Moreno, A. B. Wegrzyn, D. van der Es, N. Karu, R. M. T. Fleming, A. C. Harms and T. Hankemeier, *J. Chromatogr. A*, 2019, **1608**, 460413.
36. Z. Li, J. Tatlay and L. Li, *Anal. Chem.*, 2015, **87**, 11468–11474.
37. N. Drouin, M. van Mever, W. Zhang, E. Tobolkina, S. Ferre, A.-C. Servais, M.-J. Gou, L. Nyssen, M. Fillet, G. S. M. Lageveen-Kammeijer, J. Nouta, A. J. Chetwynd, I. Lynch, J. A. Thorn, J. Meixner, C. Lößner, M. Taverna, S. Liu, N. T. Tran, Y. Francois, A. Lechner, R. Nehmé, G. Al Hamoui Dit Banni, R. Nasreddine, C. Colas, H. H. Lindner, K. Faserl, C. Neusüß, M. Nelke, S. Lämmerer, C. Perrin, C. Bich-Muracciole, C. Barbas, Á. L. Gonzálvez, A. Guttman, M. Szigeti, P. Britz-McKibbin, Z. Kroezen, M. Shanmuganathan, P. Nemes, E. P. Portero, T. Hankemeier, S. Codesido, V. González-Ruiz, S. Rudaz and R. Ramautar, *Anal. Chem.*, 2020, **92**, 14103–14112.
38. G. Pendyala, E. J. Want, W. Webb, G. Siuzdak and H. S. Fox, *J. Neuroimmune Pharmacol.*, 2007, **2**, 72–80.
39. M. Wilm and M. Mann, *Anal. Chem.*, 1996, **68**, 1–8.
40. A. D. Southam, T. G. Payne, H. J. Cooper, T. N. Arvanitis and M. R. Viant, *Anal. Chem.*, 2007, **79**, 4595–4602.
41. T. Vasiljevic, V. Singh and J. Pawliszyn, *Talanta*, 2019, **199**, 689–697.

42. A. D. Southam, R. J. M. Weber, J. Engel, M. R. Jones and M. R. Viant, *Nat. Protoc.*, 2017, **12**, 255–273.
43. A. J. Chetwynd, A. David, E. M. Hill and A. Abdul-Sada, *J. Mass Spectrom.*, 2014, **49**, 1063–1069.
44. S. R. Wilson, C. Olsen and E. Lundanes, *Analyst*, 2019, **144**, 7090–7104.
45. K. Mejía-Carmona, J. Soares da Silva Burato, J. V. B. Borsatto, A. L. de Toffoli and F. M. Lanças, *TrAC, Trends Anal. Chem.*, 2020, **122**, 115735.
46. E. Vasconcelos Soares Maciel, A. L. de Toffoli, E. Sobieski, C. E. Domingues Nazário and F. M. Lanças, *Anal. Chim. Acta*, 2020, **1103**, 11–31.
47. Y. Liu, X. Wang, Z. Chen, D. H. Liang, K. Sun, S. Huang, J. Zhu, X. Shi, J. Zeng, Q. Wang and B. Zhang, *Anal. Chim. Acta*, 2019, **1062**, 147–155.
48. M. P. Ladd, R. J. Giannone, P. E. Abraham, S. D. Wullschleger and R. L. Hettich, *Sci. Rep.*, 2019, **9**, 1–13.
49. G. D'Orazio, *TrAC, Trends Anal. Chem.*, 2020, **125**, 115832.
50. W. Zhang and R. Ramautar, *Electrophoresis*, 2021, **42**(4), 381–401.
51. K. Segers, W. Zhang, N. Aourz, J. Bongaerts, S. Declerck, D. Mangelings, T. Hankemeier, D. De Bundel, Y. Vander Heyden, I. Smolders, R. Ramautar and A. Van Eeckhaut, *Talanta*, 2020, **217**, 121107.
52. W. Zhang, F. Guled, T. Hankemeier and R. Ramautar, *J. Chromatogr. B: Anal. Technol. Biomed. Life Sci.*, 2019, **1105**, 10–14.
53. W. Zhang, F. Guled, T. Hankemeier and R. Ramautar, *Electrophoresis*, 2020, **41**, 360–369.
54. M. van Mever, K. Segers, N. Drouin, F. Guled, Y. Vander Heyden, A. Van Eeckhaut, T. Hankemeier and R. Ramautar, *Microchem. J.*, 2020, **156**, 104949.
55. H. W. Liao, S. S. Rubakhin, M. C. Philip and J. V. Sweedler, *Anal. Chim. Acta*, 2020, **1118**, 36–43.
56. J. M. Wilkins and E. Trushina, *Front. Neurol.*, 2018, **8**, 1.
57. M. M. Niedzwiecki, D. I. Walker, J. C. Howell, K. D. Watts, D. P. Jones, G. W. Miller and W. T. Hu, *Ann. Clin. Transl. Neurol.*, 2020, **7**, 36–45.
58. L. Hu, M.-X. Dong, Y.-L. Huang, C.-Q. Lu, Q. Qian, C.-C. Zhang, X.-M. Xu, Y. Liu, G.-H. Chen and Y.-D. Wei, *Front. Mol. Neurosci.*, 2020, **13**, 80.
59. D. Miura, Y. Fujimura and H. Wariishi, *J. Proteomics*, 2012, **75**, 5052–5060.
60. J. Ferey, F. Marguet, A. Laquerrière, S. Marret, I. Schmitz-Afonso, S. Bekri, C. Afonso and A. Tebani, *Anal. Bioanal. Chem.*, 2019, **411**, 3891–3903.
61. Y. Magnusson, P. Friberg, P. Sjövall, F. Dangardt, P. Malmberg and Y. Chen, *Clin. Physiol. Funct. Imaging*, 2008, **28**, 202–209.
62. J. Balog, L. Sasi-Szabó, J. Kinross, M. R. Lewis, L. J. Muirhead, K. Veselkov, R. Mirnezami, B. Dezso, L. Damjanovich, A. Darzi, J. K. Nicholson and Z. Takáts, *Sci. Transl. Med.*, 2013, **5**, 194ra93.
63. S. Guenther, L. J. Muirhead, A. V. M. Speller, O. Golf, N. Strittmatter, R. Ramakrishnan, R. D. Goldin, E. Jones, K. Veselkov, J. Nicholson, A. Darzi and Z. Takats, *Cancer Res.*, 2015, **75**, 1828–1837.

64. S. Schadt, S. Kallbach, R. Almeida and J. Sandel, *Drug Metab. Dispos.*, 2012, **40**, 419–425.
65. K. D. Duncan and I. Lanekoff, *Anal. Chem.*, 2019, **91**, 7819–7827.
66. S. Azab, R. Ly and P. Britz-Mckibbin, *Anal. Chem.*, 2019, **91**, 2329–2336.
67. K. Faserl, B. Sarg, L. Sola and H. H. Lindner, *Proteomics*, 2017, **17**, 1700310.
68. C. G. Vasilopoulou, K. Sulek, A. D. Brunner, N. S. Meitei, U. Schweiger-Hufnagel, S. W. Meyer, A. Barsch, M. Mann and F. Meier, *Nature communications*, 2020, **11**(1), 1–11.
69. E. Sánchez-López, G. S. M. Kammeijer, A. L. Crego, M. L. Marina, R. Ramautar, D. J. M. Peters and O. A. Mayboroda, *Sci. Rep.*, 2019, **9**, 1–9.
70. M. Saoi, K. M. Kennedy, W. Gohir, D. M. Sloboda and P. Britz-McKibbin, *Sci. Rep.*, 2020, **10**, 1–10.
71. K. Nakatani, Y. Izumi, K. Hata and T. Bamba, *Mass Spectrom (Tokyo)*, 2020, **9**(11), A0080.
72. T. Kawai, N. Ota, K. Okada, A. Imasato, Y. Owa, M. Morita, M. Tada and Y. Tanaka, *Anal. Chem.*, 2019, **91**, 10564–10572.
73. T. Kawai, N. Ota, A. Imasato, Y. Shirasaki, K. Otsuka and Y. Tanaka, *J. Chromatogr. A*, 2018, **1565**, 138–144.
74. F. Meier, S. Beck, N. Grassl, M. Lubeck, M. A. Park, O. Raether and M. Mann, *J. Proteome Res.*, 2015, **14**, 5378–5387.
75. F. Meier, A. D. Brunner, S. Koch, H. Koch, M. Lubeck, M. Krause, N. Goedecke, J. Decker, T. Kosinski, M. A. Park, N. Bache, O. Hoerning, J. Cox, O. Räther and M. Mann, *Mol. Cell. Proteomics*, 2018, **17**, 2534–2545.
76. G. S. M. Kammeijer, I. Kohler, B. C. Jansen, P. J. Hensbergen, O. A. Mayboroda, D. Falck and M. Wuhrer, *Anal. Chem.*, 2016, **88**, 5849–5856.

CHAPTER 11

More with Less: Single-cell Metabolomics by Mass Spectrometry

V. L. VERSCHOOR,[a] Y. SHIMIZU,[b] S. EMARA,[c] A. ALI*[a,d] AND T. HANKEMEIER*[a]

[a] Leiden Academic Center for Drug Research (LACDR), Leiden University, Einsteinweg 55, 2333C, The Netherlands; [b] RIKEN Center for Biosystems Dynamics Research (BDR), RIKEN, 6-2-3, Furuedai, Suita, Osaka, 565-0874, Japan; [c] Pharmaceutical Chemistry Department, Faculty of Pharmacy, Misr International University, Km 28 Ismailia Road, Cairo 44971, Egypt; [d] Research Center, Misr International University, Km 28 Ismailia Road, Cairo 44971, Egypt
*Emails: Ahmed.ali@outlook.jp; hankemeier@lacdr.leidenuniv.nl

11.1 Introduction

The cell is the smallest basic functional unit of all life. As such, the study of the cell is essential to understand any biological process of interest. The phenotype and behavior of a cell are a function of the sum of all molecular processes ongoing within it: many different approaches have been developed to study a subclass of cellular molecules, often with the goal of linking the state of those molecules to the cell's phenotype. Historically, given the technical limitations owing to low cell volume and relatively low number of molecules to be analyzed, cells have been pooled by thousands or millions and analyzed in bulk. While these bulk analyses allow for collection of vast amounts of data on cell

New Developments in Mass Spectrometry No. 10
Advanced Mass Spectrometry-based Analytical Separation Techniques for Probing the Polar Metabolome
Edited by Rawi Ramautar

Published by the Royal Society of Chemistry, www.rsc.org

populations, the information on an individual cell is lost.[1] The averaging of signal inherent to this approach will reduce any outlying data to noise, thus removing a lot of information on what makes any single particular cell 'tick'. This is especially relevant for any situation where a researcher is interested in a rare cell population, or in a cell population that is highly heterogenous. In either case, pooling multiple cells would lead to an averaged measurement that would not represent the state of the cell of interest (Figure 11.1). To fully understand the underlying mechanisms leading to a cell's behavior, we thus need to collect our data on a single-cell scale.

Single-cell analysis methods include studying the genome, transcriptome, proteome, and the metabolome, with metabolomics being the most challenging, especially on the single-cell level. Hence, it is the focus for this chapter. Metabolomics is the study of all metabolites, which covers molecules with a relatively low molecular weight. These molecules include amino acids, nucleotides, lipids, and carbohydrates which act as precursors, intermediates, and products in the metabolic network. By measuring as many metabolites as possible, we get an indication of the ongoing processes in the cell. Metabolomics has an interesting role as an intermediate between the genome and the actual phenotype of the cell.[2,3] More than the genome, transcriptome, and proteome, the metabolome is extremely dynamic, responding very quickly to cellular and extracellular impulses. While this is advantageous in studying a cellular response to a stimulus, it is difficult to perform metabolomics measurements on a cell without perturbing its environment. This can quickly

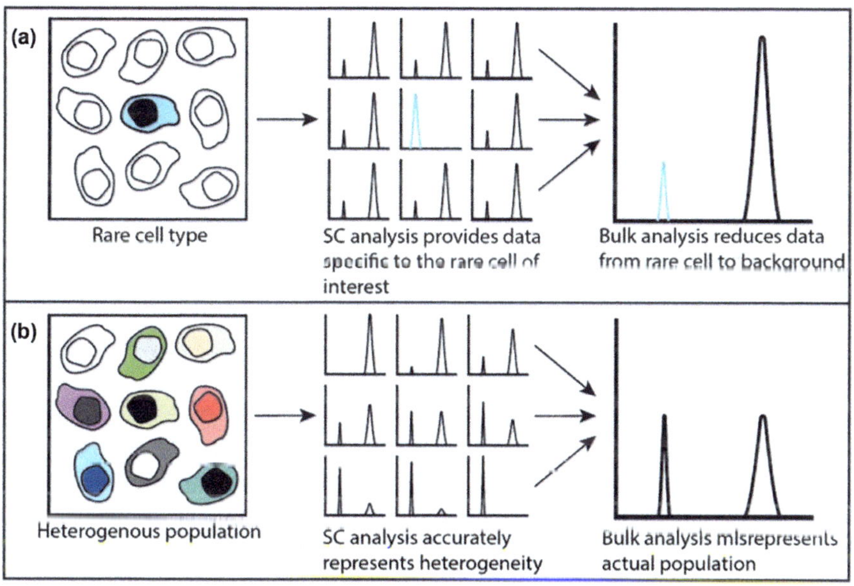

Figure 11.1 Advantages of single-cell analysis: when analyzing either a rare cell type or heterogeneous cell population, bulk analysis will average out the individual signals and cell specific data will be lost.

affect the metabolome and the validity of the analysis. Therefore, the sampling strategy is of extreme importance in order to gain trustworthy data, and we will discuss some sampling strategies and considerations in this chapter.

As mentioned, another challenge in single-cell metabolomics is the extremely small volume and low amounts of measurable molecules that constitute a single cell. With a volume in the femtoliter range, manipulation and treatment of isolated single cells are complex. Addition of any volume of derivatization agent can quickly reduce the concentration of metabolites to below the limit of detection of a mass spectrometer. While single-cell genomics and transcriptomics have similar problems, the genome and transcriptome can be amplified. Since the metabolome cannot, single-cell metabolomics requires highly sensitive analytical techniques and limited sample manipulation to avoid dilution. In this chapter, we seek to introduce readers with experience in mass spectrometry to the possibilities and challenges of live single-cell metabolomics and discuss strategies to sample, handle, and analyze single-cell samples. Toward this end, several strategies for sampling and isolating single cells will be highlighted, as well as several mass spectrometry-based techniques that are especially suited for single-cell metabolomics. Despite the utility of other single-cell mass spectrometry techniques, we chose to emphasize methods that minimize disruption to the cell's microenvironment, and by extension, their metabolome. Therefore, we focused mainly on probe-based sampling methods and ambient ionization techniques. A more generalized overview of single-cell metabolomics is excellently summarized in ref. 4–6.

11.2 Sampling Strategies

Since a typical single-cell metabolomics experiment usually starts with sampling, a correct sampling strategy is essential for the success of any single-cell experiment, especially in the case of live cell metabolomics. Reliably isolating a live single cell from its environment while avoiding disturbance of the metabolome is challenging, and the sampling approach will need to be optimized for both the cell source and method of analysis available. In this section, we discuss some sampling considerations and several single-cell isolation strategies.

11.2.1 Sampling Considerations

Single-cell sample preparation is perhaps the largest source of unwanted technical variability in single-cell analysis.[7] As such, optimization of the steps involved in isolation and preparation of single-cell samples is required. Sampling preparation is usually dependent on the source of live cells and on the planned analysis technique. Samples for live single-cell metabolomics will generally be from cell culture, blood samples, or freshly excised tissue. While blood samples require little pre-treatment to get a single-cell suspension, cell culture samples are often trypsinized, and tissues often require enzymatic digestion.[8] However, these enzymatic perturbations can lead to metabolomic changes. When possible, reducing metabolic activity in the sample can help mitigate these changes: by working at lower temperatures, metabolic activity will decrease. Another option is to manually isolate single cells of interest using microneedles or capillaries,[9] at the cost of lower throughput and higher labor.

Another aspect to consider when designing a live single-cell sampling approach is the analytical method that will be used. Several analytical methods will be discussed in depth later in this chapter, but it is important to mention here that the analytical method will dictate the state the cell has to be in at the start of analysis. For example, certain approaches, such as MALDI and DESI, require the cell to be present on a specific type of surface.[10] Other methods can require the cell to be in suspension at the start of analysis. Defining the sample preparation protocol thus requires considering the state in which sample cells start and what state they have to be in in order to start the analysis of choice.

Once these factors have been defined, it is important to consider if, when, and how the cell will be quenched. By quenching a cell, all metabolic processes are stopped. This is generally done either by adding organic solvent – methanol is commonly used – to the cells,[11] or by snap-freezing the cell.[12] By quenching as early as possible, the disruption of the cell metabolome is kept to a minimum.

As discussed, the sampling strategy is to be considered thoroughly when planning any metabolomics approach, especially in single-cell resolution. Broadly speaking, isolation approaches fall into 3 categories: microfluidics, fluorescence activated cell sorting (FACS), and micro-sampling, the three methods will be discussed in the following section, along with their advantages and possible limitations.

11.2.2 Microfluidics

Microfluidics-based approaches are a great complement to single-cell research: the dimensions (micrometers) and volumes (pico- to microliters) involved in microfluidics techniques are inherently more suited for single-cell manipulation than traditional biological tools such as petri dishes or well-plates.[13] The small sample volumes involved carry a lower risk of dilution of the sample, and several strategies have been developed which allow a great amount of control over a cell's location and environment. In microfluidics-based approaches, single cells are usually isolated either by capturing them in droplets or microstructures.

Droplets or slugs containing single cells can be produced by combining a flow of aqueous single-cell suspension with a flow of carrier oil at a microfluidic structure. Examples of suitable structures are T-junctions[14] or co-flow channels.[15] Once cells are encapsulated in these volumes, they can be manipulated: droplets can be sorted, inspected, incubated, and released from the chip. Picoinjectors that can specifically inject small volumes of reagent into passing droplets or slugs have also been designed.[16]

Several microfluidic microstructures have been designed to capture single cells: Di Carlo *et al.* designed a device containing several parallel trapping arrays, each consisting of a number of U-shaped polydimethylsiloxane (PDMS) structures designed to capture and retain single HeLa cells that are introduced into the device.[17] Shalek *et al.* use a device containing chambers that can be closed and opened on demand using pressure regulation.[18] Lohr *et al.* use micropatterned nano wells, which capture cells with gravity and can be closed off using a glass slide.[19]

All the mentioned methods have been extensively applied and reviewed.[13,20] However, not all approaches are suitable for single-cell metabolomics by mass spectrometry. Microfluidic devices are commonly produced using PDMS, which can be problematic for mass spectrometry analyses due to the chemical background signal it produces.[21] Furthermore, removing the isolated single cells from the microfluidic device can be challenging. Suitability of a specific microfluidic method will depend on the source of cells and analytical technique a researcher plans to use.

11.2.3 FACS

FACS is a flow-cytometry method that allows for sorting of single cells based on characteristics such as cell size, density distribution, as well as fluorescently labeled proteins.[22] In short, a single-cell suspension is stained with fluorescently labeled antibodies against proteins of interest. The stained cells then flow through a flow cell one by one and are exposed to a laser. Based on scattering of the laser and on emission from the fluorescent label, a cell will be sorted according to pre-set characteristics. While the sorting approach is generally used to isolate multiple cells of a certain phenotype, modern FACS systems are capable of sorting and isolating single cells as well. In fact, in a 2015 article, FACS was found to be the most commonly used technology for single-cell isolation.[23]

As one of the most used methods for cell identification, sorting and single-cell isolation, the appeal of FACS as part of live single-cell sample preparation is clear. The method is robust, high throughput (up to 70 000 events per second can be analyzed[23]) and commonly available. The number of labels that can be detected by a FACS system has increased from 3–5 to 14,[24] and many fluorescent antibodies have been developed over the years. Interestingly, a recent article even devised a computational automated gate-setting method where cells were not fluorescently stained, instead relying purely on cell-intrinsic parameters such as cell size, granularity, and mitochondrial content.[25] FACS can be especially suitable for the isolation of rare cell types from a mix of cells. For example, Hiyama *et al.* used FACS to isolate $CD45^-/GD2^+$ circulating tumor cells from a patient blood sample, which were then collected with a capillary, sonicated, and electrosprayed into an LTQ Orbitrap.[26] The FACS selection enabled identification of several amino acids and lipids in a cell population which would have been hard to isolate using other approaches.

Despite the discussed positives, FACS is not commonly used for single-cell metabolomics studies.[27] This might be due to the additional steps required for FACS analysis. Cells are already in a single-cell suspension before FACS and have to be stained and sorted. During these processes, cells are not quenched, which, along with the fact that they undergo mechanical stress within the FACS system,[28] might lead to changes in the metabolome during isolation. Another disadvantage is the starting material required: for most reliable FACS analyses, at least 10 000 recorded events will be required. Taking into account cell loss and cells required for setting up FACS gates, one FACS run will require a number of cells that might not be available, depending on the cell source. These concerns should be considered when devising a single-cell metabolomics approach using FACS.

11.2.4 Micro-sampling

Micro-sampling techniques aim to reduce perturbations to the cell's microenvironment, and by extension, its metabolome. They usually involve minimal, or no, sample pre-treatment, and the sampling process is done in ambient conditions, with some methods even sampling the cells in their native microenvironment *i.e.*, cell media. They can be easily integrated into the aforementioned isolation techniques, as well as ambient ionization mass spectrometry-based methods. The working principle of micro-sampling approaches is simple. A pulled glass pipette is used to sample the whole cell, or part of it, under constant microscopic observation. The pipette is usually controlled by means of a 3-axis micromanipulator. Suction of the cell, or its contents can be done by attaching a pump (manual or automatic) to the pipette holder and applying negative pressure. Afterwards, the sampled cellular contents are then lysed, extracted, and fed into the mass spectrometer, usually *via* nanospray ionization (NSI). Since micro-sampling methods are usually integrated with other techniques, they will be discussed at more detail in the analytical methods section.

Despite their obvious utility, micro-sampling techniques have limitations that should be kept in mind when designing a given single-cell metabolomics experiment. Since the sampling capillary is controlled by a 3D micromanipulator, the sampling process is often labor intensive, skill dependent and has low throughput; automation addresses most of these issues, however, it's often expensive and difficult to implement.

11.2.5 Discussion

It's important to mention that the sampling strategy used in single-cell metabolomics experiments is often dictated by the experimental goal itself. There is no perfect technique that possesses single-cell resolution, high throughput, and minimal perturbations to the cell, or its microenvironment. More often than not, any given sampling strategy will possess one, or two of the preceding qualities, therefore, the needs of the experiment itself and the biological or clinical phenomenon to be investigated usually decides which method is best.

If the goal is to investigate rare subpopulations that are mixed in with a large number of other cells, then the benefits of using high throughput techniques such as FACS and microfluidics offset the potential downsides of large perturbations to the cells in question. One typical example of this is circulating tumor cells analysis (CTCs). CTCs are present in the peripheral blood of cancer patients in concentrations less than 10 cells per mL of blood,[29] mixed with 4–6 million other, naturally occurring, cells.[30] Therefore, FACs and microfluidics are ideal for isolating these rare cells. On the other hand, if the main purpose of the experiment is investigating cellular heterogeneity, or if a strong emphasis is directed at minimizing external changes to the cell state (*e.g.*, drug discovery studies), then the minimal perturbations caused by micro-sampling offset any potential limitations in throughput and skill requirements.

11.3 Analytical Methods

After successful cell isolation and metabolic quenching, the next step is analyzing the isolated cells. The method most often used for single-cell metabolic studies is mass spectrometry. Hence, methods based on mass spectrometry will serve as the main focus in this section. A multitude of methods was developed to analyze single cells: as with any analytical technique, each comes with its own solutions to the balancing act between throughput, robustness, resolution, sensitivity, and finally, reducing perturbations to the cell's microenvironment. Since the aim is to study the metabolome of the cell, only methods that measure live cells are included in this section. These methods either integrate capillary micro-sampling, or ablation techniques that ablate the cellular contents into the mass spectrometer by means of nanospray ionization (Figure 11.2). Each will be discussed at length in the following section, along with their possible applications, uses, and limitations.

11.3.1 Live Single-cell Mass Spectrometry (LSC–MS)

Live single-cell mass spectrometry (LSC–MS) is a method pioneered by the late Masujima and his group in RIKEN, Japan.[31] It exemplifies the ethos of single-cell metabolomics, by performing the sampling, lysis, extraction, and subsequent introduction into the MS instrument in the same capillary/pipette. Therefore, reducing the possibility of sample dilution and maximizing sensitivity, while minimizing disruption to the cell's microenvironment, and allowing for sampling cells under microscopic observation in their native conditions. Masujima's team successfully applied this technology to showcase the potential of single-cell metabolomics in biology,[32–34] medical,[35,36] and pharmaceutical applications.[37,38] This was done by using a platinum coated pipette to sample the cell, then,

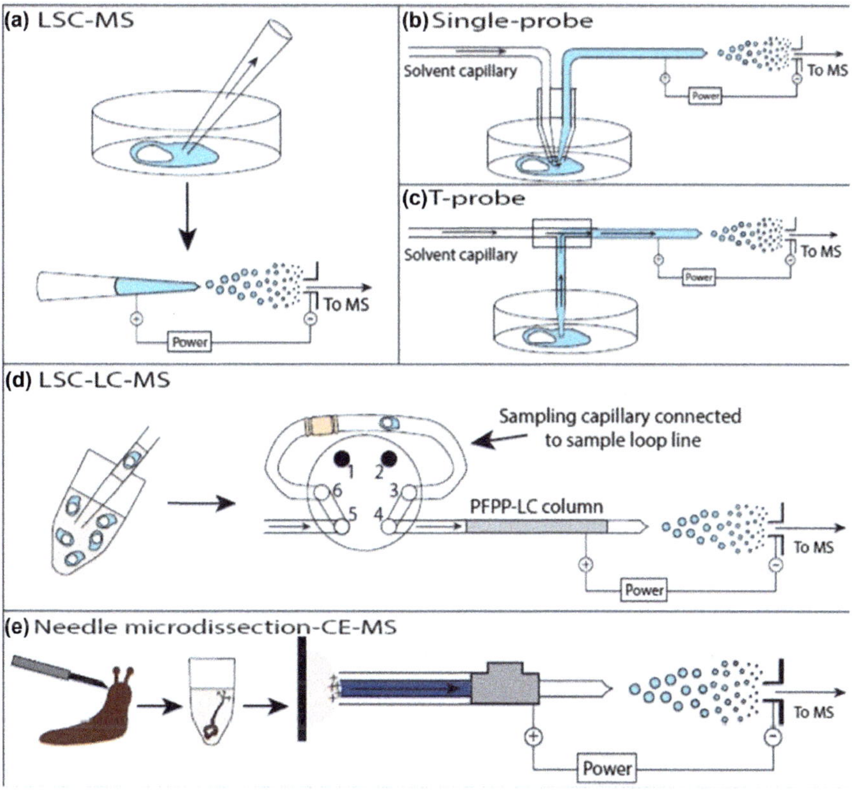

Figure 11.2 Overview of capillary sampling-based single-cell analysis methods.

adding the ionization solvent from the other end of the pipette that lyses and extracts the cellular contents. The pipette is then attached to the NSI source of the MS instrument; thanks to the conductivity afforded by the platinum coating, it can then act as an NSI emitter which introduces the contents of the capillary into the MS.

In traditional LSC–MS, the capillary contents are directly sprayed into the MS, without chromatographic separation, which limits specificity. Zhang *et al.* and Fuji *et al.* attempted to circumvent this by incorporating a separation step based on ion mobility into the protocol (Figure 11.1b),[39,40] which gave LSC–MS the ability to distinguish between isometric compounds, as well as improve its general specificity. Vertes group applied this technique in the subcellular analysis of single identified Type 1 and Type 2 F group (Fgp) neurons from *Lymnaea stagnalis* CNS.[41] Their study showcased subcellular distributions of FMRFamide-like neuropeptides between the cytoplasm and nucleus in individual Type 2 Fgp. In addition, a 28-residue neuropeptide was found for the first time and sequenced by tandem MS in a single Type 2 Fgp neuron. These results demonstrate the feasibility of detecting peptide localization in single identified neurons with sub-cellular resolution as well as facilitate the analysis of sub-cellular level changes underlying memory function and dysfunction in both invertebrate and vertebrate organisms.

Furthermore, the mismatch between the spray rate (10–100 nL per min) and typical mammalian cells volumes (pico-liter range) means that there is usually not enough time to perform detailed MS/MS studies to confirm the structure of analytes of interest. An interesting approach that attempts to address this was suggested by Xinrong Zhang's group: they incorporate droplet micro-extraction to LSC–MS. It involves using the sampling capillary/pipette to introduce a small amount (∼nL) of an extraction solvent to the cell in question, after waiting for 10 seconds, the extract was aspirated using the sample capillary. The solvent is then evaporated, and the cellular extract is reconstituted with 2 nL of ionization solvent followed by MS/MS analysis. Using this technique, it was possible to identify more than 300 phospholipids according to their MS/MS profile.[42]

A limitation that is shared across virtually all MS-based methods is that MS is an exhaustive technique *i.e.*, it consumes the sample while measuring it. The impact of this limitation on any given experiment is inversely proportional to the volume of the sample measured. While in traditional experiments, this is hardly an issue, in single-cell measurements, where a typical cell volume is measured in the low picolitres, the exhaustiveness of MS measurements limits what can be measured in a single cell. To circumvent this, several groups tried to integrate relatively non-invasive techniques with single-cell MS measurements, mainly, microscopic[43] or spectroscopic imaging.[44] One recent example is the work done by Ali *et al.*, where 3D holography and tomography were successfully integrated into LSC–MS. Owing to this unique combination of measurements, the heterogeneity of several metabolites in subcellular space was investigated, as well as the possibility of performing absolute quantitation in subcellular compartments by mapping out the cell in 3D before and after sampling it with a small capillary.[32] Another possible integration was also proposed by Ali *et al.*, where LSC–MS was coupled with Raman measurements.[38] This was achieved by mounting a micromanipulator to the Raman microscope, then, under microscopic observation, single drug-treated HepG2 cells were measured by Raman, then immediately sampled and analyzed using LSC–MS. Raman measurements were used to build a predictive model, that could predict with a high degree of accuracy whether a given cell was affected by tamoxifen or not, while mass spectrometry measurements could detect, then quantify, the amount of the drug in the cell, as well as its metabolites. The two datasets were also correlated, and several potential biomarkers of the drug activity could be isolated from the complex Raman spectra using mass spectrometry data.[38]

11.3.2 Single Probe and T-probe

Another, more integrated technique is the single probe, pioneered by Yang's group at the University of Oklahoma. The single-probe brings the possibility of performing real-time, *in-situ* metabolic measurements of the cell in question.[45]

This is achieved by using a dual-bore quartz needle with a pulled tip, into the bores of which two smaller capillaries are inserted. One of the capillaries delivers a flow of ionization solvent, while the other capillary functions as a nano-ESI emitter. The combination of solvent flow and suction from the nano-ESI emitter produces a liquid junction at the tip of the dual-bore needle. When this liquid junction comes in contact with a given cell, metabolites are extracted and are carried away through the second channel to the MS inlet, where they are sprayed by NSI into the MS instrument and analyzed.

The tip of the single-probe is small enough that it can be inserted into a single cell to probe subcellular localization of metabolites, and analysis can be performed even on cells covered with a layer of liquid (Pan *et al.* performed their experiments with cells covered by PBS[45]). This mitigates possible metabolic disturbance caused by cells being dry, which is required for some of the techniques that ionize directly from a surface. Using the single-probe to analyze HeLa cells, Pan *et al.* were able to identify several energy metabolites, lipids, and anticancer drugs that the cells were treated with. The single-probe has also been used for mass spectrometry imaging of mouse tissue slices and cultured HeLa cells.[46]

An updated design named the T-probe was developed, which involves a single-bore capillary for cellular or subcellular sampling connected to a T-junction. In the T-junction, aspirated cellular contents are mixed with the ionization solvent and directed toward the emitter capillary to be sprayed into the mass spectrometer (Figure 11.2).[47] Again, the continuous depletion of solvent from the ESI emitter is the driving force behind the suction used to aspirate cellular content.

Both the single-probe and the T-probe are very convenient and robust methods of single-cell sampling. The small size of the probe tip allows for sampling at a subcellular resolution, and analysis is near-real time due to the direct link between the sampling probe and the ESI. This direct link also has its downsides: using the current design of both probe types, it is not possible to do any sample treatment or analyte separation. These are tradeoffs that will have to be considered when working with either type of probe.

11.3.3 Nano-LC

The aforementioned methods directly introduce the extracted cellular contents into the mass spectrometer, however, incorporating a separation step, such as the earlier mentioned ion mobility separation, can aid in specificity of the analysis and distinction between isometric compounds. Nakatani *et al.* attempted to address this by using micro-sampling to feed single cells directly into a nano-LC system.[48] To showcase their technique, HeLa cells were cultured and trypsinized into a single-cell suspension at 10 cells per uL. 100 nL of this solution, corresponding to 1 cell on average, was picked up using a capillary connected to a nanopipette device. Collection of a single cell in this volume was confirmed by microscopy, at which point the capillary was connected to a sample loop line. The single cell was then introduced into the nano-LC system by switching a sample loop valve. In the LC system, a combination of the acidic mobile phase, high system back pressure, and an applied voltage combine to lyse the cell and release the intracellular metabolites into the mobile phase. These metabolites are then separated on a PFPP-Nano-LC column. Using this method, Nakatani *et al.* were able to identify 18 hydrophilic metabolites in 22 single cells analyzed. These metabolites include GSH, dTMP, adenosine, and 15 amino acids. By comparing average peak areas per cell for each of these metabolites, the authors show heterogeneity between HeLa cells from the same sampled population, perhaps due to differences in cell cycle or microenvironment.

Using this technique, cells to be measured could be loaded into the LC system within 25 minutes of being trypsinized and within 5 minutes from being sampled from the single-cell suspension. While this is relatively quick, especially for a single-cell technique that includes a separation step, care should be taken that the trypsinization and time spent unquenched as a single-cell suspension affect the metabolome as little as possible.

11.3.4 Manual Needle Microdissection and CE

Another approach combining manual capillary sampling and a separation technique comes from the Sweedler group at the University of Illinois. Liao *et al.* micro-dissected sensory neurons from *A. californica* (a species of sea slug) using sharp tungsten needles.[49] Metabolic reactions were quenched by placing neurons in methanol and storing samples at −20 °C until analysis. Metabolites were extracted in several rounds of organic solvent addition and sample drying. Samples were separated using Field-Amplified Sample Injection Capillary Electrophoresis (FASI-CE). FASI is a technique where sample analytes are stacked on the boundary of the injected sample matrix and a sample buffer, which ensures a much higher sensitivity compared to injection without sample stacking.[50] 37 metabolites could be identified from pleural sensory neurons with an average cell body diameter of ∼50 μm. Most of these metabolites were amino acids, and functionally relevant metabolites such as tyramine, arginine, and citrulline were detected. Using two dipeptides –alanine–alanine and proline–leucine as internal standards, 21 metabolites could be quantified.

CE is very well suited for separation in single-cell metabolomics analyses, due to the high sensitivity and the wide dynamic range of the method. Furthermore, stacking techniques like FASI allow a degree of control over the analytes, which can help ‘concentrate’ the analytes in a relatively diluted sample before separation.[51] However, it should be noted that CE use is often limited to polar compounds only, which should be kept in mind when choosing this method.

11.3.5 (Nano-)DESI

Desorption Electrospray Ionization (DESI), developed by Takáts *et al.*, is one of the first described ambient ionization strategies for mass spectrometry.[52]

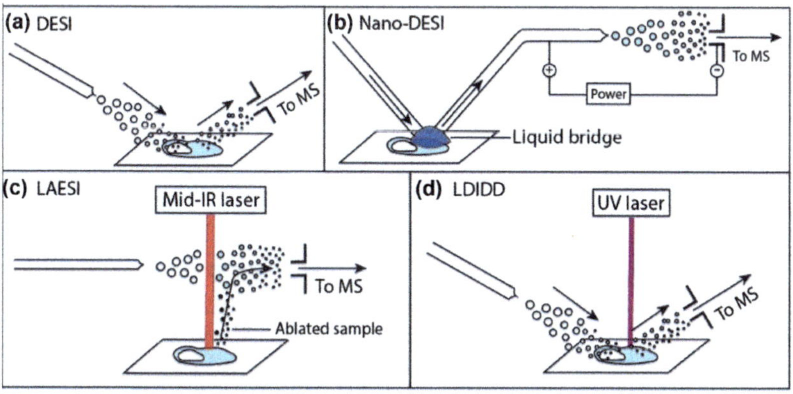

Figure 11.3 Overview of surface ionization-based single-cell analysis methods.

In DESI, a flow of electrosprayed solvent droplets is targeted at a sample surface, where analytes are desorbed and carried with secondary charged droplets into the MS inlet (Figure 11.3).[53] DESI is extensively used for ambient ionization of tissue samples and is the most widely recognized ambient ionization method for mass spectrometry imaging (MSI).[54] However, its average spatial resolution of ~100 μm[54] is not ideal for single-cell analysis.

An analog of DESI with a higher spatial resolution is nanospray desorption electrospray ionization. Nano-DESI is an ionization technique that in principle also resembles the single-probe and T-probe discussed above. It was developed in the labs of Julia Laskin and Alexander Laskin at Purdue University.[55] In this approach, a continuous flow of solvent in the form of a liquid bridge is used to extract metabolites from a surface. This liquid bridge of solvent is formed by two capillaries, the first delivering the solvent, and the second a nanospray ionization capillary (Figure 11.3). By positioning this liquid bridge on a single cell, the contents of the cell are extracted, ionized, and analyzed by a mass spectrometer. This extraction is performed under ambient conditions and as such allows for near real-time analysis of cells without major perturbation to their metabolism.

Bergman *et al.* used Nano-DESI to analyze single human cheek cells deposited on glass slides and were able to analyze them within 2 minutes of collection.[56] This allowed identification of several metabolic compounds without pretreatment, derivatization, or separation. In total, 11 amino acids could be consistently detected with a reliable signal-to-noise ratio, as could 6 metabolites and several lipid species in the lipid classes PE, PC, sphingomyelin, and plasmalogen. By infusing a PC 46 : 0 internal standard in the DESI solvent, Bergman *et al.* were even able to perform quantification of endogenous PC species in these single cells, an approach that can be used for any single compound of interest. To increase throughput, Nano-DESI can be used for MSI. By automatically scanning the Nano-DESI probe over a sample surface containing single cells, Bergman *et al.* mapped the abundance of creatine and proline on a slide seeded with single cells, in lines spaced 150 μm apart.

The Julia Laskin lab also uses Nano-DESI for MSI, with higher resolution: by using a smaller capillary with an outer diameter of 15–25 μm, their Nano-DESI MSI approach reaches line spacing of 15 μm and an average spatial resolution of 9.8 μm.[57] A shear force microscopy (SFM) probe, in the form of a third capillary with an outer diameter of 10 μm, is included in their protocol. This SFM probe enables precise control over the distance – within 1 μm - between the Nano-DESI probe and sample. By controlling this distance during scanning, samples with an uneven surface can be imaged with high spatial resolution. Like Bergman *et al.*, Yin *et al.* added internal standards – LPC 19 : 0, PI, and arachidonic acid-d8 – to their solvent for absolute quantification. While this approach has only been used for MSI of high-resolution frozen murine uterine tissue slices,[57] the proposed improvements can conceivably be applied to single-cell analysis as well.

Aside from the discussed upsides, a downside of Nano-DESI as an ambient single-cell metabolomics technique is that cells cannot be sampled directly from their native environment: for the liquid bridge of solvent to function, cells need to be positioned on a relatively flat surface, with no liquid layer (*e.g.*, medium) which would disrupt the liquid bridge flow. Indeed, Bergman *et al.* simply spread collected cells on a glass slide with no liquid and state in their article that no care was taken to keep cells alive until analysis.[56] While the time from collection to analysis is short, monitoring and promoting cell survival during analysis will be required when metabolic perturbations are to be kept to a minimum.

11.3.6 LAESI

LAESI is an ambient ionization method where electrospray ionization (ESI) is combined with laser ablation (LA). The energy from a mid-IR laser is used to ablate a targeted area containing a cell, producing a plume of ejected cell contents. A spray of charged droplets from an ESI source is intersected with the plume of cell contents, which ionizes the neutral molecules from the ablated cell and introduces them into the mass analyzer (Figure 11.3). This method is well suited to ionization of compounds with a broad range of molecular masses.

The method was developed in the lab of Akos Vertes at George Washington University and was adapted to a single-cell scale by Bindesh Shrestha.[58,59] The limiting factor for this adaptation of LAESI to a single-cell analysis technique was the size of the ablation spot left by the laser. In the first paper describing the technique, the ablation spot had a diameter of 350–400 μm,[58] far larger than the average size of a single mammalian cell.[60] Reducing this ablation size was made possible by etching the tip of an optical fiber to reduce its diameter. By etching the tip to a radius of ~15 μm, the size of the ablation spot can be decreased to a diameter of ~30 μm.

LAESI has been extensively used to analyze single plant cells,[59,61] and has also been applied to a broad range of animal cell types, such as sea urchin eggs,[59] *Xenopus laevis* eggs,[62] rat brain tissue,[63] and human cells.[64]

In the article introducing the method, Shrestha *et al.* were able to detect 32 compounds from a single A. cepa bulb epidermal cell (10 of which were monosaccharides or polysaccharides with varying degrees of polymerization). They identified several compounds only present in pigmented A. cepa bulb epidermal cell compared to unpigmented cells. A number of detected nominal *m/z* peaks from analyzed sea urchin eggs are provided, but these were not further identified in the article.[59] In a later article, Shrestha *et al.* also used LAESI on *X. laevis* eggs to identify 52 small metabolite ions and at least 92 various types of lipids, from lipid classes including fatty acids, phosphatidylcholines, and triacylglycerol.[62] The authors were even able to distinguish metabolomic profiles of the different poles of the polarized *X. laevis* egg and compared lipid profiles of unfertilized and fertilized eggs. While the metabolic coverage of these results is impressive, it is important to consider that a single *X. laevis* egg has an average diameter of ∼1.4 mm, compared to an average diameter of ∼20 μm for a single mammal cell. The limited amount of sample material available in single-cell metabolomics is a less relevant concern for this model system. LAESI has also been used for analysis of mammal and human cells. These analyses have typically not been performed at a single-cell resolution, however. Rat brain tissue slices were subjected to LAESI–MSI with an ablation spot size of ∼200 μm, rat insulinoma cells were analyzed ∼100 cells at a time,[65] human buccal epithelial were analyzed as low as 6 at a time,[65] and a recent protocol for LAESI combined with ion mobility separation in human adherent cell lines is not for single-cell analysis and does not indicate the expected ablation spot size for this technique.[64]

While LAESI has been used effectively to profile single plant cells, single mammal cell analysis has been limited. As such, when pursuing LAESI as the ionization method for a single-cell metabolomics experiment in mammal cells, some optimization might be required.

11.3.7 LDIDD

Laser Desorption Ionization Droplet Delivery (LDIDD) is a technique that combines laser desorption-based sampling with ionization by electrosprayed droplets, in a similar approach as LAESI (Figure 11.3). It was developed by Lee *et al.* in Richard Zare's lab at Stanford.[66] While similar to LAESI in principle, there are some practical differences between the two approaches. First, the laser type is different: LAESI uses a mid-IR laser with a wavelength of 2940 nm,[59] while LDIDD uses a UV laser with a wavelength of 266 nm.[66] Second, in LAESI the stream of electrosprayed droplets is perpendicular to the laser orientation, while the electrospray in LDIDD is aimed directly at the laser-irradiated sample region.

The mentioned differences in laser type affect both the process of ablation and the ablation spot size. The mid-IR laser ablation used in LAESI functions by introducing energy into water molecules in the sample. These absorb laser energy strongly in the mid-IR range due to symmetric stretching of their O–H bonds, which leads to phase explosion into a plume of desorbed material.[58] The UV laser used in LDIDD is more energetic due to its lower wavelength and is capable of photoionizing sample molecules:[66] generally, ablated LAESI materials will be neutral, while LDIDD ablation leads to more ionized compounds. In both methods, the electrospray aids in ionization of any neutral compounds present.

Another consequence of the choice of laser is the resolution of the technique: the smallest possible focused laser ablation spot size is directly related to wavelength. The high-wavelength mid-IR laser can reach a spot size of ~30 μm, while the ablation spot of a low-wavelength UV laser can get as small as 3 μm. This resolution difference is especially relevant in case these techniques are used for MSI.

Using LDIDD, Lee *et al.* were able to profile the levels of different phosphatidyl choline lipids during apoptosis of single HEK (Human Embryonic Kidney) 293T cells.[66] They identified that several PC lipids with shorter chains (32 : 0, 33 : 1, 34 : 1) were significantly more abundant in an apoptotic cell compared to a healthy one, while PC lipids with longer chains (36 : 1) became less abundant. The authors were also able to perform LDIDD–MSI on mouse brain tissue slices, producing high resolution (with a maximal spatial resolution of 3 um) images of lipids such as sulfatide C24 : 1, PS (34 : 1), PS (36 : 1), and PI (38 : 4).

Lastly, LDIDD is capable of detecting molecules in aqueous solutions, which Lee *et al.* used to detect secreted molecules from live cells. From the medium of a bulk culture of PC12 neuronal cell compounds such as phenethylamine, tyramine, and dopamine could be detected. Scaling down this approach to allow for detection of secreted molecules from single live cells would be an interesting added dimension for single-cell metabolomics. Indeed, the Zare lab has expressed an interest in further development of this approach, with the goal of measuring secreted molecules from a single live neuronal cell.

11.4 Data Considerations

The data generated from typical single-cell metabolic studies typically involve thousands of signals; drawing any biological interpretation from these complex datasets is wholly dependent on the ability to accurately identify metabolites. To this end, several publicly available databases can be used to putatively identify metabolites, such as the human metabolome database[67] (HMDB), METLIN,[68] lipid maps,[69] and the Kyoto encyclopedia of gene and genomes[70] (KEGG). It is also important to state the degree of confidence associated with any metabolic identification, a facet that is often overlooked in metabolic studies. There are typically four levels of metabolite identification that were defined by the metabolomics initiative:[71] positively identified compounds, putatively annotated compounds, putatively characterized compound classes, and unknowns. To positively identify a compound of interest, a minimum of two orthogonal and independent data should be used, as well as a reference standard of the compound of interest. Typically, mass to charge ratio, isotopic pattern, ion mobility profile, and fragmentation pattern are used in single-cell metabolic studies. If no reference compound was used, and the analyte was identified by spectral libraries, then the compound is said to be putatively annotated, which should affect any biological or clinical interpretations based on this compound.

11.5 Discussion

After establishing the need for single-cell metabolomics to answer biological/clinical questions, the next question that comes to mind is: Which analytical method is the best? No analytical method is perfect, this is doubly-so in the case of nascent fields such as single-cell metabolomics. Therefore, the answer to this question often depends on the experimental design itself. The first question that should be answered is: How big are the cells in question? If the experiment is designed around typical mammalian cells (few microns in diameter) then perhaps capillary micro-sampling techniques, such as the single probe, or LSC–MS are best suited for the task. The resolution in micro-sampling techniques is only limited by how small the diameter of the sampling capillary is, which ranges from tens of microns, to few hundred nanometers, allowing for sub-cellular analysis as well. If the cells in question are larger in size, then ablation-based methods, or nano-DESI are better equipped for the task at hand, since the capillaries used in micro sampling methods are liable to be plugged by large cells.

Furthermore, micro-sampling techniques often involved micromanipulation to sample the cells, a labor and skill intensive, not to mention, operator dependent step, which can adversely affect reproducibility and robustness. Throughput is also adversely affected by the lack of automation and is very operator sensitive, for example, the number of cells sampled using LSC–MS can vary between 5 and 30 cells per hour depending on the experimental setup and the skill of the operator in micromanipulation. Another thing that should be kept in mind is the sample itself, for cells embedded in tissues, the single-probe, nano-DESI, and other ablation techniques might be better than LSC–MS for example, where the very fragile capillary tips used in traditional LSC–MS are prone to breakage when sampling from tissue.[72] On the other hand, LSC–MS involves a less complicated setup, and separated sampling and MS analysis steps, which allows for the sampling to be done offsite. This improves throughput and makes it easier to perform interdisciplinary collaborations, where sampling can be done in one lab, frozen, and then shipped off to be analyzed by another laboratory. Finally, if more sensitivity is needed, and there is less emphasis on non-invasive sampling and sample preparation, Nano-DESI, and other imaging-based MS approaches might be better suited for the task.[4,73,74] In conclusion, despite the extra steps involved in single-cell metabolomic studies, it is important to remember that the data generated from said studies cannot be obtained from regular, population-based assays, and that, in combination with genomics, transcriptomics, and proteomics, this uncovered dimension will lead to new discoveries and insights in the fields of medicine, biology, and drug discovery.

References

1. M. Fessenden, Metabolomics: Small molecules, single cells, *Nature*, 2016, **540**, 153–155.
2. H. Br, *et al.*, The metabolome as a link in the genotype-phenotype map for peroxide resistance in the fruit fly, Drosophila melanogaster, *BMC Genomics*, 2020, **21**(1), 341.
3. P. P. Handakumbura, B. Stanfill, A. Rivas-Ubach, D. Fortin, J. P. Vogel and C. Jansson, Metabotyping as a Stopover in Genome-to-Phenome Mapping, *Sci. Rep.*, 2019, **9**(1), DOI: 10.1038/s41598-019-38483-0.
4. A. Ali, *et al.*, Single-cell metabolomics by mass spectrometry: Advances, challenges, and future applications, *TrAC, Trends Anal. Chem.*, 2019, **120**, 115436, DOI: 10.1016/j.trac.2019.02.033.
5. K. D. Duncan, J. Fyrestam and I. Lanekoff, Advances in mass spectrometry based single-cell metabolomics, *Analyst*, 2019, **144**(3), 782–793.
6. R. Zenobi, Single-Cell Metabolomics: Analytical and Biological Perspectives, *Science*, 2013, **342**(6163), 1243259.
7. P.-Y. Tung, *et al.*, Batch effects and the effective design of single-cell gene expression studies, *Sci. Rep.*, 2017, **7**(1), 39921.
8. A. Reichard and K. Asosingh, Best Practices for Preparing a Single Cell Suspension from Solid Tissues for Flow Cytometry, *Cytometry*, 2019, **95**(2), 219–226.
9. M. V. Rapallino and A. Cupello, Holger Hydén's technique of preparation of single Deiters' neurons and study of permeability characteristics of their plasma membranes, *Brain Res. Protoc.*, 2001, **8**(1), 58–67.
10. J. Krismer, J. Sobek, R. F. Steinhoff, R. Brönnimann, M. Pabst and R. A. Zenobi, Methodology for Studying Metabolic Heterogeneity of Single Cells in a Population, *Methods Mol. Biol.*, 2020, **2064**, 113–124.
11. Q. Teng, W. Huang, T. W. Collette, D. R. Ekman and C. Tan, A direct cell quenching method for cell-culture based metabolomics, *Metabolomics*, 2009, **5**(2), 199–208.
12. M. J. Sáez and R. Lagunas, Determination of intermediary metabolites in yeast. Critical examination of the effect of sampling conditions and recommendations for obtaining true levels, *Mol. Cell. Biochem.*, 1976, **13**(2), 73–78.
13. T. Luo, L. Fan, R. Zhu and D. Sun, Microfluidic Single-Cell Manipulation and Analysis: Methods and Applications, *Micromachines*, 2019, **10**(2), 104.
14. Q. Zhang, *et al.*, Development of a facile droplet-based single-cell isolation platform for cultivation and genomic analysis in microorganisms, *Sci. Rep.*, 2017, **7**(1), 41192.
15. E. Z. Macosko, *et al.*, Highly Parallel Genome-wide Expression Profiling of Individual Cells Using Nanoliter Droplets, *Cell*, 2015, **161**(5), 1202–1214.
16. A. R. Abate, T. Hung, P. Mary, J. J. Agresti and D. A. Weitz, High-throughput injection with microfluidics using picoinjectors, *Proc. Natl. Acad. Sci. U. S. A.*, 2010, **107**(45), 19163–19166.
17. D. D. Carlo, L. Y. Wu and L. P. Lee, Dynamic single cell culture array, *Lab Chip*, 2006, **6**(11), 1445.

18. A. K. Shalek, *et al.*, Single-cell RNA-seq reveals dynamic paracrine control of cellular variation, *Nature*, 2014, **510**(7505), 363–369.
19. J. G. Lohr, *et al.*, Whole-exome sequencing of circulating tumor cells provides a window into metastatic prostate cancer, *Nat. Biotechnol.*, 2014, **32**(5), 479–484.
20. S. M. Prakadan, A. K. Shalek and D. A. Weitz, Scaling by shrinking: empowering single-cell 'omics' with microfluidic devices, *Nat. Rev. Genet.*, 2017, **18**(6), 345–361.
21. K. Huikko, *et al.*, Poly(dimethylsiloxane) electrospray devices fabricated with diamond-like carbon–poly(dimethylsiloxane) coated SU-8 masters, *Lab Chip*, 2003, **3**(2), 67–72.
22. A. Cossarizza, *et al.*, Guidelines for the use of flow cytometry and cell sorting in immunological studies: Guidelines for the use of flow cytometry and cell sorting in immunological studies, *Eur. J. Immunol.*, 2017, **47**(10), 1584–1797.
23. A. Gross, J. Schoendube, S. Zimmermann, M. Steeb, R. Zengerle and P. Koltay, Technologies for Single-Cell Isolation, *Int. J. Mol. Sci.*, 2015, **16**(8), 16897–16919.
24. S. C. De Rosa, J. M. Brenchley and M. Roederer, Beyond six colors: A new era in flow cytometry, *Nat. Med.*, 2003, **9**(1), 112–117.
25. C. S. Baron, *et al.*, Cell Type Purification by Single-Cell Transcriptome-Trained Sorting, *Cell*, 2019, **179**(2), 527–542.e19.
26. E. Hiyama, *et al.*, Direct Lipido-Metabolomics of Single Floating Cells for Analysis of Circulating Tumor Cells by Live Single-cell Mass Spectrometry, *Anal. Sci.*, 2015, **31**(12), 1215–1217.
27. P. Hu, W. Zhang, H. Xin and G. Deng, Single Cell Isolation and Analysis, *Front. Cell Dev. Biol.*, 2016, **4**, 116.
28. A. Binek, *et al.*, Flow cytometry has a significant impact on the cellular metabolome, *J. Proteome Res.*, 2019, **18**, 169–181.
29. B. Gold, M. Cankovic, L. V. Furtado, F. Meier and C. D. Gocke, Do circulating tumor cells, exosomes, and circulating tumor nucleic acids have clinical utility? A report of the association for molecular pathology, *J. Mol. Diagn.*, 2015, **17**(3), 209–224.
30. L. Dean, *Blood and the Cells It Contains*, National Center for Biotechnology Information (US), 2005.
31. T. Fujii, *et al.*, Direct metabolomics for plant cells by live single-cell mass spectrometry, *Nat. Protoc.*, 2015, **10**(9), 1445–1456.
32. A. Ali, *et al.*, Quantitative Live Single-cell Mass Spectrometry with Spatial Evaluation by Three-Dimensional Holographic and Tomographic Laser Microscopy, *Anal. Sci.*, 2016, **32**(2), 125–127.
33. K. Yamamoto, *et al.*, Cell-specific localization of alkaloids in Catharanthus roseus stem tissue measured with Imaging MS and Single-cell MS, *PNAS*, 2016, **113**(14), 3891–3896.
34. S. Enya, *et al.*, Dual Roles of Glutathione in Ecdysone Biosynthesis and Antioxidant Function During Larval Development in Drosophila, *Genetics*, 2017, **207**(4), 1519–1532.

35. E. Hiyama, *et al.*, Direct Lipido-Metabolomics of Single Floating Cells for Analysis of Circulating Tumor Cells by Live Single-cell Mass Spectrometry, *Anal. Sci.*, 2015, **31**(12), 1215–1217.
36. Y. Abouleila, *et al.*, Live single cell mass spectrometry reveals cancer-specific metabolic profiles of circulating tumor cells, *Cancer Sci.*, 2019, **110**(2), 697–706.
37. Y. Fukano, N. Tsuyama, H. Mizuno, S. Date, M. Takano and T. Masujima, Drug metabolite heterogeneity in cultured single cells profiled by pico-trapping direct mass spectrometry, *Nanomedicine*, 2012, **7**(9), 1365–1374.
38. A. Ali, *et al.*, Single-Cell Screening of Tamoxifen Abundance and Effect Using Mass Spectrometry and Raman-Spectroscopy, *Anal. Chem.*, 2019, **91**(4), 2710–2718.
39. L. Zhang, L. L. Allworth and A. Vertes, Identification of Metabolites in Single Cells by Ion Mobility Separation and Mass Spectrometry, in *Single Cell Metabolism: Methods and Protocols*, ed. B. Shrestha, Springer New York, New York, NY, 2020, pp. 9–18.
40. T. Fujii, *et al.*, Direct metabolomics for plant cells by live single-cell mass spectrometry, *Nat. Protoc.*, 2015, **10**(9), 1445–1456.
41. L. Zhang, *et al.*, Subcellular Peptide Localization in Single Identified Neurons by Capillary Microsampling Mass Spectrometry, *Sci. Rep.*, 2018, **8**(1), 1–10.
42. X.-C. Zhang, *et al.*, Combination of Droplet Extraction and Pico-ESI-MS Allows the Identification of Metabolites from Single Cancer Cells, *Anal. Chem.*, 2018, **90**(16), 9897–9903.
43. T. D. Do, *et al.*, Optically Guided Single Cell Mass Spectrometry of Rat Dorsal Root Ganglia to Profile Lipids, Peptides and Proteins, *ChemPhysChem*, 2018, **19**(10), 1180–1191.
44. E. K. Neumann, *et al.*, Multimodal Chemical Analysis of the Brain by High Mass Resolution Mass Spectrometry and Infrared Spectroscopic Imaging, *Anal. Chem.*, 2018, **90**(19), 11572–11580.
45. N. Pan, W. Rao, N. R. Kothapalli, R. Liu, A. W. G. Burgett and Z. Yang, The Single-Probe: A Miniaturized Multifunctional Device for Single Cell Mass Spectrometry Analysis, *Anal. Chem.*, 2014, **86**(19), 9376–9380.
46. W. Rao, N. Pan and Z. Yang, Applications of the Single-probe: Mass Spectrometry Imaging and Single Cell Analysis under Ambient Conditions, *J. Visualized Exp.*, 2016, **112**, 53911.
47. R. Liu, N. Pan, Y. Zhu and Z. Yang, T-Probe: An Integrated Microscale Device for Online In Situ Single Cell Analysis and Metabolic Profiling Using Mass Spectrometry, *Anal. Chem.*, 2018, **90**(18), 11078–11085.
48. K. Nakatani, Y. Izumi, K. Hata and T. Bamba, An Analytical System for Single-Cell Metabolomics of Typical Mammalian Cells Based on Highly Sensitive Nano-Liquid Chromatography Tandem Mass Spectrometry, *Mass Spectrom.*, 2020, **9**(1), A0080.
49. H.-W. Liao, S. S. Rubakhin, M. C. Philip and J. V. Sweedler, Enhanced single-cell metabolomics by capillary electrophoresis electrospray

ionization-mass spectrometry with field amplified sample injection, *Anal. Chim. Acta*, 2020, **1118**, 36–43.

50. G. Hempel, Strategies to improve the sensitivity in capillary electrophoresis for the analysis of drugs in biological fluids, *Electrophoresis*, 2000, **21**(4), 691–698.
51. D. M. Osbourn, D. J. Weiss and C. E. Lunte, On-line preconcentration methods for capillary electrophoresis, *Electrophoresis*, 2000, **21**(14), 2768–2779.
52. Z. Takáts, J. M. Wiseman, B. Gologan and R. G. Cooks, Mass Spectrometry Sampling under Ambient Conditions with Desorption Electrospray Ionization, *Science*, 2004, **306**(5695), 471–473.
53. F. M. Green, P. Stokes, C. Hopley, M. P. Seah, I. S. Gilmore and G. O'Connor, Developing Repeatable Measurements for Reliable Analysis of Molecules at Surfaces Using Desorption Electrospray Ionization, *Anal. Chem.*, 2009, **81**(6), 2286–2293.
54. L.-E. Lin, *et al.*, Precision biomarker discovery powered by microscopy image fusion-assisted high spatial resolution ambient ionization mass spectrometry imaging, *Anal. Chim. Acta*, 2020, **1100**, 75–87.
55. P. J. Roach, J. Laskin and A. Laskin, Nanospray desorption electrospray ionization: an ambient method for liquid-extraction surface sampling in mass spectrometry, *Analyst*, 2010, **135**(9), 2233.
56. H.-M. Bergman and I. Lanekoff, Profiling and quantifying endogenous molecules in single cells using nano-DESI MS, *Analyst*, 2017, **142**(19), 3639–3647.
57. R. Yin, K. E. Burnum-Johnson, X. Sun, S. K. Dey and J. Laskin, High spatial resolution imaging of biological tissues using nanospray desorption electrospray ionization mass spectrometry, *Nat. Protoc.*, 2019, **14**(12), 3445–3470.
58. P. Nemes and A. Vertes, Laser Ablation Electrospray Ionization for Atmospheric Pressure, in Vivo, and Imaging Mass Spectrometry, *Anal. Chem.*, 2007, **79**(21), 8098–8106.
59. B. Shrestha and A. Vertes, In Situ Metabolic Profiling of Single Cells by Laser Ablation Electrospray Ionization Mass Spectrometry, *Anal. Chem.*, 2009, **81**(20), 8265–8271.
60. M. B. Ginzberg, R. Kafri and M. Kirschner, On being the right (cell) size, *Science*, 2015, **348**(6236), 1245075.
61. S. A. Stopka, *et al.*, Laser-ablation electrospray ionization mass spectrometry with ion mobility separation reveals metabolites in the symbiotic interactions of soybean roots and rhizobia, *Plant J.*, 2017, **91**(2), 340–354.
62. B. Shrestha, *et al.*, Subcellular Metabolite and Lipid Analysis of Xenopus laevis Eggs by LAESI Mass Spectrometry, *PLoS One*, 2014, **9**(12), e115173.
63. P. Nemes, A. S. Woods and A. Vertes, Simultaneous Imaging of Small Metabolites and Lipids in Rat Brain Tissues at Atmospheric Pressure by Laser Ablation Electrospray Ionization Mass Spectrometry, *Anal. Chem.*, 2010, **82**(3), 982–988.

64. *Ion Mobility-Mass Spectrometry: Methods and Protocols*, ed. G. Paglia and G. Astarita, Springer US, New York, NY, 2020, vol. 2084.
65. J. A. Stolee and A. Vertes, Toward Single-Cell Analysis by Plume Collimation in Laser Ablation Electrospray Ionization Mass Spectrometry, *Anal. Chem.*, 2013, **85**(7), 3592–3598.
66. J. K. Lee, E. T. Jansson, H. G. Nam and R. N. Zare, High-Resolution Live-Cell Imaging and Analysis by Laser Desorption/Ionization Droplet Delivery Mass Spectrometry, *Anal. Chem.*, 2016, **88**(10), 5453–54.
67. Human Metabolome Database, https://hmdb.ca/ (accessed Feb. 28, 2020).
68. https://metlin.scripps.edu/landing_page.php?pgcontent=mainPage (Accessed Feb. 28, 2020).
69. LIPID MAPS Lipidomics Gateway, https://lipidmaps.org/ (accessed Feb. 28, 2020).
70. KEGG: Kyoto Encyclopedia of Genes and Genomes, https://www.genome.jp/kegg/ (accessed Feb. 28, 2020).
71. R. Goodacre, *et al.*, Proposed minimum reporting standards for data analysis in metabolomics, *Metabolomics*, 2007, **3**(3), 231–241.
72. W. Rao, N. Pan and Z. Yang, High Resolution Tissue Imaging Using the Single-probe Mass Spectrometry under Ambient Conditions, *J. Am. Soc. Mass Spectrom.*, 2015, **26**(6), 986–993.
73. M. E. Dueñas, J. J. Essner and Y. J. Lee, 3D MALDI Mass Spectrometry Imaging of a Single Cell: Spatial Mapping of Lipids in the Embryonic Development of Zebrafish, *Sci. Rep.*, 2017, **7**(1), 14946.
74. M. K. Passarelli, *et al.*, The 3D OrbiSIMS—label-free metabolic imaging with subcellular lateral resolution and high mass-resolving power, *Nat. Methods*, 2017, **14**(12), 1175–1183.

CHAPTER 12

Chiral Metabolomics

CARLOS CALDERÓN[a,b] AND MICHAEL LÄMMERHOFER*[a]

[a] Institute of Pharmaceutical Sciences, Pharmaceutical (Bio-)Analysis, University of Tübingen, Auf der Morgenstelle 8, 72076 Tübingen, Germany; [b] Escuela de Química, Universidad de Costa Rica, San José 11501-2060, Costa Rica
*Email: michael.laemmerhofer@uni-tuebingen.de

12.1 Introduction

Metabolomics aims at the comprehensive analysis of metabolites in a biological system. Nowadays, advanced hyphenated separation techniques such as LC, GC, SFC, and CE with MS for detection enable the routine analysis of hundreds to thousands of compounds in a single run.[1,2] Still, a comprehensive coverage of the metabolome is challenging and illusive because of the high structural diversity and wide range of physicochemical properties of metabolites as well as their concentrations.[3] Ultimately, it must comprise structural isomers (constitutional isomers and stereoisomers). Unfortunately, the differentiation of isomers is mass spectrometry's Achilles heel, and this selectivity must be provided by the prior separation step. In general, common achiral separation technologies like RP and HILIC can resolve constitutional isomers and also diastereomers. Yet, oftentimes the selectivity may still be critical, as for instance typically observed for the various sugar phosphates (*i.e.* hexose and pentose phosphate constitutional isomers). On the other hand, enantiomers cannot be separated at all on achiral separation systems nor can MS distinguish them; therefore, they need specific methods. This is the focus of the current chapter – to outline the possibilities of

New Developments in Mass Spectrometry No. 10
Advanced Mass Spectrometry-based Analytical Separation Techniques for Probing the Polar Metabolome
Edited by Rawi Ramautar

Published by the Royal Society of Chemistry, www.rsc.org

implementing separation technologies in metabolomics approaches in order to allow enantioselective analysis of metabolites.

Within the structural diversity that metabolites can possess, chirality is a very critical aspect to be considered. For a long time the predominant occurrence of a particular enantiomer in some families of compounds (*e.g.* D-sugars, L-amino acids, *etc.*) has been well known, and it was believed that only these enantiomers have a biological function. However, since the 1980s the occurrence of uncommon enantiomers in organisms has been increasingly described, which has occurred along with the development of techniques for chiral analysis. Furthermore, some of these uncommon enantiomers have been linked to critical metabolic functions and thus have been analyzed as biomarkers of disease.

Considering the comprehensive description which metabolomics tries to achieve, it is therefore mandatory as far as possible to determine correctly the stereochemistry of metabolites which are present in a certain system and from that point to establish unequivocal correlation with their biological role within an organism.

Chiral analysis of metabolites can be achieved by two main types of approaches (Figure 12.1): The direct method employs a chiral stationary phase CSPs (which are obtained by immobilization of an enantiomerically pure chiral selector, SO, on a chromatographic support) to separate the enantiomers (selectants, SA) based on formation of transient diastereomeric associates on the surface of the CSP while the analyte enantiomers travel through the column. The separation originates from different complex stability constants for *R*- and *S*-enantiomers of the analytes with the chiral selector. This is the preferred approach in LC, GC, and SFC enantiomer separation. The indirect method avoids the need for a chiral stationary phase by derivatizing the enantiomers with an enantiomerically pure chiral derivatizing agent (CDA) producing diastereomers which can be analyzed by achiral methodologies (Figure 12.1). Besides, direct enantiomer separation can be also accomplished by adding the chiral selector to the mobile phase in combination with an achiral separation system such as RPLC (chiral mobile phase additive, CMPA). While this approach has no relevance for LC, it is the method of choice in CE enantiomer separation. Its main drawback with regard to metabolomics is the presence of the CMPA in the effluent making it essentially incompatible with ESI–MS unless specific techniques such as countercurrent or partial filling techniques are used.[4–6]

In this chapter, the specific utility and peculiarities of the above two approaches for chiral metabolomics with primary focus on LC will be discussed.

12.2 Direct Chiral Metabolite Analysis by HPLC

The direct analysis of enantiomers is based on the reversible transient formation of diastereomeric complexes between the SAs (enantiomers to be analyzed) and the chiral SO. Because of the chirality of the selector, different molecular interactions can be established with each enantiomer, and therefore the stability of their resultant complexes can be different. Under adequate

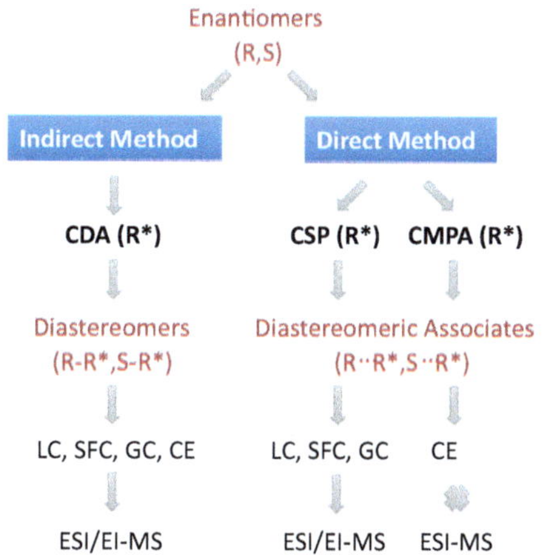

Figure 12.1 Approaches for enantiomeric separations. CDA, chiral derivatizing agent (enantiomerically pure; here indicated by *R**); CSP, chiral stationary phase; CMPA, chiral mobile phase additive; * indicates the absolute configuration of the chiral selector which is arbitrarily shown here in *R* configuration (it can be also *S* or have multiple stereogenic centers in single enantiomeric form). If *R**-CDA contains *S**-CDA as an enantiomeric impurity also *R*–*S** and *S*–*S** stereoisomers are formed, which are enantiomers to *S*–*R** and *R*–*R** stereoisomers having the consequence that they comigrate with them in achiral chromatographic methods. In the CDA approach, stationary and mobile phases are achiral. In the CSP approach, the stationary phase is chiral while the mobile phase is achiral. In the CMPA approach, the mobile phase (background electrolyte in CE) is chiral (through the achiral additive) while, if it is applied to LC, the stationary phase is achiral.

conditions, this difference in energy enables the separation of the enantiomers. Complexes between SOs and SAs are produced by different kinds of interactions such as van der Waals, π–π interaction, ionic interaction, dipole–dipole, or ion–dipole amongst others. Considering the structural diversity that a metabolite can possess, it is not surprising that different selectors and conditions have shown good results for enantioseparation of specific groups of metabolites but not for others. It means that in most of the cases enantiomeric analyses are focused only on a specific type of metabolite. In other words, there is no unique CSP and chiral column, respectively, which could suffice the proclamation of global enantioselective metabolomics. Which means a set of chiral columns may be required for a broad coverage in enantioselective metabolomics.

12.2.1 Chiral Stationary Phases for Direct Analysis

A series of chiral stationary phases have become available for direct analysis of enantiomers during the last 40 years. These stationary phases are obtained by immobilization of chiral selectors on chromatographic supports by adsorption, coating, or covalent linkages. The ones obtained by coating do

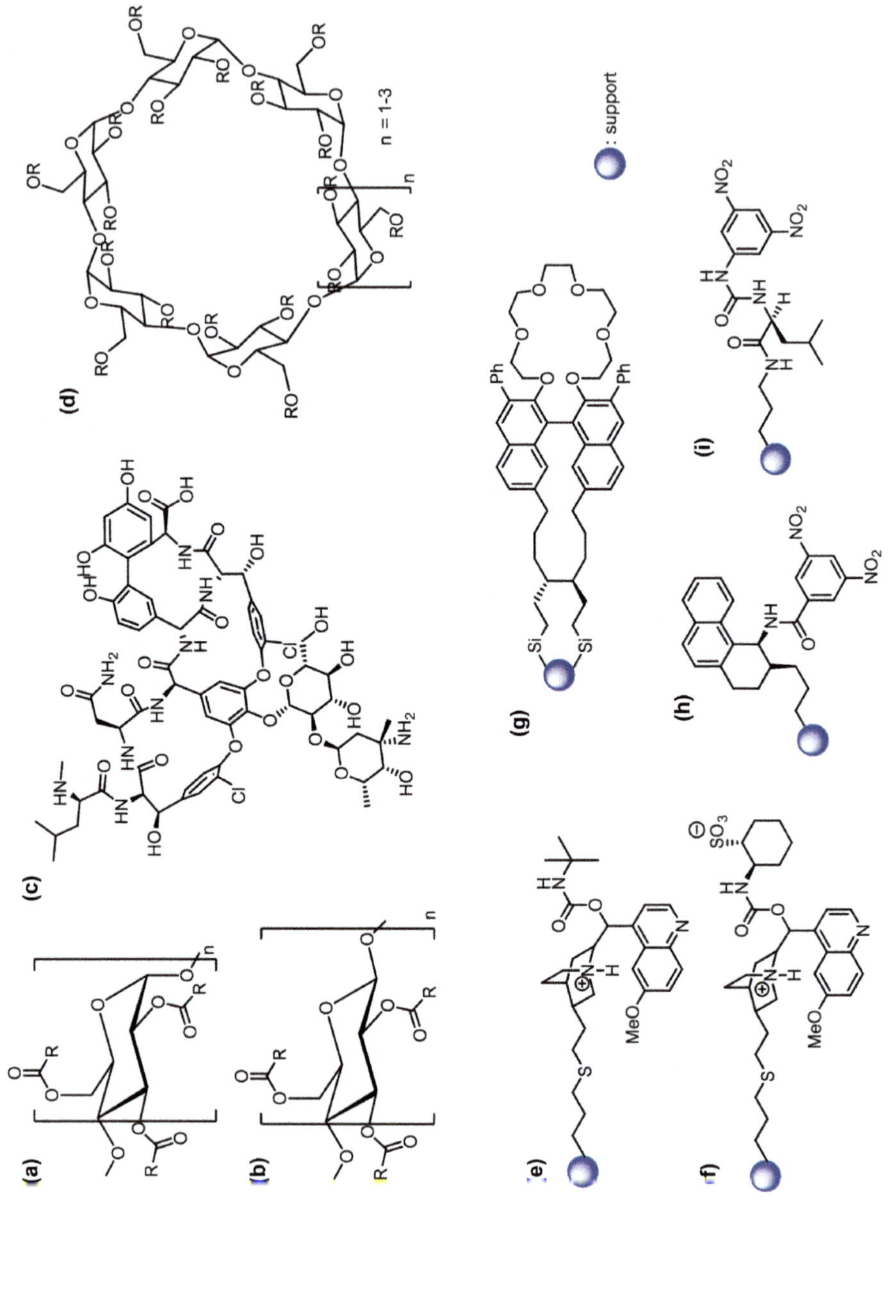
(a)
(b)
(c)
(d)
n = 1-3
(e)
(f)
(g)
(h)
(i)
: support

not tolerate all kinds of mobile phases, and users have to carefully read the column instruction manuals from suppliers. In general, chiral columns that can be used under MS-compatible elution conditions seem to be preferred for metabolomics studies unless other detectors such as fluorescence detection (Fl) are used. Most of the predominantly employed chiral columns are actually multimodal and can be employed in several elution modes such as RP, HILIC, polar organic mode, NPLC, and SFC.

Diverse chiral stationary phases have been developed for direct analysis of enantiomers. The most commonly utilized CSPs in enantioselective LC can be summarized into three groups: i) Coated and immobilized polysaccharide-based CSPs with cellulose and amylose esters and carbamates bearing distinct aryl substituents, ii) glycopeptide antibiotics (vancomycin, teicoplanin, teicoplanin aglycone and ristocetin) as well as other macrocyclic selectors (cyclofructans, cyclodextrins, chiral crown ethers), and iii) low molecular synthetic chiral selectors (Pirkle type, donor–acceptor CSPs) and cinchonan carbamate-based chiral ion-exchange selectors (anion-exchange type and zwitterionic quinine/quinidine carbamates) (Figure 12.2).[7] A few examples of the applicability of these CSPs for chiral metabolomics will be described briefly.

12.2.1.1 Polysaccharide-based CSPs

These CSPs constitute the most important chiral selector class employed nowadays (tradenames of CSPs Chiralpak, Chiralcel, Lux, *etc.*) as they have the broadest scope of applicability. Their broad application spectra are the result of the deep chiral cavities formed by the pendant arylester/arylcarbamate groups at the (slightly) helically wound glucopyranose polymer backbone. The helical structure creates a dense and regular arrangement of grooves along the cellulose backbone. The grooves act as enantioselective binding sites, offering hydrogen bonding environments at the interior, as well as sterically discriminative and π-stacking interactions at the exterior regions. The highly dynamic character of the pendant side chains allows flexible adjustment to different chiral compound structures. Chiral analytes can be stereoselectively inserted into the grooves and dynamically interact with the chiral selector moieties leading to stereodiscrimination. These CSPs are available as coated and immobilized versions. The coated polysaccharide CSPs can only be used

Figure 12.2 Structures of commonly employed CSPs. Polysaccharide-based CSPs: (a) amylose carbamates (R = –NH–Ar) (Chiralpak AD and IA as well as Lux Amylose-1, Ar = 3,5-dimethylphenyl) and (b) cellulose esters (R = Ar) (Chiralcel OJ, Ar = 4-methylbenzoate) and carbamates (R = –NH–Ar) (Chiralcel OD and Chiralpak IB as well as Lux Cellulose-1, Ar = 3,5-dimethylphenyl), (c) Macrocyclic glycopeptide (vancomycin, Chirobiotic V), (d) cyclodextrin-based (Cyclobond), (e) chiral anion-exchangers (Chiralpak QN-AX and QD-AX), (f) zwitterionic ion-exchangers (Chiralpak ZWIX(+) and ZWIX(−)), (g) crown ether bonded CSPs (Crownpak CR-I(−) and CR-I(+)) and (h and i) donor–acceptor CSPs with Whelk O1 In (h) and KSAACSP-001 in (i).

with a restricted solvent range, and often separate columns for RPLC and NPLC modes are recommended. This means that these columns should be dedicated to one mode and not switched back and forth between RP and NP modes. The immobilized versions (often indicated by the I or i in the tradename) are preferred as they can be employed with an extended solvent range, are more stable (against injection solvents), and can be switched between modes. For HPLC, 5 µm or 3 µm fully porous particle-based polysaccharide CSPs have been most popular, however, also sub-2 µm particle-based polysaccharide CSPs (Chiralpak IA-U, IB-U, IC-U, and IG-U) are now available for fast UHPLC enantiomer separations (for selectors of IA and IB see Figure 12.2a–b; of IC, cellulose tris(3,5-dichlorophenylcarbamate), and of IG, amylose tris (3-chloro-5-methylphenylcarbamate)). In order not to lose performance, they should be operated with dedicated UHPLC equipment. The distinct polysaccharide backbones (amylose *vs.* cellulose), the different functional groups (ester *vs.* carbamate), and various substituents at the phenyl group (dimethyl, dichloro, or mixed chloro/methyl) are basis for altered enantioselectivity profiles through changes in 3-D structure, hydrogen donor/acceptor strengths, and electron densities in the aromatic moieties. These structural modifications generate CSPs with complementary enantioselectivity profiles.

For metabolomics the RP elution mode is of first choice. The convenience is that the polysaccharide columns can be used with the eluent system commonly selected for achiral RPLC, *e.g.* formic acid, acetic acid, ammonium formate and acetate, as well as ammonium bicarbonate along with isocratic or gradient elution using acetonitrile or methanol as an organic modifier. Hence, elution order follows the lipophilicity trend, and the system works under common RP paths which makes it straightforward to work with. If retention is too strong or too weak, the polar organic or HILIC mode can be tested. If this remains unsuccessful, NPLC or SFC are other options, with some inconvenience if it comes to MS detection. NPLC needs a make-up flow admixed post-column before ESI detection to get a stable spray, or other ion sources such as APCI or APPI. SFC–MS may also need some modifications, *i.e.* make-up flow, to alleviate problems from precipitation upon CO_2 evaporation, and freezing of transfer line.

From a practical viewpoint, the perfect compatibility of RPLC mode with ESI–MS further contributes to the popularity of these columns in LC–MS based bioanalysis. Due to the polymeric nature and immobilization, there is no background noise from selector bleeding which is favorable for the S/N and sensitivity of such LC–MS separations.

In metabolomics these CSPs have been employed particularly for the chiral analysis of oxylipins. Multiple examples can be found in Table 12.1 and 12.2.

12.2.1.2 *Macrocyclic Glycopeptide, Cyclodextrin and Cyclofructan-based CSPs*

Macrocyclic glycopeptide columns are based on macrocyclic antibiotics such as vancomycin (V) (Figure 12.2c), teicoplanin (T), teicoplanin aglycone (TAG), and

Table 12.1 Direct chiral metabolomics by 1D HPLC.

Metabolites[c]	Sample	Reagent	CSP type	Column	Detection	Ref.
Proteinogenic amino acids, allo-Thr, homoserine, allo-Ile and homocysteine		Underivatized and AQC	ZWIX	Chiralpak ZWIX(+) and Chiralpak ZWIX(−)	SIM	23
Proteinogenic amino acids	Black vinegar		CCE	Crownpak CR-I(+)	SIM	14
D-Ser	Mouse brain		CCE	Crownpak CR(+)	SRM	13
Proteinogenic amino acids, DL-allo-Ile, DL-Ile			CCE	Crownpak CR-I(+)	MRM	61
84 chiral meatabolites including proteinogenic amino acids	Cheese		CCE, ZWIX	Crownpak CR-I(+)/CR-I(−) and Chiralpak ZWIX (−)	MRM	12
Amino acids and amines	Tobacco cells	AQC	MGP	Chirobiotic T	MRM	9
Proteinogenic amino acids		AQC, AC[a] and APC[a]	ZWIX	Chiralpak ZWIX(+)	SIM	22
D-Ala, D-Asp, D-Ser	Rat plasma and tissues	AQC, TAHS	ZWIX	Chiralpak ZWIX(+)	SRM	24
Proteinogenic amino acids	Peptide hydrolysates	NBD-F	MGP	Chirobiotic TAG	MRM	8
D-Ser, D-Ala	Mouse lumbar spinal cords	NBD-F	Pirkle	Sumichiral OA-2500S	Fl	18
Proteinogenic amino acids	Human urine	SFP	AX	Chiralpak QN-AX and QD-AX	SRM	20
HpOME(*E*)s, HpOME(*Z*)s and HOMEs			Pirkle and PS	Reprosil Chiral NR and NR-R, Chiralcel OB-H	MRM and UV 210 nm	68
HpOME(E)s, HpOME(Z)s			Pirkle	Reprosil Chiral NR	MS3	69
12S-HETE and 15S-HETE	Mice skin		PS	Chiralpak-IA	MRM and UV	70
18-HEPE, 15-HEPE, 12-HEPE and 5-HEPE, RvE1 and RvE2	Human serum		PS	Chiralpak AD- RH	UV and MRM	71
5,6-EET, 8,9-EET, 11,12-EET and 14,15-EET	Rat liver		PS	Chiralcel OJ	SIM	72
5-, 12- and 15-HETE	Mice colon mucosa, blood serum		PS	Chiralpak AD-RH	SRM	73
15-, 8-, 12- and 5-F2-iPs	Urine from control, MCI and severe AD patients		PS	Chiralpak AD-H	MRM	74

Table 12.1 (*Continued*)

Metabolites[c]	Sample	Reagent	CSP type	Column	Detection	Ref.
Lipoxin A-4	Human macrophages and MCF-7 breast carcinoma cells		PS	Lux Amylose-2[b]	MRM	75
14-HDoHE, 13,14-diHDoHE, MaR1 and MaR2	Human peripheral blood mononuclear cells		PS	Chiralpak AD-RH	MRM	76
5-, 9-, 11-, 12-, 13-, 15-HETE, 14-HDoHE, 17-HDoHE	Human lung tissue and guinea pig trachea		PS	Chiralpak AD-RH	MRM	77
8R-HEPE, 8R-HETE, 10R-HDoHE	Euphausia pacific krill		PS	Chiralpak ID[b]	MRM	78
9-HODE, 13-HODE and 12-HETE	Psoriatic skin scales		PS	Chiralpak AD	DAD, MS2 for identification	79
ATL, 6-epi-LXA4, RvD1, ATR, PDX and NPD1, LTB4, 12-epi-LTB4, 5(*S*)-HETE and 12(*S*)-HETE, LXA4, LXA5, LXB4, RvD2, MaR1, dinor-NPD1, tetranor-NPD1, PGE2, PGD2, PGJ2, TXB2, 11-dh-TXB2, 6-keto-PGF1a, PGF2a, 10(*S*),17(*S*)-DiHDPAn-3, 10(*S*)17(*R*)-DiHDPAn-6, 10(*S*),17(*R*)-DiHAdA, 15(*R*)-HETE, 15(*S*)-HETE, 20-HETE, 17(*R*)-HDoHE, 17(*S*)-HDoHE, (±) 18-HEPE and 14(*S*)-HDoHE	Human plasma		PS	Lux Amylose-1	MRM	80
HODEs, HETEs, HEPEs, HDoHEs, resolvins D1, HETrE, HOTrE and DiHETE	Platelet releasates		PS	Chiralpak IA-U	SRM	81
(9S,10R)- and (9R,10S)-epoxystearic acids	Malignant human colorectal cancer		PS	Chiralpak AD	SIM	82
Hydroperoxy fatty acids: 5-, 8-, 9-, 11-, 12-, 14-, 15-HPETE and 8-, 9-, 10-, 11-, 12-, 13-HPODE			PS and Pirkle	Chiralcel OB-H, Reprosil Chiral-NR, Chiralcel OC[b]	UV and MS3	83
3-hydroxyfatty acids (from C4 to C16)	Bacterial cultures	3,5-DMPI[a]	PS	Lux Cellulose-1	MRM	84
9-HODE, 13-HODE, 11-HETE, and 15-HETE	Human atherosclerotic plaque	Diazomethane	PS	Chiralcel OB[b]	UV	85

5,6-EET, 8,9-EET, 11,12-EET and 14,15-EET	Mouse epithelial hepatoma (Hepa) cell line	PFB	PS	Chiralpak AD-H	MRM	86
5,6-EET, 8,9-EET, 11,12-EET and 14,15-EET		PFB, diazomethane	PS	Chiralcel OB, Chiralcel OD	UV	87
5-, 9-, 10-, 12-, 13-PAHSA (palmitic acid ester of hydroxystearic acid)	Mice breast milk		PS	CHIRAL ART Amylose-SA	MS3	88
Glyceric acid	Human urine		MGP	Chirobiotic R	MRM	89
2-hydroxyglutarate	CD8 + T-lymphocytes		MGP	Chirobiotic R	MRM	90
Lactate and 3-hydroxybutyrate	Mice serum and urine	NBD-PZ[a]	PS	Chiralpak AD-RH and Chiralcel OD-RH	UV/Vis	91

[a]AC, aminophenyl-*N*-hydroxysuccinimidyl carbamate; APC, 3-aminopyridyl-*N*-hydroxysuccinimidyl carbamate; 3,5-DMPI, 3,5-dimethylphenyl isocyanate; NBD-PZ, 4-nitro-7-piperazino-2,1,3-benzoxadiazole.

[b]Lux Amylose-2, amylose tris (3-chloro-6-methylphenylcarbamate); Chiralpak ID, amylose-tris(3-chlorophenylcarbamate); Chiralcel OC, cellulose-tris (phenylcarbamate); Chiralcel OB, cellulose-tribenzoate.

[c]ATL, 15-epi-Lipoxin A4; ATR, 17-epi- resolvin D1; EET, epoxy-eicosatrienoic acid; HAdA, hydroxy-docosatetraenoic acid; HDoHE, hydroxy-docosahexaenoic acid; HDPA, hydroxy-docosapentaenoic acid; HEPE, hydroxy-eicosapentaenoic acid; HETE, hydroxy-eicosatetraenoic acid; HETrE, hydroxy-eicosatrienoic acid; HODE, hydroxy-octadecadienoic acid; HOME, hydroxy-octadecenoic acid; HOTrE, hydroxy-octadecatrienoic acid; HPETE, hydroperoxy-eicosatetraenoic acid; HPODE, hydroperoxy-octadecadienoic; HpOME, hydroperoxy-octadecenoic acids; iP, isoprostane; LT, leukotriene; LX, lipoxin; MaR, maresin; NPD1, neuroprotectin D1; PDX, protectin DX; PG, prostaglandin; Rv, resolvin.

Table 12.2 Direct chiral metabolomics by 2D HPLC (^{1}D: RP).

Metabolites	Sample	Reagent	CSP type	Column	Detection	Ref.
Proteinogenic amino acids		6-MOQ-EtOCOOSu[a]	AX, ZWIX and Pirkle	^{2}D: Chiralpak QN–AX, Chiralpak ZWIX(+), and KSAACSP-001S (*N*-3,5-dinitrophenylaminocarbonyl-L-leucine bonded aminopropylsilica)	SRM	49
Proteinogenic amino acids, Ile, allo-Ile, allo-Thr	Blood serum	AQC	AX and ZWIX	^{1}D and ^{2}D: Chiralpak QN-AX + Chiralpak ZWIX (+) or Chiralpak QD-AX + Chiralpak ZWIX(−)	MRM	21
D-Asp	Thymus and spleen of mice	NBD-F	AX	^{2}D: Chiralpak QD-AX	Fl	19
Val, allo-Ile, Ile, and Leu	Cerebrum, cerebellum, liver, kidney, plasma and urine of rats	NBD-F	AX	^{2}D: Chiralpak QN-AX and Chiralpak QD-AX	Fl	42
Pro and OH-Pro	Serum and collagen-rich skin tissue	NBD-F	AX and Pirkle	^{2}D: QN-2-AX, QD-2-AX, Sumichiral OA-2500S and OA-2500R	Fl	92
Ala, Val, 2-aminobutyric acid, Nva, *N*-MeAla, isoVal, 3-aminoisobutyric acid	Meteorite	NBD-F	Pirkle	^{2}D: Sumichiral OA-4600S[b] and KSAACSP-001S	Fl	93
Proteinogenic amino acids	Plasma from chronic kidney disease patients	NBD-F	Pirkle	^{2}D: Sumichiral OA-3200[b]	Fl and MRM	50
D-Ala	Blood of Wistar rat	NBD-F	Pirkle	^{2}D: Sumichiral OA-2500S	Fl	16
His, Ser, Asn, Gln, Arg, Asp, allo-Thr, Glu, Thr	Rat cerebrum, plasma, urine	NBD-F	Pirkle	^{2}D: narrowbore-Sumichiral OA-2500S	Fl	17
D-Leu and D-Pro	Brains of mice and rats	NBD-F	Pirkle	^{2}D: Sumichiral OA-2500S or Sumichiral OA-2500R	Fl	94
D-Ser and D-Ala	Brain tissues, peripheral tissues, serum and urine of mice	NBD-F	Pirkle	^{2}D: Sumichiral OA-2500S	Fl	43

D-Ser and D-Ala	Brain, periphery, plasma and urine of mutant rats	NBD-F	Pirkle	^{2}D: Sumichiral OA-2500S	Fl	44
D-Asp	Pineal glands of rodents	NBD-F	Pirkle	^{2}D: Sumichiral OA-2500S	Fl	95
D-Ser	Brain tissues, peripheral tissues, CSF, blood serum and urine of mice	NBD-F	Pirkle	^{2}D: Sumichiral OA-2500S	Fl	96
D-Asp and D-Glu	Brain tissues, peripheral tissues, plasma and urine of mice	NBD-F	Pirkle	^{2}D: Sumichiral OA-2500S	Fl	97
Citrulline and ornithine	Urine of mice	NBD-F	Pirkle	^{2}D: KSAACSP–105S[b]	Fl and SIM	32
Ala, Asp, Glu, Leu, Pro and Ser	Human plasma and urine	NBD-F	Pirkle	^{2}D: KSAACSP-001S	Fl and MRM	31
Proteinogenic amino acids, Ile, allo-Ile, allo-Thr	Mice with acute kidney injury	NBD-F	Pirkle	^{2}D: KSAACSP-001S	Fl	45
Asn, Ser, Ala, and Pro	Plasma of patients with chronic kidney disease	NBD-F	Pirkle	^{2}D: anion exchange; ^{3}D: KSAACSP-001S	Fl	30
D-Ser	Blood serum and urine of mice	NBD-F	Pirkle	^{2}D: KSAACSP-001S	Fl	98
D-Ser	Blood serum and urine of CKD patients	NBD-F	Pirkle	^{2}D: KSAACSP-001S or Sumichiral OA-3200	Fl	99
Ala, 2-aminobutyric acid (2AB), Val, nVal and iVal	Murchison meteorite and the Antarctic meteorites	NBD-F	Pirkle	^{2}D: anion exchange; ^{3}D: Singularity CSP-001S	Fl	29
Proteinogenic amino acids, allo-Ile	Black vinegar (Kurozu)	NBD-F	Pirkle and AX	^{2}D: Sumichiral OA-2500S, and Chiralpak QN-AX	Fl	67
D-Asp and D-Glu	Rats (brain and peripheral tissues, plasma, urine)	NBD-F	Pirkle and AX	^{2}D: Sumichiral OA-2500S, Chiralpak QN-AX, Chiralpak QD-AX	Fl	54
N-Me-D-Asp and its analogs	Wistar rats	NBD-F	Pirkle and AX	^{2}D: Sumichiral OA-2500S, Sumichiral OA-2500R and QN-2-AX	Fl	46
	Lung and kidney tissue		PS	^{2}D: Chiralcel OD-H	MRM	51

Table 12.2 (*Continued*)

Metabolites	Sample	Reagent	CSP type	Column	Detection	Ref.
5,6-EET, 8,9-EET, 11,12-EET and 14,15-EET						
EETs, HETEs, EEQs, HEPEs, EDPs and HDoHEs	Plasma and liver samples from mice, human plasma		PS	^{2}D: Lux-Cellulose-3[b], Lux- Amylose-1	MRM	35
PGE 2 and entPGE 2	Mice brain		PS	^{2}D: Lux Amylose-2	MRM	52
17-HDoHE	Cell culture supernatants		PS	^{2}D: Lux Amylose-2	MRM	53
Lactate and 3-hydroxybutyrate	Human plasma, urine, and saliva	NBD-COCl	Pirkle and AX	^{2}D: Chiralpak QD-AX and KSAACSP- 001S	Fl	47
Lactate and 3-hydroxybutyrate	Human plasma and urine	NBD-PZ	PS	^{2}D: Chiralpak AD-H	Fl	48

[a]6-MOQ-EtOCOOSu, 2,5-dioxopyrrolidin-1-yl(2-(6-methoxy-4-oxoquinolin-1(4*H*)-yl)ethyl) carbonate.
[b]Sumichiral OA-4600S, *N*-(*S*)-1-(α-naphthyl)aminocarbonyl-(*S*)-tert-leucylamidopropyl bonded silica (urea type); Sumichiral OA-3200, *N*-3,5-dinitroanilinocarbonyl-(*S*)-tert-leucylamidopropyl bonded silica (urea type); KSAACSP–105*S*, *N*-(3,5-dinitrobenzoyl)-3-(2-naphthyl)-L-alanylamidopropyl bonded silica; Lux-Cellulose-3, cellulose tris (4-methylbenzoate).

ristocetin (R) which are available immobilized on 5 μm silica gel (Chirobiotics columns). Lately, they became available as superficially porous particle columns (Poroshell 120 Chiral-V and -T) that exhibit performance comparable to sub-2 μm columns and afford fast enantiomer separations (down to the sub-minute time scale). They also have relatively broad applicability spectra and are often first choice when polysaccharide CSPs fail, in particular for very polar analytes like underivatized amino acids (T, TAG), peptides (T), and hydroxy acids (R). The macrocyclic structure is formed from a peptide backbone with bridged aromatic side chains forming a shallow basket for enantioselective inclusion of chiral analytes driven by polar interactions such as hydrogen bonding or dipole–dipole interactions. These columns have shown multimodal applicability (RP, HILIC, polar organic, NP, SFC) but are most promising when utilized in the polar organic mode (non-aqueous with polar organic solvents like MeOH and ACN or mixtures thereof), HILIC or RP elution mode which are very compatible with ESI–MS. These CSPs have been employed in metabolomics, for example for the analysis of amino acids in peptide hydrolysates[8] and tobacco cells[9] (see Table 12.1).

Chiral columns with a long tradition are cyclodextrin (CD)-bonded CSPs (Figure 12.2d) obtained by alkylether (Cyclobond) or carbamate linkage (ChiraDex). These oligomeric macrocyclic selectors possess a toroidal cone built from 6, 7, and 8 glucopyranose units connected *via* α-1,4-glycosidic bonds in α-, β-, and γ-CD, respectively. They discriminate between enantiomers through stereoselective inclusion of aromatic or bulky aliphatic substituents of chiral compounds under aqueous conditions into the hydrophobic cavity driven by hydrophobic interactions which makes them generally suitable for enantioselective ESI–MS. Since a majority of endogenous metabolites do not have aromatic interaction sites for enantioselective inclusion in the CD cavity, they have been of minor importance in LC enantiomer separation in the context of metabolomics. However, it has to be stressed that these selectors, either in native but most often in derivatized form, are first choice for CE as well as GC enantiomer separation, for the latter bonded *via* linkers to polysiloxanes that are coated to the FS wall of the GC column. Similarly, cyclofructan (CF)-based CSPs with macrocyclic ring have shown great potential to separate enantiomers with primary amines, usually aromatic ones but also with alicyclic rings.[10] They are cyclic oligosaccharides made of β-2,1-linked fructofuranose units and usually derivatized to yield carbamates. Both CD- and CF-based CSPs can be purchased based on superficially porous particle columns such as Poroshell 120 Chiral-CD (hydroxypropylated-β-cyclodextrin) and -CF (derivatized cyclofructan) for fast LC enantiomer separation, but were not widely used in enantioselective metabolomics related work.

12.2.1.3 Chiral Crown-ether CSPs

Chiral crown-ether (CCE)-based CSPs (Figure 12.2g) are popular tools for the enantioseparation of chiral ammonium compounds, in particular amino acids. The CCEs consist of a series of ether groups in a cyclic manner with at least one integrated chiral moiety as a bridge between two ethers; essentially, they are derived from 18-crown-6 with one or two ethylene bridges replaced by a chiral element (binaphthyl in Crownpak CR and tartaric acid in ChiroSil). They are used in combination with strongly acidic aqueous eluents (pH between 1 and 3.5) which ensures the formation of ammonium ions. They can bind enantioselectively to the macrocyclic ring through inclusion complexation driven by triple hydrogen bond formation between the ammonium ion and three oxygens of the crown-ether. While they are typically used with low organic content in RP mode (*e.g.* 20%), immobilized versions (Crownpak CR-I, ChirоSil) are stable against organic solvents and can be used in HILIC[11] or polar organic mode as well. In the latter, secondary amines could also be separated. For amino acids, it was reported that both retention and enantioselectivity increased when the ACN content of the mobile phase was above 90%. Eighteen (18) proteinogenic amino acid enantiomer pairs (all chiral ones except Pro) were resolved within 10 min using LC–ESI–QTOF–MS.[12] The strong acids (*e.g.* TFA) may be less appealing for MS, however, it has been documented that such eluent does not preclude ESI–MS hyphenation. It is worthwhile mentioning that these columns are available in two enantiomeric forms, *e.g.* Crownpak CR-I(+) and (−) which show opposite elution orders for the respective enantiomers. It is a useful strategy to reverse enantiomer elution orders for verification of results of enantiomeric excess determination. CSPs with selectors from the natural chiral pool, mentioned above, do not have this option as they exist only in one enantiomeric form. CCE columns have been employed in metabolomics for analysis of amino acids in mouse brains,[13] black vinegar,[14] and cheese.[12]

12.2.1.4 Donor–Acceptor Phases (Brush-type or Pirkle-type CSPs)

CSPs of this type are obtained by bonding a low-molecular neutral synthetic chiral selector, having aromatic moieties, amide, and urea groups combined with bulky elements, *via* an achiral spacer to silica (Figure 12.2h–i). They are actually less suitable for endogenous metabolites because their operation principle for chiral recognition is strongly based on aromatic interactions (π–π interactions) supported by hydrogen bonding and steric interactions. However, tagging of metabolites (*e.g.* amino acids, hydroxy acids) through precolumn derivatization makes them suitable for enantiomer separations of metabolites. Whelk O1 (Figure 12.2h) is one of the most popular columns of this class, and Sumichiral columns are also frequently used, *e.g.* Sumichiral OA 2500 based on *N*-3,5-dinitrobenzoyl-1-naphthylglycine bonded aminopropylsilica. In numerous examples, Hamase and co-workers demonstrated the usefulness of this approach for amino acid and hydroxy acid enantiomer separations after NBD labeling or other tagging strategies. In order to be able to resolve all amino acid enantiomers on the same column, a custom designed selector (KSAACSP-001S, *N*-3,5-dinitrophenylaminocarbonyl-L-leucine bonded aminopropylsilica, Figure 12.2i) was developed to fulfill this requirement which is still challenging. Some examples of the application of this kind of selector in metabolomics are the analysis of amino acids in plasma,[15,16] urine,[17] and some mice tissues.[18]

12.2.1.5 Chiral Anion-exchangers and Zwitterionic Ion-exchangers

These CSPs employ selectors which are derived from quinine (QN) and quinidine (QD) alkaloids (Figure 12.2e–f). Their enantioselectivity can be effectively tuned by their carbamate residues (*tert*-butyl in Chiralpak QN-AX/QD-AX, Figure 12.2e, and 2-sulfohexyl in Chiralpak ZWIX(+) and (−), Figure 12.2f). The former are weak chiral anion-exchangers (AX) capable of separating chiral acids, and the latter have extended applicability spectra; they allow separation of chiral acids, bases, and zwitterionic compounds through merging chiral WAX and chiral SCX moieties providing a zwitterionic ion-exchanger (ZWIX). Efficient elution requires some counterions, *i.e.* buffer ions in the mobile phase to disrupt the driving ionic interactions causing strong retention. Through the intramolecular counterion effects on the ZWIX phases they show weaker retention. As a consequence, elution becomes possible with lower buffer (additive) concentrations which is favorable for MS compatibility. For the enantioseparation of amino acids, both chiral anion exchangers[19–21] and zwitterionic ion-exchangers[22–24] have been employed, in the former case after precolumn derivatization of the amino group (with DNP, AQC, *etc.*) in the latter both free and derivatized amino acids can be separated.[25] Separations of isobaric amino acids (Leu/Ile/allo-Ile as well as Thr/allo-Thr) have been documented on a ZWIX(+) column.[25] Additionally, side chain derivatization by *S*-alkylation with iodoacetamide was recommended for thiol-containing amino acids such as cysteine and homocysteine.[25] Furthermore, this kind of selector has been employed for the chiral separation of short chain hydroxy carboxylic acids and hydroxy fatty acids, respectively.[26–28]

12.2.2 Tagging

Although the above outlined direct approach tacitly implies that the sample can be directly analyzed without derivatization prior to injection to the HPLC system, for several reasons achiral precolumn derivatization may often be favorable or even necessary for direct HPLC enantiomer separations. Such analyte tagging with achiral derivatizing agents is frequently performed for the following reasons. Some metabolites are very small and may lead to undetectable fragment ions in the course of fragmentation in MS/MS detection schemes. Introducing a retention tag to induce interaction with the chiral selector and thus increase retentivity as well as enantioselectivity gives more favorable chromatographic properties. This is a concept often realized for amino acids and their chiral separation on Pirkle-type and chiral ion-exchanger CSPs.[29–32] Alkylation tags are required for thiol functionalities or other groups which give poor analytical performance. For example, the enantiomer separation of the amino acid cysteine (Cys) with its mercaptomethyl residue is redox active and subject to uncontrolled oxidation in the course of sample preparation but also during the ESI process. In order to avoid artifacts and preclude target analyte loss by oxidation, protection of this side chain by an alkylation tag is recommended. Iodoacetamide is a convenient alkylation reagent, well known from proteomics protocols, as it allows selective alkylation of thiols besides amino groups under carefully controlled pH conditions.[23] Its combination with an amino tag such as AQC can provide the resulting *S*-carbamidomethylated Cys (Cys-IAA-AQC) not only with stabilized analyte properties but also improved peak shape, increased enantioselectivity, and enhanced detection sensitivity in direct HPLC enantiomer separation.[23] If total (oxidized and reduced) Cys should be analyzed, a reduction step with 1,4-dithio-DL-threitol (DTT) or tris(2-carboxyethyl)phosphine (TCEP) prior to S-alkylation and N-derivatization is required. Finally, tagging is most popular for the introduction of a charge tag for MS detection or a chromophore/fluorophore for UV and fluorescence detection, respectively. The main goal for this step is to increase the detection sensitivity for MS detection (charge tag) and UV or fluorescence detection (chromophore/fluorophore) due to lack of such properties in the underivatized analyte. The charge tag for MS introduces a stable charge to analytes such as a fixed charge quaternary amine (*e.g.* *p*-*N*,*N*,*N*-trimethylammonioanilyl *N*′-hydroxysuccinimidyl carbamate iodide, TAHS, see Figure 12.3b)[24,33] or a tertiary amine with high proton affinity (*e.g.* AQC, Figure 12.3b)[24] which are favorable for a positive ESI mode. Other strategies of tagging, such as chemical isotope tagging/labeling (giving access to stable isotope labeled internal standards of analytes for compensation of matrix effects), isobaric tagging (tandem mass tags which use a specific labeling arrangement of isotopes in the tag and charges to yield different isotopes of the product ion in MS/MS)[34] are also amenable to enantioselective metabolomics (*vide infra*), but so far less popular in this field. One important aspect to consider during this step of tagging is to avoid (partial) racemization of the analyte; in fact, this is a critical issue that must be considered during method validation. The tagging reagent is usually primarily selected based on the detection technique which is available for the analysis.

A useful strategy for metabolomics could be multiple tagging.[34] The above tagging reagents are primarily aiming at a particular metabolite group with a specific functional group. For broad coverage, it might be desirable to derivatize different metabolite groups with complementary reagents. In this context, sequential tagging (one sample is subsequently derivatized with several derivatizing agents) or parallel tagging (one sample is aliquoted and tagged with different reagents) can be carried out. The latter may be advantageous as it avoids double derivatization with different labels and other effects of the labeling reaction on some target analytes. Parallel tagging is a convenient method to increase coverage yet at the expense of total analysis time (several runs have to be performed subsequently for the analysis of the differently tagged samples). It has been used for the indirect method (*vide infra*).

Some examples of enantioselective metabolomics studies which involved a derivatization step can be found in Table 12.1 in which specification of a

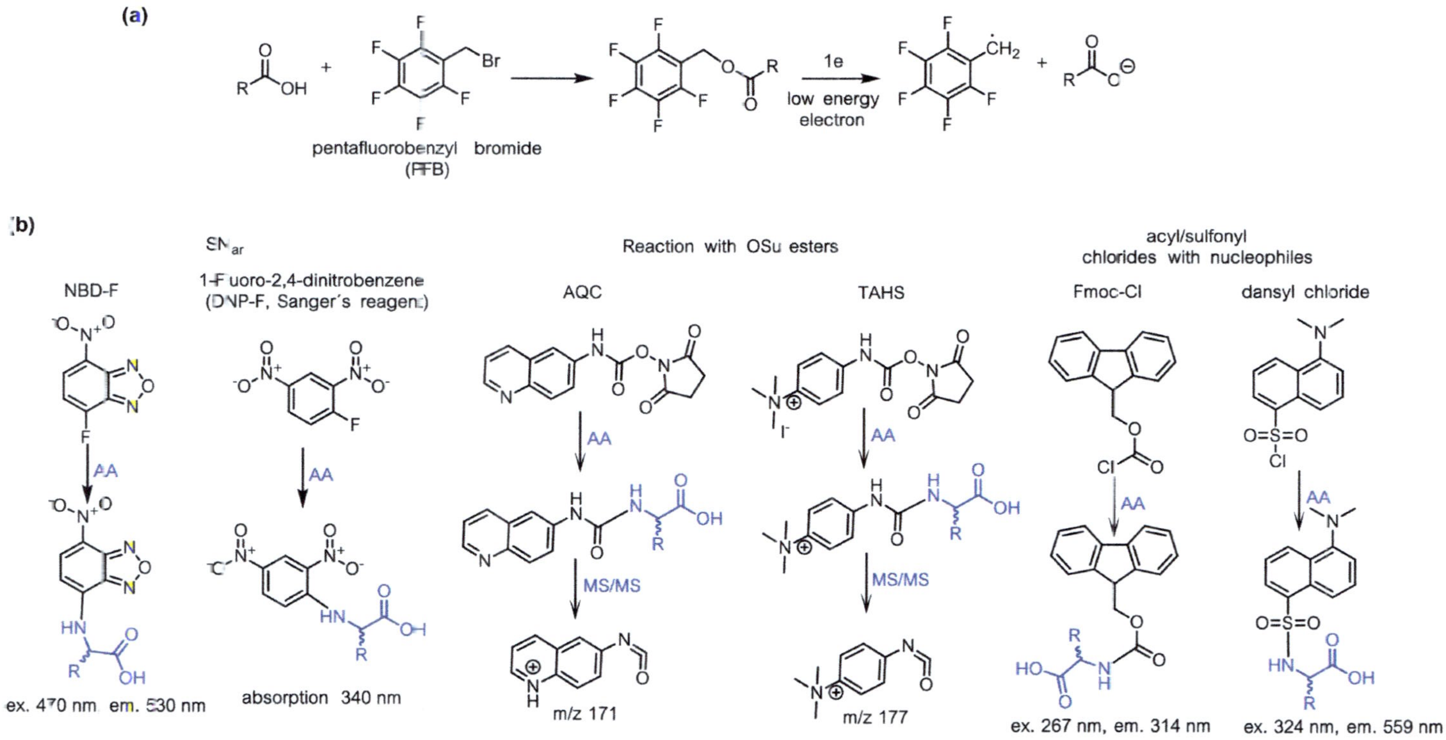

Figure 12.3 Achiral tagging strategies (a) Mechanism for the formation of carboxylate anions by dissociative electron capture APCI/MS analysis of PFB-ester derivatives. (b) Common achiral tagging reactions for amino acids.

reagent means that the analytes were tagged. For example, since chiral-LC was developed initially for normal phase conditions, and apolar eluent systems were less favorable for ESI as they affect ionization with ESI sources, electron capture atmospheric pressure chemical ionization (ECAPCI)–MS became a popular choice for chiral studies of oxylipins. In order to take advantage of the high sensitivity that this kind of ionization brings about, oxylipins were derivatized with pentafluorobenzyl bromide (PFB).[35,36] Figure 12.3a shows the mechanism for formation of PFB derivatives.

Another well-known and often utilized achiral tagging procedure is the derivatization of amino acids with 4-fluoro-7-nitrobenzofurazan (NBD-F), which is an excellent chromo- and fluorophore for fluorescence detection at 530 nm after excitation at 470 nm. Many more reagents such as dansyl chloride, 9-fluorenylmethyl-chloroformate (Fmoc-Cl), 6-aminoquinoyl-*N*-hydroxy-succinimidyl carbamate (AQC), *p*-*N*,*N*,*N*-trimethylammonioanilyl *N*′-hydroxysuccinimidyl carbamate iodide (TAHS), and 1-fluoro-2,4-dinitrobenzene (DNP-F, Sanger's reagent) are just a few examples of tags for the analysis of amino acids (Figure 12.3b).

Despite the extra work and time that this derivatization step costs, it is more than compensated by the manifold benefits from more favorable chromatographic behavior, enhanced enantioselectivity, increased detection sensitivity and/or lower LOQ, assay specificity, and others.[37]

12.2.3 Direct 1D-LC *versus* 2D-LC

One-dimensional LC with chiral stationary phases in combination with highly sensitive (fluorescence, mass spectrometry) and highly selective (mass spectrometry) detection are state-of-the-art enantiomer separation technologies with favorable sample throughput. However, it has been highlighted that 1D HPLC chiral separations lack the selectivity and sensitivity required for the analysis of trace amounts of metabolite enantiomers in complex biological samples. (Matrix) Interferences, (enantioselective) ion-suppression, limited selectivity for diastereomers and constitutional isomers are just some of the problems that may arise. Under this scenario 2D HPLC methods which combine typically the use of reversed-phase in the first dimension and enantioselective columns in the second dimension represent an outstanding alternative to achieve highly accurate results with excellent assay specificity in chiral metabolomics.[38] The down sides are a more complex instrumental setup and often excessive run times of more than 1 h up to 4 h per analysis.

2D separations can be carried out either in full comprehensive or multiple heart cutting mode. In the full comprehensive mode the entire eluate of the first dimension separation is transferred and separated by a complementary chromatographic principle in the second dimension; it has been realized for instance for amino acids by a chiral (quinine bonded CSP)×achiral (RPLC) 2D-LC separation after their derivatization with Sanger's reagent[39] and by a chiral×chiral 2D-LC setup after their precolumn derivatization to FMOC-amino acids.[40] In the latter study, a quinine carbamate bonded CSP was coupled with a quinidine carbamate CSP that has identical chemistry but opposite stereochemistry and allowed the generation of characteristic peak patterns with the corresponding D- and L-enantiomers as cross peaks along a diagonal parity line in the 2D-separation space. More popular have been multiple heart cutting 2D-LC separations, also in particular for amino acids but also hydroxy acids, in which peaks of interest were transferred from an achiral first dimension separation into a second chiral separation dimension *via* multiloop switching valves[41] (Figure 12.4). Enantioselective 2D-HPLC systems have been particularly promoted by Hamase *et al.* in multiple studies for enantiomeric analysis of proteinogenic amino acids,[17,42–45] for *N*-methyl-D-aspartic acid (NMDA) and its analogs[46] and for some short hydroxy acids such as lactate and 3-hydroxybutyrate.[47,48] In most of these cases, Pirkle type CSPs have been employed for the separation (in the second dimension), and NBD-F derivatization has been used for tagging of amino acids with the subsequent detection by fluorescence. However, some more recent studies[31,32,49,50] by Hamase *et al.* have also employed MS for detection while keeping the tagging with NBD-F (Figure 12.4a).

The achiral separation in the first dimension not only allows separation of the target amino acids from each other as well as from matrix interferences, but also the separation of isobaric amino acids (Ile/Leu/allo-Ile; Thr/allo-Thr). If 1D-LC with a chiral column is employed, the enantiomers are often well separated but the elution times of the different amino acids are similar leading to many overlapping peaks. With UV and fluorescence detection this leads to many interferences. If MS detection is used, it may possibly lead to enantioselective matrix effects in MS detection. An accurate determination of trace D-levels besides large quantities of L-amino acids is therefore usually not possible with standard 1D-LC methods, due to matrix effects this is probably also not possible with MS detection unless proper correction with isotope labeled internal standards for both enantiomers is performed. 2D-LC or even 3D-LC has been suggested as a solution. In order to separate all proteinogenic amino acids on a single chiral stationary phase, the group of Hamase developed a tailor-made donor–acceptor phase based on *N*-3,5-dinitrophenylcarbamoyl-leucylamide KSAACSP-001S (Figure 12.2i). The advantage of such synthetic phases is their availability in two enantiomeric forms. The CSP with opposite configuration has been frequently used to validate the results by reversal of elution order. It confirms that the right peak has been selected for quantitation and not wrongly a chemical impurity peak which may be present as well. During the 2D separation, the fractions of the first-dimension separation are stored in loops until the second dimension separation system is ready for the next analysis. For the "heart cutting" in enantioselective LC usually the loop size is selected such that the entire peak can be transferred to the second-dimension column. It ensures the highest possible sensitivity. Generally, a real heart cut (transferring only the center of the peak to the second-dimension column) is also adequate because the achiral column in the first dimension does not discriminate between enantiomers, and the peak profiles of the opposite enantiomers are

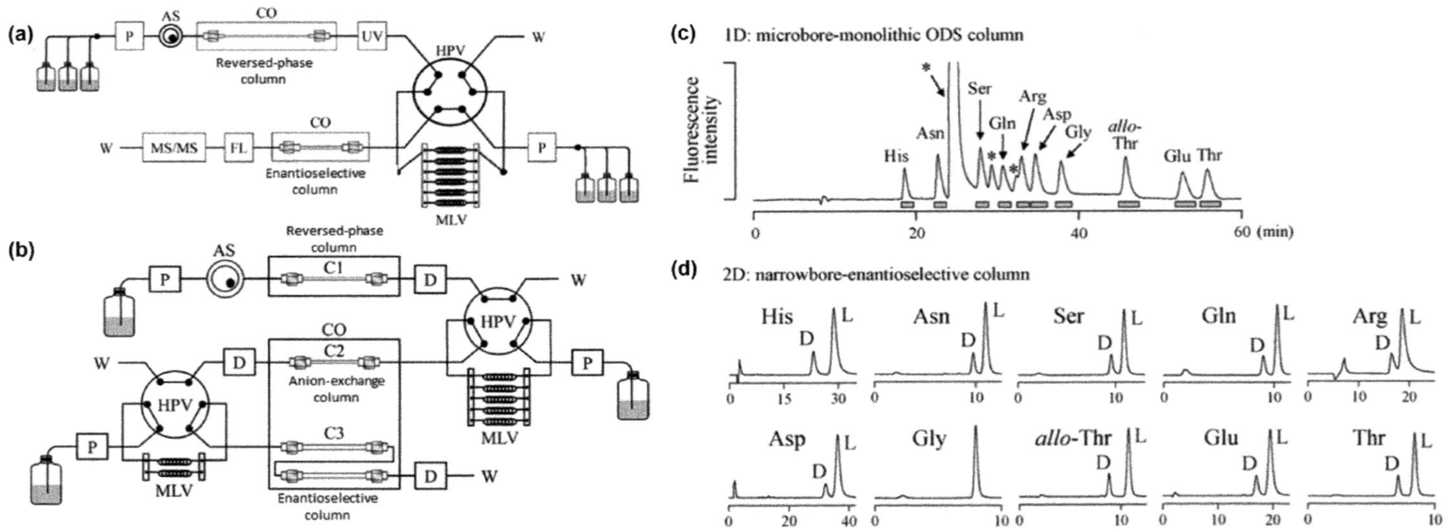

Figure 12.4 (a) 2D and (b) 3D HPLC system diagrams (reproduced from ref. 31 with permission from Elsevier, Copyright 2018), employed by the group of Hamase, as well as 2D HPLC separation of amino acids in the (c) 1st dimension (reproduced from ref. 29 with permission from Elsevier, Copyright 2020), and (d) 2nd dimension (reproduced from ref. 17 with permission from Elsevier, Copyright 2010).

totally superimposed *i.e.* in each point of the chromatographic zone the D/L-ratio corresponds to the one in the sample.

Another problem besides the limited chemoselectivity of chiral columns is their limited applicability range. There is no single chiral column which can separate the enantiomers of all relevant metabolites on one column under one mobile phase conditions applicable in a single run as typically required for a comprehensive analysis. For this reason, Kimura *et al.* recently reported a 2D HPLC chiral metabolomics method which employs different CSPs in one method, although for two analytes only. The same 2D-LC system shown in Figure 12.4a was equipped with an additional six-port/2-position valve (behind the multiple loop modulator) that allows two distinct chiral columns to be coupled (Chiralpak QD-AX for NBD-lactic acid and KSAACSP-001S for NBD-3-hydroxybutyric acid) as the second dimension chiral separation with RPLC in the first dimension. It has been utilized for the simultaneous detection of lactate and 3-hydroxybutyrate enantiomers after pre-column-derivatization with the fluorescent reagent 4-(*N*-chloroformylmethyl-*N*-methylamino)-7-nitro-2,1,3-benzoxadiazole (NBD-COCl) in human clinical samples.[47] Studies employing 2D HPLC systems have also been reported for the analysis of oxylipins. As in most of the cases for this kind of compound, polysaccharide-type CSPs have been employed in the second dimension, and MRM experiments have been used for detection.[35,51–53]

These multi-loop, multiple heart cutting RP-chiral 2D-LC methods turned out to be powerful tools for enantioselective metabolomics studies. New commercial 2D-LC instruments are equipped with UHPLC pumps, robust valves, and advanced valve topologies including multiple heart cutting valves (with up to 2×6 sampling loops as an interface between the two dimensions). They may further allow active solvent modulation. In addition, along with user-friendly software such MHC 2D-LC systems may become the future in enantioselective metabolomics studies. With such instruments run-to-run repeatability of <0.4% RSD can be readily achieved, and this excellent retention time precision is important for the time-based sampling of the first dimension eluates. In many literature reports not more than 10 cuts were taken, however, it has been documented that up to 30 cuts are also easily possible,[47] and if the second dimension separations are very fast even more can become routine.

Biological samples have an extremely complex matrix. Many metabolites of interest are present in trace levels only, such as the D-enantiomers of amino acids. Hamase *et al.* have shown that even after a 2D-LC separation there may be interferences coeluting with the trace D-amino acid levels. To overcome this problem, 3D HPLC methods (Figure 12.4b) consisting of RP column in the 1st dimension, anion-exchange stationary phase in the 2nd dimension, and an enantioselective column in the 3rd dimension have been designed and developed for the determination of the asparagine, serine, alanine, and proline enantiomers in the plasma of patients with chronic kidney disease.[30] This setup contains two loop modulators for sampling and fraction transfer between two dimensions (Figure 12.4b). Clean trace

D-enantiomer peaks resulted after undergoing 3 dimensions of complementary selectivity and allowed their detection and quantitative analysis without any interferences.

12.2.4 Metabolites Analyzed by a Direct Approach

Direct chiral metabolomics studies have been primarily focused on two important classes of chiral metabolites: amino acids and oxylipins. Chiral separations for short hydroxy fatty acids, *e.g.* 2- and 3-hydroxy fatty acids, have also been reported but to a minor extent by LC. Hydroxy acidurias on the other hand is an important topic in metabolic diseases, and thus enantioselective analysis also has some relevance in the screening of inborn errors of metabolic diseases.

12.2.4.1 Amino Acids and Analogs

For a long time, it was considered that only L-amino acids are present in biological systems. However, the occurrence of D-amino acids in organisms has been reported widely since the 1980s.[54] It has also been noted that both D- and L-enantiomers are involved in different biological functions, distributions, and metabolic pathways.[55] For several D-amino acids biological functions have been described, in mammals,[56] microorganisms,[56] but also in plants.[57] Their enantioselective analysis has therefore propagated, in particular with the goal to use them as robust biomarkers for various diseases.

Detailed reviews about the direct and indirect analysis of amino acids have been reported elsewhere,[2,37,58–60] and a summary of direct methods is given in Table 12.1. The analysis of free amino acids can be performed on teicoplanin and teicoplanin aglycone (TAG) CSPs,[8] crown-ether bonded CSPs,[12–14,61] zwitterionic ion-exchange CSPs[12] using MS (single ion monitoring, SIM) and MS/MS (selected reaction monitoring (SRM) or multiple reaction monitoring (MRM)) for detection. Each of these CSPs can separate a wide range of free amino acids but also has some shortcomings. Crown-ether CSPs require strongly acidic eluents to form ammonium ions and fail to separate secondary amino acids such as Pro which is not capable for inclusion complexation *via* triple hydrogen bonding with the ether groups. The ZWIX phases have no or limited enantioselectivity for Cys, Lys, and Ser. Teicoplanin and TAG have sometimes somewhat broader or tailing peaks due to affinity like binding mechanism, and a reversal of the elution order by switching to the enantiomeric phase is not possible as it does not exist for these *ex natural chiral pool* CSPs. A comparison of these CSPs can be found in the amino acid database on the webpage of Chiral Technologies (chiraltech.com). For this reason, the combination of two types of chiral columns, *i.e.* on the one hand a binaphthyl-based crown ether (Crownpak CR-I) and on the other a cinchona alkaloid-derived zwitterionic stationary phase (Chiralpak ZWIX(+)) was considered and found to be successful.[12] It enabled the highly sensitive analysis of 115 chiral and non-chiral metabolites using optimized MS/MS parameters in the range between 0.001–50 nmol mL^{-1}

without derivatization. Its application to food metabolomics revealed trace levels of D-amino acids in food samples such as cheese. The simultaneous analysis of 15 underivatized chiral amino acids in 25 min by liquid chromatography–electrospray tandem mass spectrometry was accomplished using a teicoplanin chiral stationary phase under isocratic conditions (acetonitrile–water, 75 : 25). Post-column addition of formic acid allowed it to boost detection sensitivity.[62]

In most of the cases chiral analysis is performed after tagging free amino acids to introduce properties favorable for their analysis. AQC- and NBD-derivatization turned out to be amongst the most popular tagging strategies. Both have favorable performance in MS-detection and are also fluorescing derivatives providing high sensitivity with fluorescence detection. For the chiral analysis of tagged amino acids macrocyclic glycopeptide CSPs have been employed in some studies. Gholami *et al.*, for example, employed a Chirobiotic T column for the analysis of AQC-proteinogenic amino acids in tobacco cells[17]. The idea of the study was to determine the effects of L- and D-ornithine (L-Orn and D-Orn) on the subsequent metabolic pathways in *Nicotiana tabacum*. Another study reported the use of a Chirobiotic TAG column for the analysis of NBD-tagged proteinogenic amino acids in a peptide hydrolysate.[8] A variety of other N-tagged amino acids such as FMOC-derivatives can also be separated on Teicoplanin (now also available on core–shell particle column as Poroshell 120 Chiral-T revealing enhanced efficiencies and faster separations) and Teicoplanin aglycone.

Anion exchange-type CSPs, QN-AX and QD-AX, have been reported for analysis of proteinogenic amino acids derivatized with succinimidyl ferrocenyl propionate (SFP) in human urine and detected by selected reaction monitoring (SRM).[20] They provide enantioselectivity for virtually all kinds of *N*-derivatized amino acids (including FMOC, DNP, NBD, and AQC-amino acids). Tagging with AQC allows the separation of all proteinogenic amino acids on quinine carbamate CSPs[63] and has, for instance, been employed to determine D- and L-amino acid levels in the whole brain of wild-type and mutant mice lacking D-amino acid oxidase activity using high-performance liquid chromatography–tandem mass spectrometry (HPLC–MS/MS).[64] Significant levels of D-Ser (25% related to L) could be detected, and significant differences in D-Leu, D-Ala, D-Ser, D-Arg, and D-Ile levels in mutant and wild type mice were found.

Furthermore, corresponding zwitterionic ion-exchangers derived from cinchona alkaloids enable enantiomer separations of *N*-derivatized amino acids with excellent selectivity profiles. Due to the intramolecular counter-ion effect of the sulfonic acid moiety of the selector, retention times are much lower than on corresponding chiral anion-exchanger columns mentioned above, and interaction with Arg and His enhanced which is favorable for their resolution.[25] Karakawa *et al.* have reported the use of the zwitterionic CSP produced from these cinchona alkaloid selectors (Chiralpak ZWIX(+)) for the analysis of D-Ala, D-Asp, and D-Ser in plasma and tissues from rats.[24] The enantiomer separation of all isobaric leucines (in particular

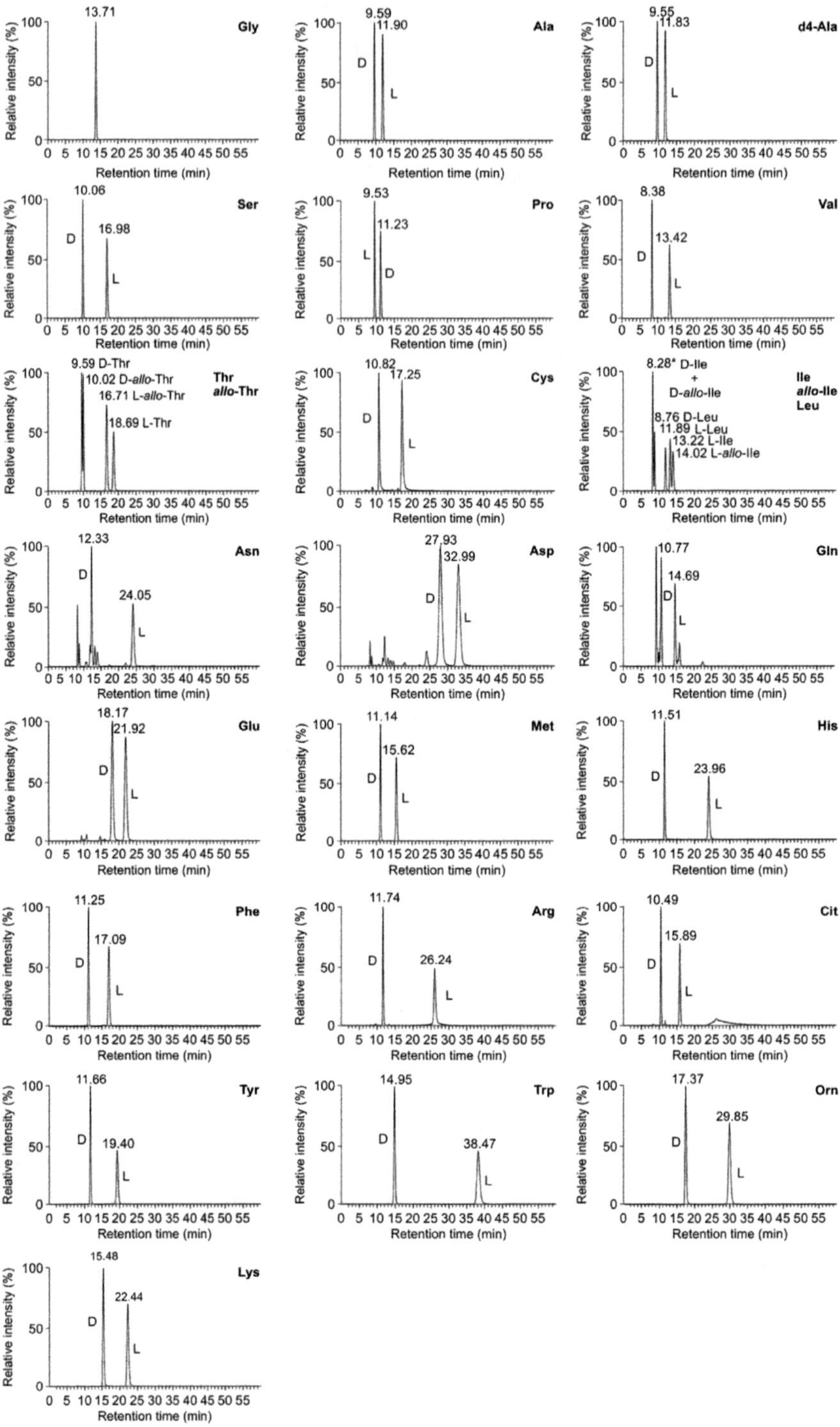

Relative intensity (%)
Retention time (min)
0 5 10 15 20 25 30 35 40 45 50 55
Gly
13.71
Ala
9.59 D
11.90 L
d4-Ala
9.55 D
11.83 L
Ser
10.06 D
16.98 L
Pro
9.53 L
11.23 D
Val
8.38 D
13.42 L
Thr
allo-Thr
9.59 D-Thr
10.02 D-allo-Thr
16.71 L-allo-Thr
18.69 L-Thr
Cys
10.82 D
17.25 L
Ile
allo-Ile
Leu
8.28* D-Ile
+
D-allo-Ile
8.76 D-Leu
11.89 L-Leu
13.22 L-Ile
14.02 L-allo-Ile
Asn
12.33 D
24.05 L
Asp
27.93 D
32.99 L
Gln
10.77 D
14.69 L
Glu
18.17 D
21.92 L
Met
11.14 D
15.62 L
His
11.51 D
23.96 L
Phe
11.25 D
17.09 L
Arg
11.74 D
26.24 L
Cit
10.49 D
15.89 L
Tyr
11.66 D
19.40 L
Trp
14.95 D
38.47 L
Orn
17.37 D
29.85 L
Lys
15.48 D
22.44 L

Ile/allo-Ile besides Leu) is still challenging on the before-mentioned approaches. To solve this shortcoming a tandem column obtained by coupling of chiral anion-exchange CSP and zwitterionic chiral ion-exchange CSP (Chiralpak QN-AX + Chiralpak ZWIX(+)) was used with MeOH/water (90 : 10; v/v) mobile phase containing 0.1% (v/v) formic acid/ammonium formate buffer additive to improve the separation of all proteinogenic amino acid enantiomers in a single LC–MS analysis in blood of patients with cognitive disorder (Figure 12.5). Resolutions for Pro enantiomers and between isobaric leucines were improved compared to separate ZWIX columns. D-Amino acids eluted earlier than their L-amino acids, except for the secondary amino acid proline for which elution order was reversed. To separate D-Ile and D-allo-Ile the corresponding tandem column Chiralpak QD-AX + - Chiralpak ZWIX(−) which showed the reversed elution order had to be used. In this study with tandem MS detection, the D-Pro ratio (D-pro/(D-pro + L-pro)) was suggested as a biomarker for mild-cognitive-impairment.[21] In conclusion, using this tandem column approach all proteinogenic amino acid enantiomers can be analyzed simultaneously by the use of LC–MS/MS along with the isobaric and secondary amino acids (Figure 12.5).

Amino acids can also be separated into enantiomers on donor–acceptor phases (Pirkle type phases) after precolumn derivatization. CSPs such as Sumichiral OA-2500 (*N*-3,5-dinitrobenzoyl-(1-naphthylglycylamido)propylsilica)[65] or custom synthesized *N*-3,5-dinitrophenylaminocarbonyl-leucine bonded aminopropylsilica (KSAACSP-001) (with urea type selector) (Figure 12.2i)[55] have shown great promise. They are available in both selector configurations, *R* and *S*, allowing the inversion of the elution order. The enantiomer's elution order is reversed while chemical impurities keep their relative retention on the CSP with opposite configuration(s), see for instance *N*-methyl-Asp and *N*-methyl-Glu on Sumichiral OA-2500S and OA-2500R in ref. 55. It can be used for validation of trace D-levels. The interference is then no longer co-eluting with the trace D-enantiomer so that accurate quantitative analysis becomes possible, also with fluorescence detection. Pirkle-type phases are most commonly applied in normal-phase mode. For LC–MS, ammonium acetate in methanol has been proposed as an alternative to ESI incompatible normal phase conditions in combination with Sumichiral OA 2500 and acidic analytes (similar to *N*-derivatized amino acids). Hamase and co-workers suggested the use of ACN–MeOH of varying ratios and formic acid concentrations of 0.05 to 0.5% as a mobile phase for chiral separations on KSAACSP-001S suitable for both fluorescence and ESI–MS/MS detection.[66]

Figure 12.5 Chromatograms of chiral amino acid separation after AQC-tagging on tandem column of Chiralpak QN-AX and Chiralpak ZWIX(+) by LC–ESI–MS/MS.

Enantioselective 1D-LC–ESI–MS is often not sensitive enough or due to enantioselective matrix effects not accurate. For this reason, K. Hamase and co-workers developed multiple heart cutting 2D-LC and 3D-LC both with fluorescence detection[30,55] but also tandem ESI–MS/MS detection[24,31,32,66] for routine enantioselective amino acid analysis in biomarker research, fermented food analysis, and some other applications, as discussed above (see also Table 12.2). The instrumental setup is depicted in Figure 12.4. Their work on amino acids analysis involves pre-column derivatization with NBD-F as the tagging reagent, a reversed phase separation of individual NBD-amino acids from each other and matrix interferences, multi-loop sampling from the ^{1}D eluate, and automated fraction transfer into the ^{2}D chiral column, usually Pirkle-type CSPs (especially Sumichiral OA-2500S, Sumichiral OA-2500R, KSAACSP-001S and KSAACSP-105S), for enantioseparation in the second dimension. However, some zwitterionic ion-exchange[49] and anion-exchange[46,54,67] CSPs have been also reported by this group as 2nd dimension columns. Up to 10 amino acids have typically been analyzed by this approach, and fluorescence detection was the first choice (Table 12.2).

Some of the studies of Hamase's group have been addressed to the analysis of D-Ser and D-Ala in brain, peripheral tissues, plasma, and urine of rats and mice. Some of the findings from these studies are the higher localized concentration of D-Ser and D-Ala in specific organs. For example, D-Ser was found at 300 nmol g^{-1} in the cerebral cortex and hippocampus of rats and D-Ala with concentrations of about 100 nmol g^{-1} in the pancreas of mice.[43,44] High concentrations of D-Ser and D-Ala have also been found in the central nervous system, and these metabolites are known as modulators of the *N*-methyl-D-aspartic acid (NMDA) receptor-mediated neurotransmission.[55] With a similar approach Han *et al.* reported high concentration of acidic amino acids. For D-Asp high concentrations were reported in the pineal gland and endocrine tissues while D-Glu was highly concentrated in urine samples of rats.[54] Some non-proteinogenic amino acids such as *N*-methyl aspartic acid (NMA) and *N*-methyl glutamic acid (NMG) have also been analyzed with 2D HPLC systems in some bivalve species. The D-enantiomer of NMA (NMDA) is well-known as an agonist of the NMDA subtype of the glutamate receptor, and it has also been reported as a regulator of some hormonal releases.[46] Also, D-citrulline and D-ornithine had increased concentration levels in D-amino acid oxidase-deficient mice.[32]

It was demonstrated that even with a 2D-LC setup, coelutions of matrix components with trace D-amino acids still exist in biological samples. Therefore, Hamase and co-workers developed a multiple heart cutting 3D-HPLC assay for enantioselective analysis of amino acids.[30] After separation by RP in the ^{1}D, an achiral anion-exchanger column was used as another achiral separation as ^{2}D, before finally the heart cuts from these two dimensions were transferred to the ^{3}D enantioselective column. The mobile phases of all three dimensions were compatible with each other. It was documented that clean peaks for D-amino acid trace levels only result after this 3-dimensional separation, which allows uncompromised accurate

D-amino acid analysis. D-Asn and D-Ser were found to be useful biomarkers for chronic kidney disease.

12.2.4.2 Oxylipins and Hydroxy Fatty Acids

Oxylipins are defined as oxygenated lipids derived from polyunsaturated fatty acids (PUFAs). Since they can possess several chiral centers along its chain, the number of constitutional and stereoisomers may be substantial. Owing to their critical role in the regulation of physiological and pathophysiological biological signaling pathways, in particular in pro-inflammatory and anti-inflammatory processes, it is therefore no big surprise that they have been a widely studied group of metabolites.

The generation of oxylipins can occur by enzymatic or non-enzymatic reactions. Non-enzymatic reactions such as autoxidation and oxidation by reactive oxygen species (ROS) in the course of oxidative stress in cellular compartments lead to racemic mixtures. On the contrary, enzymatic reactions occur stereoselectively. Enantioselective analysis of oxylipins is therefore required to derive information on whether they originate from autoxidation and ROS, respectively, or specific enzymes with distinct stereoselectivity. To this end, oxidative stress in cells and tissue can be monitored by analyzing oxylipins stereoselectively as biomarkers to obtain the full picture of biological processes in which oxylipins are involved.[100]

Enzymatic oxidation can occur by 3 different routes, either by cyclooxygenases (COX), lipoxygenases (LOX), or cytochrome P450 epoxygenases (CYP), which may show distinct stereoselectivity. Metabolites produced by these routes can be classified into series of compounds depending on the PUFA from which they are originated. Some of these oxylipins are for example prostaglandins, thromboxanes, leukotrienes, mono-, di-, and tri-hydroxy fatty acids (FAs), epoxy FAs, lipoxins, eoxins, hepoxilins, prostanoids, iso-prostanes, resolvins, protectins, and maresins. Chirality in oxylipins arises in most of the cases from hydroxyl groups attached to aliphatic fatty acid chains, although hydroperoxides and epoxy groups are also common.

Because of the involvement of these oxygenated PUFA in biological processes and because their biological activity is mostly linked to one specific enantiomer, chiral stationary phases (CSP) have been employed for their direct analysis. An extended review about the classification and biological functions of oxylipins derived from arachidonic acid, linoleic acid, γ-linolenic acid, dihomo-γ-linolenic acid, adrenic acid, α-linolenic acid, stearidonic acid, eicosapentaenoic acid, and docosahexaenoic acid is described by Gabbs *et al.*[101] Chiral separations of oxylipins, on the other hand, have been reviewed recently by Ianni *et al.*[100] Selected examples can be also found in Table 12.1.

Pirkle-type CSPs have been employed to some extent for the analysis of oxylipins. Hydroperoxy fatty acids HpOME(E)s, HpOME(Z)s, and HOMEs were analyzed with Pirkle-type CSPs Reprosil Chiral NR and NR-R and then detected by UV and LC–MS/MS with MRM.[68,69] Reprosil Chiral-NR had also been used previously along with polysaccharide-based CSPs Chiralcel OC and Chiralcel OB-H for the analysis of hydroperoxy fatty acids HPETEs and HPODEs.[83]

The majority of chiral separations of oxylipins have been performed on polysaccharide-based CSPs.[36,100] Polysaccharide-based CSPs have good chiral recognition toward oxygenated PUFAs, and they possess high versatility due to their compatibility with most relevant elution modes: normal-phase and reversed-phase, but also with polar organic/ionic-mode and SFC. Substituted tris(benzoates) and tris(phenylcarbamates) of cellulose and amylose are the most popular polysaccharide-based CSPs. The review by Ianni *et al.*[100] offers a useful guide of preferred polysaccharide-based CSPs and elution modes for the analysis of HETEs, HOMEs, HODEs, HOTEs, HEPEs, 17-HDoHE, 4-HNE, EETs, EEQs, EDPs, 14(15)-EpETrE, 9(10)-EpODE, 9(10)-EpOME, PGE2, and PGD2.[100]

Although polysaccharide-type CSPs have been successfully employed for oxylipins analysis as their free-structure, tagging strategies at the carboxylic group have been described as well. Mesaros *et al.* reported the chiral analysis of epoxyeicosatrienoic acids (EETs) in the mouse epithelial hepatoma (Hepa) cell line. For the analysis, PFB derivatization was employed (Figure 12.3a) along with ECAPCI and MRM for their detection. In this analysis, it was shown the enantioselective formation of EETs produced from arachidonic acid. Blum *et al.* reported the analysis of EETs, HETEs, EEQs, HEPEs, EDPs, and HDHAs in plasma from mice and humans and liver from mice, by using polysaccharide-based CSPs.[35] Recently, Mazaleuskaya *et al.* reported the analysis of PFB derivatized HETEs in human blood by UHPLC–ECAPCI–MS analysis. Parallel reaction monitoring was employed for detecting the metabolites after the separation on a Chiralpak AD-H column.[102] Other reported derivatization concepts employed for the analysis of oxylipins are the reaction with diazomethane which converts the carboxylic group to the corresponding methyl ester. This derivatization (suitable also for GC–MS and evaluated for Chiralcel OB) was reported for the analysis of 9-HODE, 13-HODE, 11-HETE, and 15-HETE in human atherosclerotic plaque using polysaccharide-based CSPs with UV detection at 235 nm.[85] 3,5-Dimethylphenyl isocyanate (3,5-DMPI) has also been reported as a tagging reagent for the analysis of 3-hydroxyalkanoates, from C6 to C16, on a polysaccharide-based CSP containing 3,5-dimethylphenyl carbamate bonded to a cellulose structure. Analytes were detected by MRM.[84]

Tandem columns consisting of a chiral column and a RP column[53] or amino phase[51] or two chiral columns[52] in series were proposed to accomplish the simultaneous separation of a larger number of analytes in one LC–MS/MS method. Better separation could be achieved with aminopropylsilica column in tandem with a Chiralcel OD-H column as a chiral column in a normal phase elution mode and then allowed the simultaneous analysis of EETs in lung and kidney tissue[51] and pro-resolving lipid mediators (lipoxin A4, 6-epi-lipoxinA4, lipoxin B4 and lipoxin A5, the D-series resolvins D1 and D2 as well as aspirin-triggered lipoxin A4 and resolvin D1, maresin along with protectin and the pathway markers 17(*S*)-hydroxy-docosahexaenoic acid and 17(*R*)-hydroxy-docosahexaenoic acid) with Lux Amylose-2 CSP as the chiral column in cell culture supernatants.[53] The separation of PGE 2 and entPGE 2 was carried out on two chiral Lux Amylose-2 columns connected in tandem to improve the separation in mouse brain samples with and without ischemia.[30] In all these cases LC–MS/MS with MRM acquisition was employed for detection.

Historically, most separations of oxylipins were performed in normal phase mode due to excellent enantioselectivities on various types of polysaccharide CSPs; a comprehensive overview about which type of CSP is most suitable for the different oxylipins can be found in the mentioned review by Ianni *et al.*[103] While in combination with UV detection it is adequate, mass spectrometry prefers RP conditions. An amylose tris(3,5-dimethylphenyl carbamate) coated CSP (Chiralpak AD-RH/Lux Amylose-1) and other polysaccharide CSPs, fortunately, also show broad enantioselectivity in reversed phase mode. Simple acetonitrile gradient with acetic acid gives both adequate enantioselectivity and high ionization efficiency in ESI–MS. The employed columns are typical standard HPLC columns based on 5 µm- or more recently also 3 µm-particle-based coated or immobilized polysaccharide CSPs. Recently, a subset of polysaccharide CSPs have become available based on sub-2 µm particles. They exhibit excellent performance for oxylipins under RP gradient conditions, and a single column (Chiralpak IA-U; with amylose tris(3,5-dimethylphenylcarbamate) selector immobilized on sub2-µm silica) provides broad enantioselectivity over the entire lipid mediator class (including both eicosanoids and pro-resolving lipids) (see Figure 12.6).[81] Generally, methanol-eluents provided better enantioselectivity but acetonitrile as a modifier was finally selected for its advantage regarding sensitivity. Acetic acid (0.1% added to both water and ACN) provided higher signal intensities than the more commonly employed formic acid. As can be seen in Figure 12.6, all analytes are eluted within 5 min under selected conditions. Even faster separations were possible with higher flow rates, but at the expense of detection sensitivity. This UHPLC method was finally applied for enantioselectively analyzing oxylipins in releasates of resting human platelets and of thrombin-activated platelets.

12.2.4.3 Short α- and β-hydroxy acids

Short α- and β-hydroxy acids have received special consideration for chiral analysis since they are part of, or connected to, important metabolic pathways (*e.g.* tricarboxylic acid cycle, TCA; lactic acid fermentation), and the presence of some enantiomers is directly linked to particular dysfunctions. Among them, lactic acid,[104] 2-hydroxy-3-methylbutyric acid,[105,106] 2-hydroxy-3-methylpentanoic acid,[105,106] 2-hydroxyisocaproic acid,[105–107] 2-hydroxyadipic acid,[108] and also 2-hydroxyglutaric acid[109–111] have been related to certain types of organic acidurias.

Other short hydroxy acids have also shown regulation in certain diseases. Some examples are 2-hydroxyisocaproic acid in Nocardia infection,[112]

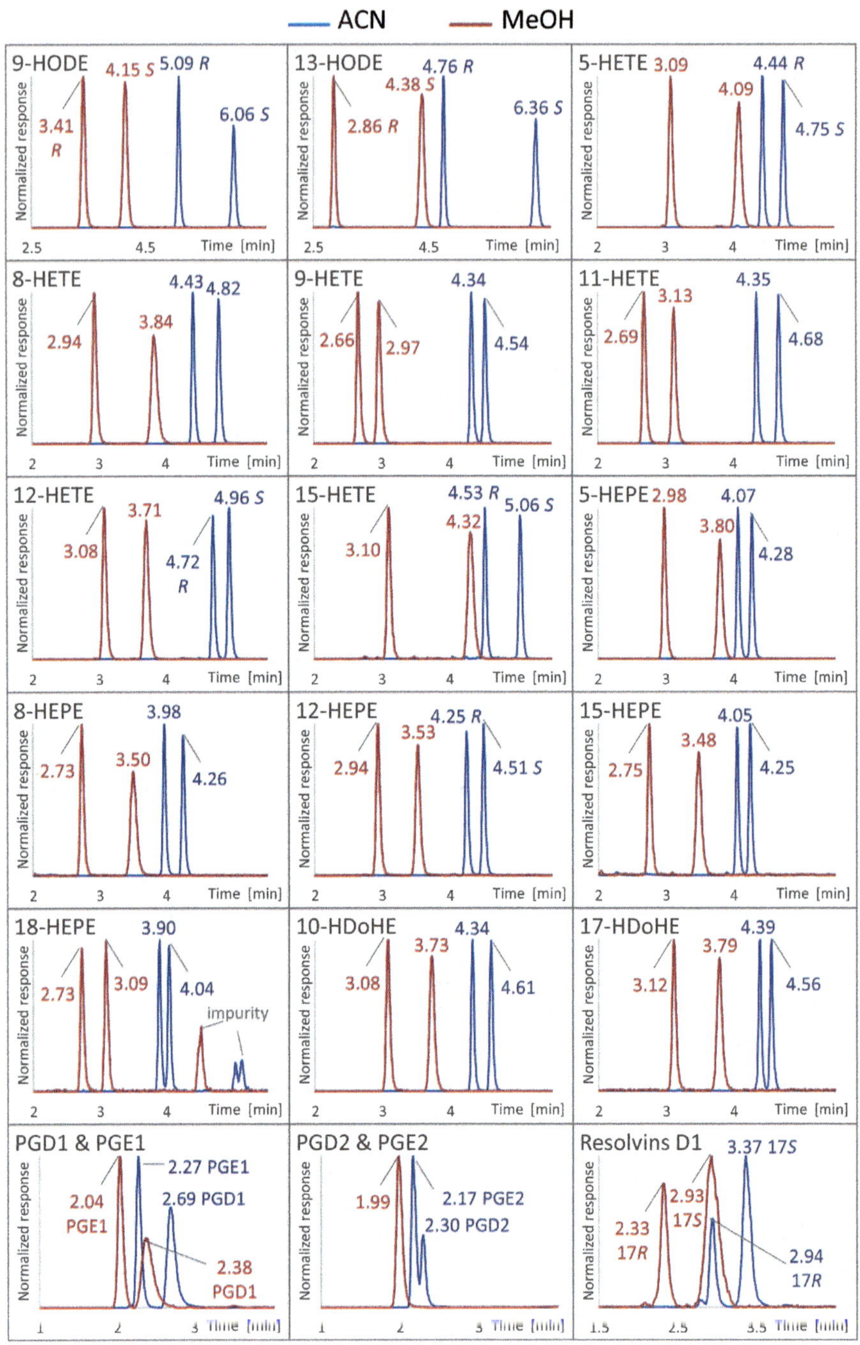
ACN
MeOH
9-HODE
3.41 R
4.15 S
5.09 R
6.06 S
13-HODE
2.86 R
4.38 S
4.76 R
6.36 S
5-HETE
3.09
4.09
4.44 R
4.75 S
8-HETE
2.94
3.84
4.43
4.82
9-HETE
2.66
2.97
4.34
4.54
11-HETE
2.69
3.13
4.35
4.68
12-HETE
3.08
3.71
4.72 R
4.96 S
15-HETE
3.10
4.32
4.53 R
5.06 S
5-HEPE
2.98
3.80
4.07
4.28
8-HEPE
2.73
3.50
3.98
4.26
12-HEPE
2.94
3.53
4.25 R
4.51 S
15-HEPE
2.75
3.48
4.05
4.25
18-HEPE
2.73
3.09
3.90
4.04
impurity
10-HDoHE
3.08
3.73
4.34
4.61
17-HDoHE
3.12
3.79
4.39
4.56
PGD1 & PGE1
2.04 PGE1
2.27 PGE1
2.69 PGD1
2.38 PGD1
PGD2 & PGE2
1.99
2.17 PGE2
2.30 PGD2
Resolvins D1
2.33 17R
2.93 17S
3.37 17S
2.94 17R
Normalized response
Time [min]

2-hydroxybutyric acid in early stages of type II diabetes,[113] 2-isopropylmalic acid in asthma,[114] and lactic acid and D-2-hydroxyglutaric acid in cancer.

In other fields, for example in wine production, the enantiomeric ratio of malic acid and lactic acid enantiomers allows the presence of microorganisms producing malolactic fermentation to be monitored. This process generates L-lactic acid, which reduces acidity and thus affects the taste of wine.[115]

In these metabolites, the proximity of the hydroxyl stereogenic center to the carboxylic group is often critical, and its location at the 2- or 3-position of the carboxylic group has typically enabled their enantioselective direct analysis by using antibiotic CSPs (*e.g.* 2-hydroxyglutaric acid on ristocetin R CSP)[116–118] and anion exchangers or zwitterionic ion-exchangers CSPs derived from quinine or quinidine alkaloids.[26–28,47,119–121] For the latter group of columns (Chiralpak QN-AX and QD-AX as well as Chiralpak ZWIX(+) and ZWIX(−)), the majority of 2- and 3-hydroxy carboxylic acids could be successfully separated.

12.2.4.4 Glycerides

Direct enantioselective analysis has been described in a few publications for glycerides. Chiral analysis of 1- and 3-mono-oleoylglyceride (1-MOG and 3-MOG) and 1- and 3-mono-linoleoylglyceride (1-MLG and 3-MLG) enantiomers has been reported by using polysaccharide-based CSPs. The developed methodology was employed to study the stereoselectivity of four microbial lipases for the esterification of glycerol with fatty acids.[122,123] García *et al.* also reported the applicability of polysaccharide-based CSPs for the separation of some 1- and 3-monoacylglycerides (1-MG and 3-MG) enantiomers, independently of the chain length and the degree of unsaturation of their fatty acids.[124]

Enantioseparation of diacylglycerides racemates of 1,2-dicapryloyl-*sn*-glycerol (1,2-DC) and 1,2-dioleoyl-*sn*-glycerol (1,2-DO) has been accomplished by using polysaccharide-based CSPs.[125]

For triacylglycerides, separation of sn1- and sn3-enantiomers of AAB-type triacylglycerides has been realized on polysaccharide-based CSPs[126–128] and cyclodextrin-based CSPs.[129,130] One disadvantage which has been highlighted for these separations is the long retention time required and the fact that the separations were performed in normal phase mode with

Figure 12.6 Comparison of enantioselective gradient RP UHPLC chromatograms obtained on sub-2μm Chiralpak IA-U column with ACN (blue) and MeOH (red) as organic modifier. Chromatographic conditions: Column: Chiralpak IA-U, flow rate: 3 mL min^{-1}; column temperature 25 °C; gradient for ACN: A: H_2O + 0.1% (v/v) acetic acid, B: ACN + 0.1% (v/v) acetic acid, 50–95% B in 5 min, 95% B for 2.5 min, 95–50% B in 0.1 min, 50% for 2.4 min; gradient for MeOH: A: H_2O + 0.1% (v/v) acetic acid, B: MeOH + 0.1% (v/v) acetic acid, 90–100% B in 5 min, 100% B for 2.5 min, 100–90% B in 0.1 min, 90% for 2.4 min.
Reproduced from ref. 81 with permission from Elsevier, Copyright 2020.

hexane–isopropanol, but also with hexane–acetonitrile–isopropanol as the eluent which renders these methods less suitable for ESI–MS.

12.3 Indirect Chiral Analysis by HPLC

Indirect chiral separation and metabolite analysis, respectively, involve a pre-column derivatization step with an enantiomeric chiral derivatizing agent (CDA) in order to transform the analyte enantiomers into diastereomers which can be separated by achiral stationary phases, usually RP columns (Figure 12.1). Generally, these achiral stationary phases have better efficiencies compared to chiral stationary phases and are more commonly found in laboratories. At the same time, this approach is promoted by the fact that many metabolites require a derivatization step anyway to improve their chromatographic properties or to improve their detection sensitivity.[131] Therefore, the indirect approach with chiral derivatization of enantiomer metabolites followed by chromatographic separation of the formed diastereomers is often preferred by researchers in the context of metabolite analysis,[58] although it is less common nowadays for enantiomer separations of pharmaceuticals. However, it must be emphasized that it requires additional validation expenditure to make sure that accurate results are obtained in metabolomics applications by the indirect approach.

In this scenario, the derivatization with CDAs has two main goals, on one hand the production of diastereomers and on the other hand the increase of sensitivity during detection through introduction of a chromophore, a fluorophore, or a moiety which can be easily ionized during MS measurements. Additionally, an increase in hydrophobicity is a side effect of derivatization which improves the separation in LC.

12.3.1 Chiral Derivatization Agents

The selection of the chiral derivatizing agent (CDA) is of prime importance in indirect chiral separations, like the selection of a chiral stationary phase in direct methods. In general, there are several features which are required for a CDA: (1) It should allow a selective reaction of the CDA with the analytes under mild conditions without side reactions. (2) A high reactivity with the target compound is desirable as it allows short reaction times and ensures quantitative reactions. (3) The derivatives must be chemically and stereochemically stable. (4) The reaction must proceed without racemization. This must be particularly considered when activation reagents are involved in the reaction, *e.g.* DCC activation of chiral carboxylic acids with the stereogenic center in the α-position is prone to racemization. (5) The CDA should have a chiral recognition moiety, *i.e.* at least one or preferentially more than one stereogenic center, close to the reactive functional group of the CDA to improve the chiral recognition and stereoselectivity of the resultant diastereomers on achiral stationary phases. The farther away the chiral element is from the stereogenic center of the analyte, the lower the separation factor is usually between the diastereomers, if not other effects which stabilize a particular conformational arrangement are present. Incorporation of the chiral center into a rigid structure is usually favorable as it reduces conformational flexibility and thereby maximizes physicochemical differences in terms of interactive hydrophobic surface area. (6) The CDA should be readily commercially available, or easy to synthesize, in both enantiomeric forms with high enantiomeric purity (>99.9% enantiomeric excess). Switching the configuration of the CDA allows a reversal of the elution order which is of practical relevance for validation of results and adjusting elution orders, *i.e.* eluting minor trace levels of metabolites first. If it is eluted as a second peak on the tailing edge of the main enantiomer peak it may lead to inaccuracies due to problems with accurate integration. (7) Besides a reactive group and an effective chiral recognition moiety, the CDA should contain a highly sensitive detection label, *i.e.* a chromophore for UV, a fluorophore for fluorescence detection, a charged moiety or group with high proton affinity for MS, or an electrochemically active moiety for ECD. (8) The molecular mass of the CDA should not be extremely large but high enough to provide characteristic fragment ions in tandem quadrupole MS/MS analysis.[131,132] To this end, efforts have been focused on the synthesis of chiral derivatizing reagents for the labeling of target metabolites with different functional groups, in particular for amino acids and carboxylic acid metabolites, that fulfill these requirements by different research groups.[132]

From these considerations it can be derived that any of the CDAs contain three essential elements: i) a reactive functional group, ii) a chiral moiety, and iii) a detection tag. The reactive functional groups in CDAs depend on the derivatizable functional moieties in the target metabolites (amino, carboxylic, carbonyl, hydroxyl, and thiol group).[131,133–135] For metabolites, indirect separations have been mainly used for amino acids, chiral amines, and hydroxy carboxylic acids. The most common reactions for chiral amines are their transformations with activated carboxylic acids (to amides), chloroformates (to carbamates), *N*-haloarylderivatives (to *N*-arylamine derivatives), isocyanates (to ureas), isothiocyanates (to thioureas), and o-phthalaldehyde (OPA) (to isoindole derivatives), for acids their reactions to amides and esters, and for alcohols their derivatizations to esters and carbamates.[55,136–138]

The number of CDAs is substantial, and a few commonly used ones are shown in Figure 12.7. The most popular approach for indirect enantiomer separation is probably based on the derivatization of primary amines, *e.g.* of amino acids, with OPA and a chiral thiol (*N*-acetyl-L-cysteine, Boc-L-cysteine, isobutyryl-L-cysteine) to isoindole derivatives.[139] The reaction is fast, yields fluorescing derivatives, and the reagent excess does not interfere with target analyte detection by FLD (because the reagent itself is non-fluorescent). The derivatization can easily be done in the autosampler just before the injection to improve repeatability and reproducibility, respectively. However, the derivatives are not stable, and reaction does not work for secondary amines and secondary amino acids, respectively. Since the above-mentioned chiral thiols (cysteine or penicillamine derived) are readily available in the

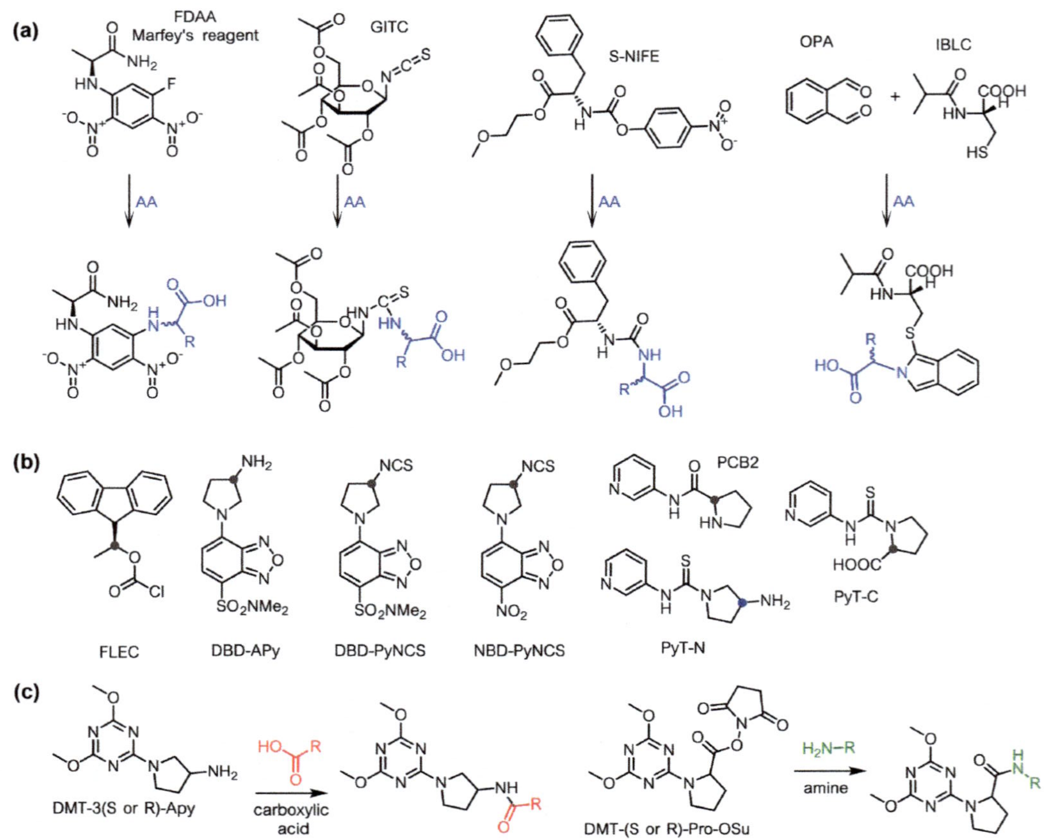

Figure 12.7 Chiral derivatizing reagents for indirect analysis by LC–UV, LC–FLD, and LC–MS.

D-configuration as well, the elution order of the corresponding diastereomeric isoindoles can be switched easily.

Another extremely popular CDA is, for example, 1-fluoro-2,4-dinitrophenyl-5-L-alanine amide, an *N*-haloarylamino acid, known as Marfey's reagent (MR).[140] Marfey's reagent has been widely used thanks to many factors such as the simplicity of the derivatization procedure, the good resolution of diastereomers using common reversed-phase stationary phases, good sensitivity in the UV region (at 340 nm), and the possibility of not only having the classical MR but also modifications with other auxiliary amino acids (as both D- and L-configured reagents) commercially available.[141] MR has been considered the most successful CDA for enantiomeric analysis of amino acids. It does not suffer many other problems observed with other CDAs (like OPA) such as an inability to react with all proteinogenic AA (it can be applied to secondary amines and amino acids as well), poor detectability, and unstable derivatives among others.[59] Its derivatives, however, are not fluorescent. Yet, Marfey's reagent was also successfully applied in combination with MS detection.[142–144] Recently, also its application to hydroxyl acids was proposed.[145] Since the tag resulting from Marfey's reagent does not contain a positive charge or a highly proton-affine nitrogen, Marfey's derivatives of amino acids are usually analyzed in negative ESI mode in which the carboxylate of the amino acids carries the negative charge.[144] For better MS compatibility of the derivatized sample, volatile amines such as triethylamine have been suggested to adjust basic conditions for derivatization instead of $NaHCO_3$.[142] Over the years, efforts have been made to improve Marfey's reagent in order to overcome limitations such as selectivity problems between allo-Ile/Ile isomers on the C18 phase. To this end, the CDA N_{α}-(2,4-dinitro-5-fluorophenyl)-L-leucinylamide (L-FDLA) has been suggested and evaluated along with ESI–MS detection.[142,146] A comparison of OPA and Marfey's reagent for LC–ESI–MS enantiomer separation has been published by S. Hess *et al.*[147]

Another frequently employed CDA is (+) or (−)-1-(9-fluorenyl)ethyl chloroformate (FLEC) (Figure 12.7b) which is a chloroformate-type CDA developed for reactions with amino groups including amino acids. Under certain conditions, FLEC can also react with hydroxyl groups. Major advantages for this CDA are its fast reaction kinetics and its simple derivatization procedure.[148] FLEC has been shown to be a versatile CDA for amino acid analysis by LC–ESI–MS[149] and very recently timsTOF without LC separation.[150]

These classical CDAs showed excellent chromatographic selectivities and have been the first choice for a long time. However, their detection properties were neither optimal in terms of fluorescence detection nor MS/MS detection. A number of new CDAs with specific focus for fluorescence and MS detection have been designed by Toyo'oka and co-workers.[132]

DBD derivatives with a chiral pyrrolidinyl moiety and isothiocyanate as the reactive group (4-(3-isothiocyanatopyrrolidin-1-yl)-7-(*N*,*N*-dimethylaminosulfonyl)-2,1,3-benzoxadiazole, DBD-PyNCS) for amino acids and amine labeling and reactive amino group (4-(3-aminopyrrolidin-1-yl)-7-(*N*,*N*-dimethylaminosulfonyl)-2,1,3-benzoxadiazole, DBD-APy) for carboxylic acid derivatization, respectively, have been originally developed for sensitive analysis by fluorescence detection.[132]

DBD-PyNCS (*R*- and *S*-forms) provided advanced performance for D-amino acid determinations. However, in biological samples peptides and other labeled compounds resulted in too many interferences with fluorescence detection.

Considering that MS has become the predominant choice for detection coupled with LC chromatographic separations, there has been a clear trend for development of CDAs designed for this technique. The work of Toyo'oka *et al.* has been of great relevance in this regard. Figure 12.7b–c shows the structures of the most relevant CDAs developed by Toyo'oka *et al.* The above DBD-labeled reagents can also be successfully utilized in (U)HPLC–MS/MS using SRM acquisition. However, CDAs with better properties for MS detection have been designed carrying a pyridyl moiety, *e.g.* (*R*)-*N*-(3-pyridylthiocarbamoyl)pyrrolidine-2-carboxylic acid ((*R*)-PyT-C) for analysis of chiral amines and amino acids. Its combination with (*S*)-3-amino-1-(3-pyridylthiocarbamoyl)pyrrolidine ((*S*)-PyT-N) for parallel analysis of carboxylic acids allowed metabolite coverage for both achiral and chiral metabolites by (U)HPLC–MS/MS-based metabolomics.[131] In both reaction batches the carboxylic acid is activated with 2-pyridyl-disulfide (DPDS) and triphenylphosphine (TPP) (60 °C, 60 min). (*R*)-PyT-C and (*S*)-PyT-N are structurally very similar CDAs which, although designed for enantiomeric analysis of distinct analyte classes *viz.* chiral amines and carboxylic acids, respectively, generate the same product ion (at *m*/*z* 137.0) upon collision-induced dissociation (CID) from protonated precursor ions in positive ESI–MS/MS.[132] LODs were in the range of 0.7–1.5 fmol for carboxylic acids and 0.5–1.9 fmol for amines (on-column).

With the idea of further increasing sensitivity, CDAs incorporating a triazine structure were synthesized (Figure 12.7c). These new CDAs, 1-(4,6-dimethoxy-1,3,5-triazine-2-yl)pyrrolidin-3-amine (DMT-3(*S*)-Apy) for analysis of chiral carboxylic acids, 1-(4,6-dimethoxy-1,3,5-triazine-2-yl)pyrrolidine-2-carboxylate (DMT-(*S*)-Pro) and the corresponding OSu-activated carboxylic reagent 2,5-dioxopyrrolidin-1-yl-1-(4,6-dimethoxy-1,3,5-triazine-2-yl) pyrrolidine-2-carboxylate (DMT-(*S*)-Pro-OSu) both for chiral amines share common product ions with *m*/*z* 195.3 and 209.3 upon CID (Figure 12.8) which is favorable for their application in broad coverage chiral metabolomics studies.[132] LODs $<$ 10 amol (on-column) have been reported for some analytes with these CDAs.

A variety of other CDAs have been favorably used for enantioselective analysis of metabolites, and more details can be found in specific reviews on this topic.[132,136,138]

12.3.2 Opportunities and Caveats of the Indirect Approach with CDAs

The indirect enantiomer separation approach using CDAs offers some advantages over the direct approach. Through the derivatization, the polar metabolites are made accessible for RPLC and with this optimally compatible with ESI–MS. This opens up the possibility to use the large variety of RP columns which are not only more efficient due to the fast exchange kinetics and lower C-term in HPLC separations, but also significantly cheaper than

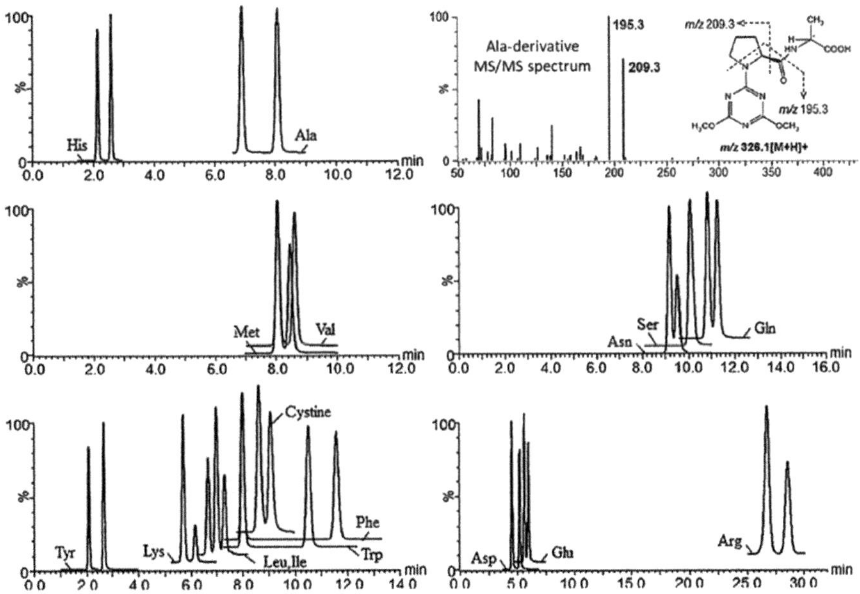

Figure 12.8 Typical SRM chromatograms of DL-amino acids derivatives. Reproduced from ref. 134 with permission from Elsevier, Copyright 2015.

chiral columns. Moreover, until recently chiral UHPLC columns (sub-2 µm fully porous particle and sub-3 µm core shell particle columns) were not available. Hence, the indirect approach by UHPLC was an attractive opportunity for highly efficient separations of multi-metabolites samples such as complex amino acid mixtures.[149] Users can apply their well-established method development strategies and even software supported simulation tools. Knowledge in the lab regarding reversed phase method development broadly exists, while it is less common with respect to the direct approach with chiral columns. The efficient separations that can be achieved on UHPLC columns are associated with the benefit that the sharper peaks exhibit improved detectability and thus enhanced sensitivity.[37]

However, users of the indirect approach must be aware of a couple of caveats which require specific attention during method development and validation. The CDA must have a high enantiomeric (stereoisomeric) purity (>99.9%). If there is a significant enantiomeric impurity in the CDA, it must be considered in quantitative analysis otherwise inaccurate results will be obtained. Consider a 1% enantiomeric impurity in both the analyte and the enantiomeric CDA (both with one stereogenic center), assuming an equal reactivity for the formation of all 4 stereoisomers, it would result in a biased enantiomeric excess (ee) of 96% instead of the true 98% due to the presence of all 4 stereoisomers (see the caption of Figure 12.1) of which the corresponding enantiomer pairs co-elute in RPLC. This means that the enantiomeric purity of the CDA must be determined and validated during method validation. It is usually neglected. Hence, the results of the indirect approach can be rarely considered accurate for trace metabolite enantiomer analysis. As a rule of thumb, if trace metabolite levels of an enantiomer should be accurately quantified, the CDA should have an enantiomeric purity of factor 10 less than the trace impurity level that should be determined. Furthermore, it has to be proven during validation that no racemization in the course of the derivatization procedure occurs. It must be also secured that no kinetic racemate resolution happens or proper measures to correct for it have to be considered, in particular when relative enantiomer quantification *via* peak areas is employed. Since reaction kinetics can be different for the two enantiomers, one has to make sure that the reaction has proceeded quantitatively by sufficient reagent excess and long enough reaction time. If the reaction is stopped too early, a different enantiomer composition in the sample would be determined than is actually present; this has to be considered and one has to make sure that no bias occurs due to this phenomenon. Last but not least, it has to be considered that diastereomers are actually entering the detector and due to distinct physicochemical properties of diastereomers their detector response might differ (often by factor of 1.1–1.2). Hence, it may not be accurate enough to derive the enantiomer ratio from the peak area ratio of diastereomers; separate calibration is required. Finally, it is needless to emphasize that the adaptation of this concept to compounds having no appropriate reactive functional group fails. Several of the problems that can occur are often properly addressed by the workflow, however, the indirect method is not without risk for bias in quantitative enantiomer analysis.

In conclusion, several disadvantages are attributed to chiral indirect analysis, and several critical factors must be properly validated in addition to common method validation, such as the enantiomeric purity of the CDA, absence of racemization or kinetic resolution, lack of interferences originating from structurally closely related side products or from excess of CDA, and if quantification by peak area ratio is performed also the response factor between diastereomers. It is clear that this represents additional validation expenditure to assure accurate quantitative results.[136] Additionally, from the perspective of comprehensive analysis, the derivatization of all chiral metabolites in the same sample seems very difficult considering the diversity of functional groups or the absence of them in the compounds. Multiple parallel derivatizations with orthogonal CDAs can be a solution. Yet there are chiral compounds that cannot be derivatized at all. Finally, complications may arise from (incomplete) multiple derivatizations of metabolites with several reactive functional groups.

12.3.3 Metabolites Analyzed by Indirect Analysis

12.3.3.1 Amines and Amino Acids

As discussed above, amino acids are chiral metabolites of primary interest as they are central metabolites and therefore of major bioanalytical interest. Levels of amino acids are already controlled in newborn screening for inborn errors of metabolism. However, this is commonly performed non-stereoselectively. D-Amino acids on the other hand have been reported to be elevated in age related diseases and are therefore becoming of increasing importance in this context. Moreover, some D-amino acids have modulating neurotransmitter bioactivity such as D-Ser at the NMDA receptor. Besides that, elevated levels of D-amino acids are commonly found in food, and their biological role or effects are of particular interest. It is therefore not surprising that a lot of research is directed to their enantioselective analysis. Indirect separations of chiral amines using chiral derivatization agents have been extensively reported, and usually these derivatizations are addressed to the analysis of amino acids (see Table 12.3). Detailed reviews about the CDAs for analysis of proteinogenic,[136] nonproteinogenic[137] and unusual amino acids[138] have been published. For practical applications in the context of enantioselective metabolomics, the coupling of indirect separations with MS is of major interest. Wide coverage, at least for all proteinogenic amino acids, is needed, and this can be accomplished by many indirect methods in combination with MS detection since stereoselectivity is usually adequate and selective detection by MS/MS (*via* EICs of specific precursor or fragment ions, respectively) does not require complete separation of all components. In addition, mass spectrometry allows specific isotopic patterns to be captured. This can be of relevance when the isotopic distribution provides additional information, such as in stable isotope labeling experiments of amino acids in cell culture (SILAC), where the incorporation of stable isotopes is used for generating quantitative information (*vide infra*).[143]

A variety of CDAs are available for enantioselective amino acid analysis, and all of the classical CDAs have proven their usefulness for LC–MS. It was reported that the popular Marfey's reagent allowed the LC–MS/MS-based separation and detection of the resulting diastereomers for quantification of the 19 common L- and D-amino acids and glycine.[144] While acidic formic acid based mobile phase did not allow full separation of all 19 enantiomer pairs of amino acids, a water/acetonitrile/ammonium acetate (pH 6.5) solvent was successful in combination with a C18 phase and a 30 min gradient. These conditions were also favorable in view of the sensitivity in negative ESI mode as the reagent itself was not designed for MS application and the label did not specifically support electrospray ionization efficiency. SRM acquisition was based on the three most sensitive Q3 fragments for each amino acid, while no universal reagent specific fragment ion was amongst the three SRMs. As a single internal standard $^{13}C_3$-Ala was used to compensate for errors such as injection variance. The method employed L-Marfey's reagent as CDA. However, due to the lack of appropriate D-amino acid standards (validated L-amino acid standards are commercially available, similar D-amino acids standards are not) the authors used D-Marfey's reagent to generate D-Marfey-L-amino acids and thus provide L-Marfey-D-amino acid standard surrogate which due to the enantiomeric relationship have the same retention time (see the caption of Figure 12.1). In this way the (separation) method could be developed. The authors also carefully investigated the reaction kinetics (with 4-fold molar

Table 12.3 Indirect chiral metabolomics.

Metabolites	Sample	Chiral derivatizing reagent	Detection	Reference
D,L-selenoMet, D,L-Met, D,LCys		(*R*)-1-Boc-2-piperidinecarbonyl chloride	MRM	161
Proteinogenic D-amino acids	Human plasma, cerebrospinal, urine	(*S*)-NIFE	MRM	162
Proteinogenic amino acids		16 CDRs	UV	163
Proteinogenic amino acids		FLEC	SIM	152
Phe, Leu, Ala, Ile, Lys	Surugamides A–E, cyclic octapeptide	GITC, FDAA	MRM	164
D-Asp, D-Ser, D-Ala	Brains of mice and rats	OPA and Boc-L-Cys	Fl	94
Proteinogenic D,L-amino acids	Serum, plasma, urine, gut	OPA/IBLC	SIM	165
N-Aβ_{1-5} peptides containing L,D-Asp, L,D-isoAsp	Human brain tissue	*R*(−)-DBD-PyNCS	MRM	166
Trp	Rat plasma	*R*(−)-DBD-PyNCS	Fl	167
Trp, kynurenine	Rat plasma	*R*(−)-DBD-PyNCS	Fl	168
18 Proteolytic amino acids	Human serum, food products	L-PGA-OSU	MRM	3
Amines and amino acid methyl esters	Rat plasma	L-PGA	MRM	169
Proteinogenic amino acids and amines	Saliva	DMT-(*S*)-Pro-OSu	SRM	134
Amines and carboxylic acids	CSF of Patients with Alzheimer's Disease	DMT-(*S*)-A and DMT-(*S*)-PO	DDA	170
Amines and carboxyls	Alzheimer's disease (AD) brain homogenates	DMT-(*S*,*R*)-Pro-Osu and DMT-3(*S*,*R*)-Apy	precursor ion scan	133
Amines and carboxylic acids	Human saliva	PyT-C and PyT-N	SRM	131
Carboxylic acids	Saliva of diabetic patients	DMT-3(*S*)-Apy, DMT-1(*S*)-Apy	MRM	135
2-hydroxy acids		(+)-FLEC	UV	155
Glyceric acid, lactic acid, 2-hydroxybutyric acid, 3-hydroxybutyric acid, mandelic acid, phenyllactic acid and *N*-acetyl-D-tryptophan	Human serums and mammalian cells	4 CAPA analogs	MRM	160
Lactic acid	CSF of children with tuberculous meningitis	DATAN	MRM	156
2-Hydroxyglutarate	Renal cell carcinoma	DATAN	SIM	157
2-Hydroxyglutarate and related metabolites	Synovium tissue, serum	TSPC	MRM	158

excess of the reagent). Indeed, as mentioned above it is a critical factor not only for sensitivity of the method but primarily for the fact that kinetic resolution effects might lead to bias of the determined enantiomer composition. Fast reaction kinetics was observed for the majority of the mono-adduct forming amino acids ($4 \times t_{\frac{1}{2}}$ was used to indicate an incubation time for 95% completion of the derivatization reaction) (95% conversion between 30 min (Pro) and 10 h). A 24 h incubation was found sufficient for all amino acids except His which required >78 h for quantitative formation of the bis-adduct. Significantly different kinetic constants have been determined for D- and L-amino acids which emphasize the risk for kinetic resolution and need careful optimization of the reaction time and temperature to avoid such bias *i.e.* reaction must be driven to completeness (Marfey's reagent in reaction mixtures with at least ~4-fold excess of CDA to afford pseudo-first order reaction kinetics). The Marfey's derivatives are stable for at least 24 h. With this RPLC–MS/MS method LLOQ between 2–8 pmol and linearity up to 800 pmol on-column could be achieved, which is not the most sensitive method but useful for a variety of applications. MRSA (multiple resistant *Staphylococcus aureus*) extracts were analyzed and the D levels ranged from several not detected D-amino acids to a majority between 1–3% and a few relatively highly abundant amino acid (His, Asp, Ser 8-10%; Glu 21% and Ala 28% D).[144] A persisting problem of the Marfey's method is the inability to differentiate between L-Ile and L-allo-Ile as well as D-Ile and D-allo-Ile. This is annoying because both Ile and also allo Ile often occur in natural products. An improved resolution for Ile/allo-Ile was reported for a so-called C3 Marfey's method whereby the C3 stands for a C3-bonded silica column (*e.g.* Agilent Zorbax SB-C3) used instead of a C18 column in the standard C18 Marfey's method.[151]

SOPs for the derivatization and a performance comparison of Marfey's reagent (FDAA) with some other classical indirect amino acid enantiomer separation methods (GITC, *S*-NIFE, and OPA-IBLC) (see Figure 12.7a) under unified conditions on a reversed phase column (Zorbax C18) using mass spectrometry (ESI–quadrupole MS detection) with MS compatible buffers (B: 5% acetic acid/B: 10% methanol in acetonitrile) were reported by S. Hess.[143] GITC derivatized amino acids are less stable than FDAA and must be analyzed within 24 h. *S*-NIFE derivatives are stable. OPA-IBLC derivatives are commonly analyzed by fluorescence, which rapidly decreases indicating the instability of these derivatives. The usually observed high enantioselectivity for OPA-IBLC could not be maintained with the mass spectrometry compatible eluent. OPA-IBLC showed the lowest enantioselectivity for most amino acids except for aromatic amino acids. A general order for the magnitudes of enantioselectivity was FDAA > S-NIFE > GITC > OPA-IBLC, but for aromatic amino acids, the order was *S*-NIFE > OPA-IBLC > GITC > FDAA. The mobile phase with 5% acetic acid in the water phase and adding 10% methanol to the acetonitrile phase generally showed lower selectivity compared to 0.1% TFA in water/acetonitrile but improved signal intensities in ESI–MS. In terms of sensitivity, the order was GITC > *S*-NIFE ~ FDAA > OPA-IBLC with typical LODs in the low picomolar range (on column).

Another common amino-reactive CDA with fluorogenic label, *viz.* FLEC, has been also successfully utilized in RP–UHPLC–MS/MS.[152] Often, reversed-phase HPLC using nonpolar stationary phases (C4, C8 and C18) and tetrahydrofuran (THF) based mobile phases were reported to be successful for a complete separation of all 19 proteinogenic amino acid enantiomer pairs after FLEC derivatization. Since THF is normally not appreciated in LC–MS, as alternatives phenyl-based stationary phases and several mobile phases consisting of ammonium acetate and different common organic solvents (acetonitrile (ACN), methanol (MeOH), 2-propanol (IPA)) were tested using gradient elution. After optimization of the separation conditions by experimental design, complete chiral separation could be achieved for 18 out of 19 FLEC-DL-AAs in less than 30 min in LC–ESI–MS. Recently, it was demonstrated that FLEC-derivatives of amino acids can also be resolved in the gas-phase without prior chromatography by trapped ion mobility spectrometry–TOF-MS (TIMS–TOFMS), as sodiated but not as protonated species.[150] Since diastereomers exhibit conformational differences, their collisional-cross sections are distinct which resulted in different gas phase mobilities in TIMS–TOF-MS. Voltage ramp, ramp time (in 255 or 510 ms), and accumulation time (in 50 or 100 ms) were optimized for each AA. The resulting TIMS resolution values ($K_0/\Delta K_0$) for the FLEC-amino acid diastereomers were on average about 115, requiring mobility (K_0) differences of about 0.009 $cm^2 V^{-1} s^{-1}$ to achieve 50%-valley separation which was accomplished for 17 out of 21 tested amino acids. Three methods were required for that because the different FLEC-amino acids required specific voltage ramps. The quantitative performance for real samples was not fully evaluated and validated but the work documented that TIMS–TOF-MS can be a fast screening method for enantiomeric compositions.

A detailed description of the powerful 4-(3-isothiocyanatopyrrolidin-1-yl)-7-(*N,N*-dimethylaminosulfonyl)-2,1,3-benzoxadiazole (DBD-PyNCS) CDA for the analysis of amino acids is given by Toyo'oka.[153] Although it was designed for fluorescence detection, it has shown reasonable utility for MS as well. However, Toyo'oka's group developed advanced CDAs, (*S* and *R*)-2,5-dioxopyrrolidin-1-yl-1-(4,6-dimethoxy-1,3,5-triazin-2-yl)pyrrolidine-2-carboxylate (DMT-(*S* and *R*)-Pro-OSu) for amines and amino acids, with tailored properties for ESI–MS/MS and improved detection sensitivities,[154] which are described in more detail in the context of enantioselective metabolomics workflows below.

12.3.3.2 Carboxylic Acids

Besides amino acids, chiral organic acids are the second most important class of metabolites for which enantioselective analysis is of relevance. A number of chiral organic acids have been described to show stereochemical alterations under certain metabolic and pathological conditions, respectively. 2- and 3-hydroxy acids as well as oxylipins are amongst the most important metabolites with carboxylic acid function for which enantioselective analysis is of biological relevance.

Hydroxy carboxylic acids can be either derivatized at the hydroxl group or carboxylic acid group. A number of hydroxy-reactive CDAs have been proposed, *e.g.* FLEC,[155] DATAAN,[156,157] and TSPC,[158] but also Marfey's reagent[145] has been used for this purpose (see Table 12.3). The reaction with the chloroformate reagent FLEC is typically performed under alkaline conditions (*e.g.* pyridine in the reaction mixture). As side reactions the carboxylate group can also be derivatized. DATAAN (*O*,*O*′-diacetyl-tartaric acid anhydride) derivatization occurs under acidic catalysis and yields the corresponding ester derivatives. *N*-(*p*-toluenesulfonyl)-L-phenylalanyl chloride (TSPC) labeling is efficient and is done in pyridine at 25 °C for 10 min. DATAAN, and TSPC have become popular CDAs for the enantioselective analysis of the oncometabolite D-2-hydroxyglutarate (2HG).[157–159]

On the other hand, a number of different carboxy-reactive CDAs for the enantioselective metabolomics analysis have been reported. Amongst them are a variety of amino group-containing chiral compounds comprising, DBD-Apy (mainly for fluorescence detection), DMT-3(*S*)-Apy (mainly for MS) and cinchona alkaloid derived primary amines (for fluorescence and MS).[160] Their application will be exemplarily discussed below.

Of particular interest in some applications can be CDAs that can react with different functional groups and allow the targeting of two distinct analyte classes. For example, the above-mentioned sulfonylchloride-type reagent TSPC which has hydroxyl- and amino reactive capability enabled the simultaneous analysis of chiral amino acids and hydroxy acids.[158] In a recent study, $^{18}O_2$-labeled-TSPC reagent was used for internal standardization to improve the accuracy in the quantification of 13 immunometabolites in a 2-HG metabolic panel.[158]

12.4 Workflows

The workflows in enantioselective metabolomics follow largely the general concepts established for achiral analytes including sampling, sample preparation, and data processing with exception of the separation step. Enantiomer discrimination during sampling and sample preparation has to be avoided to preserve the enantiomer ratio in the sample which means achiral materials have to be employed for this purpose.

12.4.1 Targeted Assays

The majority of studies on enantioselective metabolomics, both with direct and indirect approach, are employing targeted assays. However, rarely more than 20 metabolites are determined. The most comprehensive coverage is usually related to amino acid and oxylipin analysis. Coverages of 100s to 1000s of metabolites, which are quite common in achiral metabolomics, are far away in targeted enantioselective metabolomics. However, the number of studies on enantioselective targeted LC–MS of metabolites using selected reaction monitoring (SRM) is increasing. They need to be validated like achiral assays.[81] However, there is one aspect that is often critical. Matrix effects in both direct and indirect enantioselective metabolomics may occur stereoselectively. This can be compensated for by stable isotope labeled internal standards. They are often available for one enantiomer, but many times not for both. For instance, amino acids are readily available as a uniformly 13-carbon (U-^{13}C)-labeled L-enantiomer mixture. However, a corresponding U-^{13}C-labeled D-amino acid enantiomer mixture or corresponding DL-mixture cannot be purchased commercially. Since D-amino acids are usually present in low quantities besides large concentrations of L-amino acids and overlapping with the major constituents or other matrix components, the D-enantiomers might be significantly suppressed and subjected to different ion suppression than the corresponding L-enantiomer. Neither quantification by peak area ratio nor use of corresponding U-^{13}C-labeled L-amino acid internal standards can provide accurate results for the case of enantioselective matrix effects. A racemization method for a U-^{13}C-labeled L-amino acid mixture has been proposed as a potential solution of this problem to obtain by a straightforward thermal racemization procedure a U-^{13}C-labeled DL-amino acid mixture using AQC, not only as a tag for driving chromatographic separation on a Chiralpak ZWIX(+) column and boosting MS detection sensitivity, but also as a racemization tag.[22] In general, more awareness for the problem of enantioselective matrix effects and its proper correction both in direct and indirect approaches is required. On the other hand, this example shows that it must be proven that no racemization of the analytes occurs during achiral tagging, as stressed above for derivatization with CDAs in indirect methods.

For this and other reasons, in enantioselective amino acid metabolomics 2D-LC approaches with achiral fluorogenic tagging and fluorescence detection have been propagated that are currently probably the most accurate assays for the purpose of trace D-amino acid enantiomer analysis, as discussed above.[31] Multiple heart cutting (achiral–chiral) RPLC-chiral LC 2D-LC and (achiral–achiral–chiral) RPLC-weak anion exchange-chiral LC 3D-LC assays[30] have been developed for NBD-tagged amino acids in various biological samples. Usually, the assay is developed for less than 10 amino acids as targets and is barely considered as a metabolomics approach. However, these assays have been validated properly in order to ensure that also trace D-amino acid levels can be accurately analyzed. A downside for large scale clinical studies is a longer total analysis time. The method, however, was proven in several clinical studies to be a useful tool *e.g.* in biomarker studies of trace D-amino acid concentrations.

Another targeted chiral derivatization-based enantioselective metabolomics approach, somehow related to the isotope coded metabolomics methodologies, was based on a novel 4-plex chemical labeling strategy with 4 analogs of cinchona alkaloid-derived primary amines (CAPAs) which finally allowed the simultaneous determination of 16 enantiomers of 8 chiral carboxylic acids by UHPLC–ESI–MS/MS with MRM acquisition on a C18 column.[160] The CAPA CDAs are derived from cinchonan-9-amine and carry distinct substituents in position 3 of the quinuclidine (either ethyl in CAPA analogs *a* and *c* or vinyl in CAPA analogs *b* and *d*) and 6′ of the quinolone ring (either methoxy in CAPA analogs *a* and *c* or vinyl in CAPA analogs *b* and *d*). They showed exceptional selectivity for the enantiomers of 8 chiral carboxylic acids such as lactic acid, 2- and 3-hydroxybutyric acid, glyceric acids and *N*-acetyl-Trp, amongst others. The derivatization of the carboxylic group was catalyzed by TPP/DPDS, and the reaction was completed in 30 min at 40 °C. The 4 CAPA analogs can be readily distinguished by their characteristic MS fragment ions. Detection sensitivity was enhanced by up to 3 orders of magnitude compared to the non-derivatized analytes. These reagents were finally utilized in a 4-plex approach to achieve high-throughput analysis. In this 4-plex approach one CAPA analog was employed to label chiral carboxylic acid standards that served as internal standards (ISs), while the other 3 CAPA analogs were used to label endogenous chiral carboxylic acids in 3 different biological samples, which were mixed before injection like in isotope coded metabolomics. Although in this strategy matrix correction was not achieved by the internal standard, good accuracy could be proven by validation. The developed method was validated for the analysis of 16 chiral carboxylic acid enantiomers in human serum and mammalian cells and finally applied to serum samples from type 2 diabetes mellitus and colorectal cancer patients with significant changes for 5 and 4 chiral carboxylic acid enantiomers compared to the healthy control.

12.4.2 Untargeted Assays

Targeted assays have the limitation that new discoveries beyond the limited number of metabolites for which they have been developed cannot be made. To overcome this shortcoming, a few untargeted enantioselective metabolomics approaches have also been suggested. They involve chiral derivatization strategies.

One strategy is isotope coded chiral derivatization metabolomics[3,154,171,172] (Figure 12.9). This approach is used to compare the metabolite enantiomer levels in two distinct groups of samples, *e.g.* sample group A of diseased patients *versus* sample group B of healthy controls (differential profiling). Samples of group A are reacted with the light version of the CDA (light CDA) while the samples of group B are derivatized with a stable isotope labeled version of the CDA (heavy CDA). Before the analysis, the A and B samples are mixed and then analyzed by HPLC–MS/MS with high resolution MS in data dependent acquisition mode (untargeted) or triple quadrupole for dedicated sub-groups of metabolites *e.g.* amino acids (targeted). Heavy and light forms of the two diastereomers coelute in the RPLC separation, but can be distinguished by MS due to a mass shift corresponding to the respective isotope labeling. It means that the enantiomer ratio can be roughly estimated on the retention time domain and the difference in the D- and L-enantiomer levels in the distinct study groups on the *m*/*z* domain. The coeluted peaks from heavy and light version are subject to the same degree of ion suppression by the

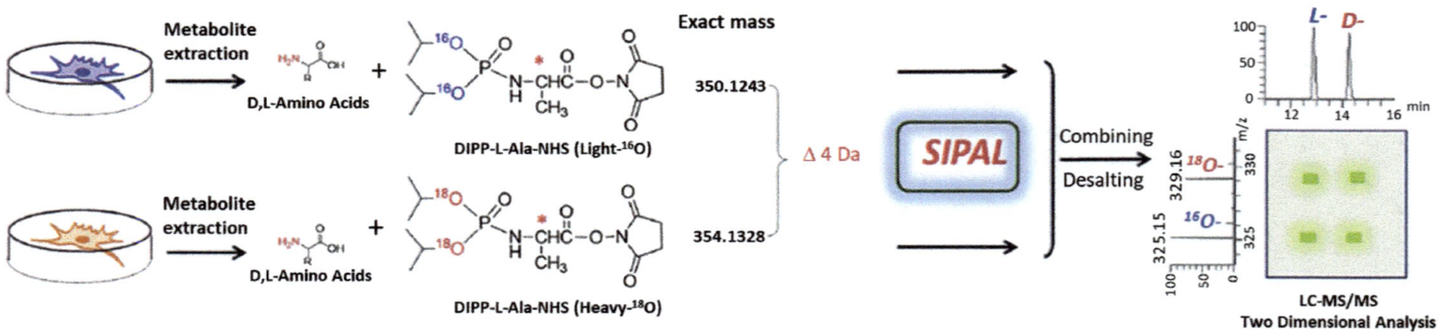

Figure 12.9 Strategy of stable isotope coded enantioselective differential metabolomics. Stable isotope *N*-phosphoryl amino acids labeling (SIPAL) for quantitative metabolomics analysis. Chemical structure of *N*-diisopropyl phosphoryl L-alanine (DIPP-L-Ala). The general structures of *N*-hydroxysuccinimide ester of DIPP-L-Ala (DIPP-L-Ala-NHS) as two-plex SIPAL tags. Experimental workflow of SIPAL strategy for quantitative analysis of chiral metabolites in biological samples using LC–MS. Reproduced from ref. 171 with permission from Elsevier, Copyright 2017.

matrix, so that the results are less biased by matrix effects. Significant up and down regulation of a particular enantiomer in the two study groups can be conveniently derived from the corresponding heavy and light versions of a fragment ion of the tag. Some caveats of this approach are the chemical and stereochemical impurities in the reagents, chromatographic isotope effects which may lead to (partial) separation of heavy and light derivatives, accompanied by imperfect compensation for matrix effects. A number of different isotope labeled CDAs have been proposed. Toyo'oka's group developed a duplex (*i.e.* two isobaric forms) NHS ester of L-pyroglutamic acid ($^2H_0/^2H_5$-L-PGA-NHS) for chiral amine and amino acid enantiomer ratio analysis.[3] Upon CID, the derivatized amino acids demonstrated characteristic product ion peaks originating from the 2-pyrrolidone moiety (*m*/*z* 84.04 and 89.04 for heavy and light versions, respectively). Although the proof of principle with this reagent was successful, the reagent suffers from some shortcomings. First, only 9 of the 19 chiral amino acids could be separated, some deuterium-labeled atoms are labile and can be back exchanged (α-^{2}H). A carboxylic acid reactive reagent for this approach was obtained with heavy and light version of $^2H_0/^2H_6$-DMT-3(*S*)-Apy.[132] Differences in D/L ratios of lactic acid in saliva samples from a healthy person and a patient could be determined. Recently, Toyo'oka's group presented a triplex form of this reagent (^{12}C-DMT-3(*S*)-Apy, $^{13}C_2$-DMT-3(*S*)-Apy, and d_6-DMT-3(*S*)-Apy) for isotope coded chiral differential metabolomics profiling.[173] It allows the comparison of three sample groups with each other. Its potential for enantioselective metabolomics is yet to be demonstrated. Another reagent for this approach, duplex isotope coded derivatization reagents, $^{18}O_0$- and $^{18}O_2$-*N*-diisopropyl phosphoryl L-alanine *N*-hydroxysuccinimide esters ($^{18}O_0/^{18}O_2$-DIPP-L-Ala-NHS) was successfully applied for amino acid metabolomics in an approach called stable isotope *N*-phosphoryl amino-acid labeling (SIPAL)[171] (Figure 12.9). The reagent cannot only be utilized in LC–MS/MS but also ^{31}P-NMR metabolomics. Using C18 RPLC with a Q-Exactive Orbitrap in data-dependent acquisition most of the 19 amino acids were well separated. The method was applied to human urine samples, and several D-amino acids, such as D-Ala, D-Arg, and D-His, were detected. Sakamoto *et al.* (2019) reported another diastereomeric isotope coded CDA, $^2H_0/^2H_5$-succinimidyl (4*S*)-(3-[(benzyloxy)-carbonyl]-5-oxo-1,3-oxazolidin-4-yl)acetate, for the chiral separation of D/L-amino acid which was utilized for the monitoring of the amino acid changes in rat plasma after administration of the alanine-serine-cysteine transporter-1 (Asc-1) inhibitor *S*-methyl-L-cysteine.[174]

While it is straightforward to determine known chiral molecules in untargeted approaches by targeted feature extraction, it is very difficult to identify and/or determine unknown enantiomer(s) in real samples by untargeted metabolomics.[132] Thus, Toyo'oka's group developed an untargeted enantioselective metabolomics strategy which they called chiral metabolomics fingerprinting and chiral metabolomics extraction.[133] The approach was based on derivatization with the CDAs DMT-(*S* or *R*)-Pro-OSu and DMT-3(*S* or *R*)-Apy (Figure 12.7c) for chiral amine and carboxy metabolites, respectively. Both DMT-Pro-OSu and DMT-3-Apy derivatives give the same characteristic product ion by CID and can be used to monitor analytes which carry the DMT label (*m*/*z* 209.2) by precursor ion scan (PIS). The enantioselective metabolomics study was performed in two steps (see Figure 12.10). In the first step (chiral metabolomics fingerprinting), the analytes (amines and carboxylic acids) of the samples were labeled with one enantiomeric form of the corresponding CDAs. The samples were then analyzed by UHPLC–ESI–MS/MS using a triple quadrupole instrument using PIS acquisition. The labeled analytes detected by PIS were then subjected to multivariate statistical analysis (PCA and OPLS-DA) to find features carrying the DMT tag and being significantly differently abundant in the different sample groups. At this stage, the selected significantly different features (potential biomarkers) of interest included both chiral (D and L forms of chiral amines and carboxylic acids) and achiral ones (achiral amines and carboxylic acids). To distinguish the chiral from the achiral features, in the second step (chiral metabolomics extraction) the samples were mixed and one aliquot derivatized with the *R*-reagents and another aliquot with the *S*-reagents of the above specified amino and carboxy-reactive CDAs. In the RPLC–MS/MS separation of the two samples, the achiral compounds are eluted at the same retention time with *R*- and *S*-reagent. In contrast, the chiral compounds still give two separated peaks, but with reversed elution order (if scalemic the reversal of elution order is obvious in the chromatogram by the different peak areas, if racemic it is not). Therefore, the peak area is also the same for the achiral compounds in the two methods. In contrast, the chiral compounds (if scalemic and not racemic) show distinct peak areas and can easily be found by multivariate statistics *e.g.* in the S-plot of OPLS-DA. The method allowed D-lactic acid to be found as a potential diagnostic marker in saliva of diabetic patients. Furthermore, it was applied for the determination of biomarker candidates in brain homogenate samples of patients suffering from Alzheimer's disease (AD). As a result, L-phenylalanine and L-lactic acid were identified as decreased and increased biomarker candidates in the AD brain, respectively.

As discussed above, the accuracy of any metabolomics approach, not only targeted but also untargeted ones, may be severely compromised by matrix effects. Trace enantiomer peaks may be more affected. To overcome this problem, a stable isotope corrected chiral and achiral nontargeted metabolomics approach involving the light version of the above CDA DMT-(*S*)-Apy for carboxylic acids and DMT-(*S*)-Pro-OSu for amines as well as their corresponding heavy ($^{13}C_2$-isotope-coded) versions, iDMT-(*S*)-Apy and iDMT-(*S*)-Pro-OSu.[170] The strategy was based on the derivatization of the analytical samples with the light version of CDAs, while mixed aliquots of all study samples (QC pool) were derivatized with heavy version of CDAs (correction samples). Derivatized analytical and correction samples were mixed and analyzed by RP–UHPLC–ESI–QTOF–MS/MS using data-dependent acquisition (DDA). The data were preprocessed by MS-DIAL which supports peak finding, background subtraction, noise filtering, deisotoping, adduct

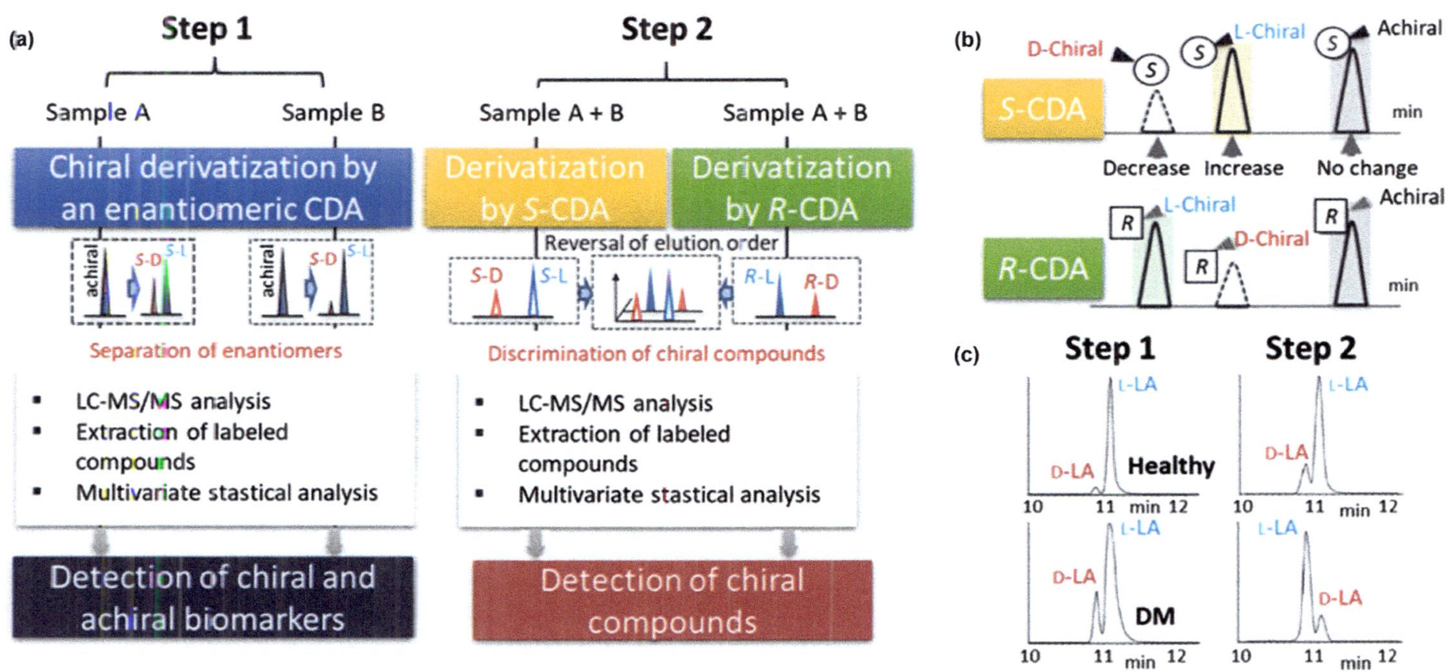

Figure 12.10 Untargeted enantioselective metabolomics strategy by chiral metabolomics fingerprinting (step 1) and chiral metabolomics extraction (step 2). (a) General workflow. Reproduced from ref. 132 with permission from The Japan Society for Analytical Chemistry, Tokyo, Copyright 2017. (b) Distinction between chiral and non-chiral metabolites by reversal of elution order through parallel analysis with *S*-CDA and *R*-CDA. Reproduced from ref. 154 with permission from Springer Nature, Copyright © 2019, Springer Science Business Media, LLC. (c) Practical example of a structurally annotated feature in step 1 showing the increase of D-lactic acid (LA) in the saliva sample of a diabetes mellitus (DM) patient compared to a healthy person and the verification by chiral metabolomics extraction by the mixed saliva sample (step 2) that the two peaks are indeed the corresponding L- and D-LA. *S* and *R* configurations always refer to the absolute configuration in the CDA and L and D to the configuration of the analyte (lactic acid).
Reproduced from ref. 133 with permission from Elsevier, Copyright 2015.

assignment, alignment and identification. Thus, the obtained peak list was further processed by an Excel macro for the discrimination of isotope coded/noncoded peak pairs (called DINA). With this program peak-picking was based on CDA-characteristic product ions (CPI) (*m*/*z* 209.1053 for both reagents, *m*/*z* 226.1315 for the carboxylic-reactive reagent and *m*/*z* 195.0987 for the amino-reactive reagent) and their calculated peak intensity ratios between light and heavy derivatives of each annotated feature with an isotope shift of 2.0067 Da used for subsequent multivariate statistical analysis. Since the heavy labeled metabolites coelute with their light labeled analogs, the ion suppression is corrected for by this approach. The precision for the peak intensities could be improved significantly as well for the 327 detected carboxylic acid and 511 amines (chiral and achiral ones) from around 70% of features having RSD below 20% without isotope correction to around 80% with isotope correction. It was demonstrated by human pooled serum spiked at different ratios of enantiomers (D/L = 0/100, 0.5/100, 1/100, and 2/100) of lactic acid (LA), 3-hydroxybutyric acid (3HA), 2-hydroxybutyric acid (2HA), and *N*-acetyl Trp (Ac-Trp) (carboxylic acids) or Ala, Val, Phe, and Trp (amines) that also differences in minor enantiomer peaks between samples can be detected. The different samples were grouped in the score plot of a PCA according to their spiking levels of D-forms. S-Plots indicated that elevated levels of Ac-D-Trp, D-LA, D-3HA and D-2HA were responsible for the grouping of the samples into 4 clusters according to the 4 different spiking levels. This method was used for biomarker discovery in the cerebrospinal fluid of patients with Alzheimer's disease (AD) (tested *vs.* non-AD samples). 24 biomarker candidates (including kynurenine, kynurenic acid and Trp) were successfully identified, including 8 chiral ones. Although it was unclear what additional information was obtained from the chiral metabolomics approach, as compared to a non-chiral one, the work showed that chiral untargeted metabolomics can be successfully realized and may have some promising aspects for future metabolomics studies.

12.5 Conclusions

Enantioselective metabolomics is a special topic within the wider field of metabolomics. Since common metabolomics approaches do not have an adequate coverage of chiral metabolites, special methods that can distinguish the distinct isomeric forms of chiral metabolites are needed. Achiral assays do not give a correct picture of the metabolic and physiological state in a variety of applications. The main application fields are amino acids, hydroxy acids and oxylipins for which enantioselective metabolomics approaches represent added value. The majority of studies reported in this field are targeted assays based on triple quadrupole MS with selected reaction monitoring acquisition focusing on a limited number of metabolites (usually <20). Chiral stationary phases with promising applicability for polar metabolites are CSPs based on polysaccharides, macrocyclic antibiotics, chiral crown ethers and cinchona-alkaloid derived chiral ion-exchangers. These phases have shown potential for the direct separation of underivatized polar metabolites, but also of their tagged derivatives. For labeled metabolites donor–acceptor phases (Pirkle type CSPs) have also shown great potential (*e.g.* for amino acid derivatives). Despite the extra effort and time that this derivatization step costs, it is more than compensated by the manifold benefits of more favorable chromatographic behavior, enhanced enantioselectivity, and increased detection sensitivity. Usually a specific class of metabolites (amino acids, hydroxy acids, oxylipins) is solely covered, yet rarely exceeds more than 20 targets which is far from the state in achiral targeted metabolomics which can cover 100s to 1000s of metabolites nowadays. Matrix interferences and matrix effects are problematic in enantioselective LC–MS/MS. Hence, 2D- and 3D-LC assays have been proposed *e.g.* for determination of trace D-levels of amino acids in biological samples, in combination with precolumn derivatization and fluorescence detection. Such targeted multidimensional assays can be regarded as the most accurate enantioselective assays currently available, however, it comes at the expense of excessive total analysis times. Since a derivatization step is usually involved anyway, and UHPLC columns were not available until recently, indirect enantiomer separation methods involving precolumn derivatization with a chiral derivatizing agent were quite popular for enantioselective metabolomics. Critical issues such as enantiomeric impurities were often not adequately considered and are a limiting factor for this approach, especially if trace enantiomers should be determined. Nevertheless, the majority of untargeted enantioselective metabolomics studies reported in the literature are based on chiral derivatizing agents and indirect enantiomer separation. Differential enantioselective metabolomics with isotope-coded CDA labeling and chiral metabolomics fingerprinting combined with chiral metabolomics extraction using CDA with opposite configurations are a few workflows that follow advanced untargeted metabolomics concepts. A broader implementation may be expected in the future. However, the entire field will remain strongly focused on D-amino acids as neuromodulators and biomarkers of disease, some hydroxy acids like 2-hydroxyglutaric acid and oxylipins for which enantioselective metabolomics has some significant importance in clinical analysis and research.

References

1. K. Segers, S. Declerck, D. Mangelings, Y. Vander Heyden and A. Van Eeckhaut, *Bioanalysis*, 2019, **11**, 2297–2318.
2. S. Tanwar and R. Bhushan, *Chromatographia*, 2015, **78**, 1113–1134.
3. T. Mochizuki, K. Todoroki, K. Inoue, J. Z. Min and T. Toyo'oka, *Anal. Chim. Acta*, 2014, **811**, 51–59.
4. B. Chankvetadze and G. Blaschke, *J. Chromatogr. A*, 2001, **906**, 309–363.
5. B. Chankvetadze, Enantioseparations by Capillary Electromigration Techniques, in *Chiral Analysis*, Elsevier, 2nd edn, 2018, pp. 565–605, https://doi.org/10.1016/B978-0-444-64027 7.00023-9.

6. B. Chankvetadze, Chiral Recognition and Enantioseparation Mechanisms in Capillary Electrokinetic Chromatography, in *Chiral Recognition in Separation Methods: Mechanisms and Applications*, ed. A. Berthod, Springer Berlin Heidelberg, Berlin, Heidelberg, 2010, pp. 97–152, ISBN 978-3-642-12445-7, https://www.springer.com/gp/book/9783642124440#aboutBook.
7. G. K. E. Scriba, *Chromatographia*, 2012, **75**, 815–838.
8. T. Miyamoto, M. Sekine, T. Ogawa, M. Hidaka, H. Homma and H. Masaki, *Chem. Biodiversity*, 2010, **7**(6), 1644–1650.
9. M. Gholami, B. A. Boughton, A. R. Fakhari, F. Ghanati, H. H. Mirzaei, L. Y. Borojeni, Y. Zhang, Z. S. Breitbach, D. W. Armstrong and U. Roessner, *Process Biochem.*, 2014, **49**, 140–147.
10. G. Hellinghausen and D. W. Armstrong, Cyclofructans as Chiral Selectors: An Overview, in *Chiral Separations*, 2019, vol. 1990, pp. 183–200, Series: Methods in Molecular Biology, 1985, 183–200, https://doi.org/10.1007/978-1-4939-9438-0_11.
11. Y. Konya, M. Taniguchi, M. Furuno, Y. Nakano, N. Tanaka and E. Fukusaki, *J. Chromatogr. A*, 2018, **1578**, 35–44.
12. Y. Nakano, M. Taniguchi and E. Fukusaki, *J. Biosci. Bioeng.*, 2019, **127**, 520–527.
13. K. Kinoshita, S. Jingu and J. I. Yamaguchi, *Anal. Biochem.*, 2013, **432**, 124–130.
14. Y. Konya, M. Taniguchi, M. Furuno, Y. Nakano, N. Tanaka and E. Fukusaki, *J. Chromatogr. A*, 2018, **1578**, 35–44.
15. T. Kimura, K. Hamase, Y. Miyoshi, R. Yamamoto, K. Yasuda, M. Mita, H. Rakugi, T. Hayashi and Y. Isaka, *Sci. Rep.*, 2016, **6**, 26137.
16. S. Karakawa, Y. Miyoshi, R. Konno, S. Koyanagi, M. Mita, S. Ohdo and K. Hamase, *Anal. Bioanal. Chem.*, 2013, **405**, 8083–8091.
17. K. Hamase, Y. Miyoshi, K. Ueno, H. Han, J. Hirano, A. Morikawa, M. Mita, T. Kaneko, W. Lindner and K. Zaitsu, *J. Chromatogr. A*, 2010, **1217**, 1056–1062.
18. J. Sasabe, Y. Miyoshi, M. Suzuki, M. Mita, R. Konno, M. Matsuoka, K. Hamase and S. Aiso, *Proc. Natl. Acad. Sci. U. S. A.*, 2012, **109**, 627–632.
19. H. Han, Q. Q. Wang, H. H. Wu and H. Wang, *Chinese J. Anal. Chem.*, 2014, **42**, 891–898.
20. R. J. Reischl, L. Hartmanova, M. Carrozzo, M. Huszar, P. Frühauf and W. Lindner, *J. Chromatogr. A*, 2011, **1218**, 8379–8387.
21. R. Kimura, H. Tsujimura, M. Tsuchiya, S. Soga, N. Ota, A. Tanaka and H. Kim, *Sci. Rep.*, 2020, **10**, 1–12.
22. J. Horak and M. Laemmerhofer, *Anal. Chem.*, 2019, **91**, 7679–7689.
23. J. Horak and M. Laemmerhofer, *J. Chromatogr. A*, 2019, **1596**, 69–78.
24. S. Karakawa, K. Shimbo, N. Yamada, T. Mizukoshi, H. Miyano, M. Mita, W. Lindner and K. Hamase, *J. Pharm. Biomed. Anal.*, 2015, **115**, 123–129.
25. J. Horak and M. Lämmerhofer, *J. Chromatogr. A*, 2019, **1596**, 69–78.
26. C. Calderón, J. Horak and M. Lämmerhofer, *J. Chromatogr. A*, 2016, **1467**, 239–245.

27. C. Calderón and M. Lämmerhofer, *J. Chromatogr. A*, 2017, **1487**, 194–200.
28. C. Calderón, C. Santi and M. Lämmerhofer, *J. Sep. Sci.*, 2018, **41**, 1224–1231.
29. A. Furusho, T. Akita, M. Mita, H. Naraoka and K. Hamase, *J. Chromatogr. A*, 2020, **1625**, 461255.
30. A. Furusho, R. Koga, T. Akita, M. Mita, T. Kimura and K. Hamase, *Anal. Chem.*, 2019, **91**, 11569–11575.
31. C. Ishii, T. Akita, M. Mita, T. Ide and K. Hamase, *J. Chromatogr. A*, 2018, **1570**, 91–98.
32. R. Koga, Y. Miyoshi, Y. Sato, M. Mita, R. Konno, W. Lindner and K. Hamase, *J. Chromatogr. A*, 2016, **1467**, 312–317.
33. K. Shimbo, A. Yahashi, K. Hirayama, M. Nakazawa and H. Miyano, *Anal. Chem.*, 2009, **81**, 5172–5179.
34. T. Huang, M. R. Armbruster, J. B. Coulton and J. L. Edwards, *Anal. Chem.*, 2019, **91**, 109–125.
35. M. Blum, I. Dogan, M. Karber, M. Rothe and W.-H. Schunck, *J. Lipid Res.*, 2019, **60**, 135–148.
36. C. Schneider, Z. Yu, W. E. Boeglin, Y. Zheng and A. R. Brash, *Methods Enzymol.*, 2007, **433**, 145–157.
37. I. Ilisz, A. Péter and W. Lindner, *TrAC, Trends Anal. Chem.*, 2016, **81**, 11–22.
38. I. Ali, M. Suhail and H. Y. Aboul-Enein, *TrAC, Trends Anal. Chem.*, 2019, **120**, 115634.
39. A. Acquaviva, G. Siano, P. Quintas, M. R. Filgueira and C. B. Castells, *J. Chromatogr. A*, 2020, **1614**, 460729.
40. U. Woiwode, R. J. Reischl, S. Buckenmaier, W. Lindner and M. Lämmerhofer, *Anal. Chem.*, 2018, **90**, 7963–7971.
41. U. Woiwode, S. Neubauer, W. Lindner, S. Buckenmaier and M. Lämmerhofer, *J. Chromatogr. A*, 2018, **1562**, 69–77.
42. K. Hamase, A. Morikawa, T. Ohgusu, W. Lindner and K. Zaitsu, *J. Chromatogr. A*, 2007, **1143**, 105–111.
43. Y. Miyoshi, K. Hamase, Y. Tojo, M. Mita, R. Konno and K. Zaitsu, *J. Chromatogr. B: Anal. Technol. Biomed. Life Sci.*, 2009, **877**, 2506–2512.
44. Y. Miyoshi, K. Hamase, T. Okamura, R. Konno, N. Kasai, Y. Tojo and K. Zaitsu, *J. Chromatogr. B: Anal. Technol. Biomed. Life Sci.*, 2011, **879**, 3184–3189.
45. Y. Nakade, Y. Iwata, K. Furuichi, M. Mita, K. Hamase, R. Konno, T. Miyake, N. Sakai, S. Kitajima, T. Toyama, Y. Shinozaki, A. Sagara, T. Miyagawa, A. Hara, M. Shimizu, Y. Kamikawa, K. Sato, M. Oshima, S. Yoneda-Nakagawa, Y. Yamamura, S. Kaneko, T. Miyamoto, M. Katane, H. Homma, H. Morita, W. Suda, M. Hattori and T. Wada, *JCI Insight*, 2018, **3**, 1–16.
46. R. Koga, Y. Miyoshi, E. Negishi, T. Kaneko, M. Mita, W. Lindner and K. Hamase, *J. Chromatogr. A*, 2012, **1269**, 255–261.

47. S. L. Liu, T. Oyama, Y. Miyoshi, S. Y. Sheu, M. Mita, T. Ide, W. Lindner, K. Hamase and J. A. Lee, *J. Pharm. Biomed. Anal.*, 2015, **116**, 80–85.
48. C. L. Hsieh, R. Koga, A. Furusho, T. Akita, M. Mita, T. Ide, J. A. Lee and K. Hamase, *J. Sep. Sci.*, 2018, **41**, 1298–1306.
49. T. Oyama, E. Negishi, H. Onigahara, N. Kusano, Y. Miyoshi, M. Mita, M. Nakazono, S. Ohtsuki, A. Ojida, W. Lindner and K. Hamase, *J. Pharm. Biomed. Anal.*, 2015, **116**, 71–79.
50. T. Kimura, K. Hamase, Y. Miyoshi, R. Yamamoto, K. Yasuda, M. Mita, H. Rakugi, T. Hayashi and Y. Isaka, *Sci. Rep.*, 2016, **6**, 26137.
51. L. Kiss, J. Bier, Y. Röder, N. Weissmann, F. Grimminger and W. Seeger, *Anal. Bioanal. Chem.*, 2008, **392**, 717–726.
52. S. A. Brose, B. T. Thuen and M. Y. Golovko, *J. Lipid Res.*, 2011, **52**, 850–859.
53. J. Homann, C. Lehmann, A. S. Kahnt, D. Steinhilber, M. J. Parnham, G. Geisslinger and N. Ferreiros, *J. Chromatogr. A*, 2014, **1360**, 150–163.
54. H. Han, Y. Miyoshi, K. Ueno, C. Okamura, Y. Tojo, M. Mita, W. Lindner, K. Zaitsu and K. Hamase, *J. Chromatogr. B: Anal. Technol. Biomed. Life Sci.*, 2011, **879**, 3196–3202.
55. R. Koga, H. Yoshida, H. Nohta and K. Hamase, *Chromatography*, 2019, **40**, 1–8.
56. G. Genchi, *Amino Acids*, 2017, **49**, 1521–1533.
57. Ü. Kolukisaoglu, *Int. J. Mol. Sci.*, 2020, **21**, 1–13.
58. Y. Miyoshi, R. Koga, T. Oyama, H. Han, K. Ueno, K. Masuyama, Y. Itoh and K. Hamase, *J. Pharm. Biomed. Anal.*, 2012, **69**, 42–49.
59. S. Batra and R. Bhushan, *Biomed. Chromatogr.*, 2018, 32, e4090.
60. G. Carenzi, S. Sacchi, M. Abbondi and L. Pollegioni, *Amino Acids*, 2020, **52**, 849–862.
61. Y. Nakano, M. Taniguchi, Y. Umakoshi, D. Watai and E. Fukusaki, in *Amino Acid Analysis: Methods and Protocols*, ed. M. A. Alterman, Springer New York, New York, NY, 2019, vol. 2030, pp. 253–261.
62. K. Petritis, A. Valleix, C. Elfakir and M. Dreux, *J. Chromatogr. A*, 2001, **913**, 331–340.
63. R. Hellinger, J. Horak and W. Lindner, *Anal. Bioanal. Chem.*, 2013, **405**, 8105–8120.
64. S. Du, Y. Wang, C. A. Weatherly, K. Holden and D. W. Armstrong, *Anal. Bioanal. Chem.*, 2018, **410**, 2971–2979.
65. M. Kato, T. Fukushima, T. Santa, K. Nakashima, R. Nishioka and K. Imai, *Analyst*, 1998, **123**, 2877–2882.
66. C. Ishii, T. Akita, M. Nagano, M. Mita and K. Hamase, *Chromatography*, 2019, **40**, 83–87.
67. Y. Miyoshi, M. Nagano, S. Ishigo, Y. Ito, K. Hashiguchi, N. Hishida, M. Mita, W. Lindner and K. Hamase, *J. Chromatogr. B: Anal. Technol. Biomed. Life Sci.*, 2014, **966**, 187–192.
68. E. H. Oliw, A. Wennman, I. Hoffmann, U. Garscha, M. Hamberg and F. Jernerén, *J. Lipid Res.*, 2011, **52**, 1995–2004.

69. E. H. Oliw and A. Wennman, Chiral Phase-HPLC Separation of Hydroperoxyoctadecenoic Acids and Their Biosynthesis by Fatty Acid Dioxygenases, *Advanced Protocols in Oxidative Stress III, Methods in Molecular Biology*, ed. D. Armstrong, Springer Science + Business Media, New York, 2015, vol. 1208, pp. 85–95, https://doi.org/10.1007/978-1-4939-1441-8_7.
70. S. Hong, B. V. Alapure, Y. Lu, H. Tian and Q. Wang, *Br. J. Dermatol.*, 2014, **171**, 30–38.
71. S. F. Oh, P. S. Pillai, A. Recchiuti, R. Yang and C. N. Serhan, *J. Clin. Invest.*, 2011, **121**, 569–581.
72. S. Wei, J. J. Brittin, J. R. Falck, S. Anjaiah, K. Nithipatikom, L. Cui, W. B. Campbell and J. H. Capdevila, *Anal. Biochem.*, 2006, **352**, 129–134.
73. A. P. Neilson, J. Ren, Y. H. Hong, A. Sen, W. L. Smith, D. E. Brenner and Z. Djuric, *Nutr. Cancer*, 2012, **64**, 163–172.
74. V. Goss, A. Cazenave-Gassiot, A. Pringle and A. Postle, *Curr. Anal. Chem.*, 2014, **10**, 121–131.
75. N. Ferreiros, J. Homann, S. Labocha, N. Grossmann, J. S. Hahn, B. Brune and G. Geisslinger, *Talanta*, 2014, **127**, 82–87.
76. B. Deng, C. W. Wang, H. H. Arnardottir, Y. Li, C. Y. C. Cheng, J. Dalli and C. N. Serhan, *PLoS One*, 2017, DOI: 10.1371/journal.pone.0102362.
77. J. Kolmert, A. Fauland, D. Fuchs, J. Säfholm, C. Gómez, M. Adner, S.-E. Dahlén and C. E. Wheelock, *Anal. Chem.*, 2018, **90**, 10239–10248.
78. H. Yamada, K. Kumagai, A. Uemura and S. Yuki, *Biosci., Biotechnol., Biochem.*, 2020, **84**, 455–462.
79. M. Bayer, A. Mosandl and D. Thaçi, *J. Chromatogr. B: Anal. Technol. Biomed. Life Sci.*, 2005, **819**, 323–328.
80. A. Toewe, L. Balas, T. Durand, G. Geisslinger and N. Ferreiros, *Anal. Chim. Acta*, 2018, **1031**, 185–194.
81. M. Cebo, X. Fu, M. Gawaz, M. Chatterjee and M. Lämmerhofer, *J. Chromatogr. A*, 2020, **1624**, 461206.
82. Y. Xiao and F. P. Guengerich, *J. Lipid Res.*, 2012, **53**, 1610–1617.
83. U. Garscha, T. Nilsson and E. H. Oliw, *J. Chromatogr. B: Anal. Technol. Biomed. Life Sci.*, 2008, **872**, 90–98.
84. A. M. Abdel-Mawgoud, F. Lepine and E. Deziel, *J. Chromatogr. A*, 2013, **1306**, 37–43.
85. E. Waddington, K. Sienuarine, I. Puddey and K. Croft, *Anal. Biochem.*, 2001, **292**, 234–244.
86. C. Mesaros, S. H. Lee and I. A. Blair, *Rapid Commun. Mass Spectrom.*, 2010, **24**, 3237–3247.
87. T. D. Hammonds, I. A. Blair, J. R. Falck and J. H. Capdevila, *Anal. Biochem.*, 1989, **182**, 300–303.
88. M. Brezinova, O. Kuda, J. Hansikova, M. Rombaldova, L. Balas, K. Bardova, T. Durand, M. Rossmeisl, M. Cerna, Z. Stranak and J. Kopecky, *Biochim. Biophys. Acta, Mol. Cell Biol. Lipids*, 2018, **1863**, 126–131.

89. M. S. Rashed, H. Y. Aboul-Enein, M. AlAmoudi, M. Jakob, L. Y. Al-Ahaideb, A. Abbad, S. Shabib and E. Al-Jishi, *Biomed. Chromatogr.*, 2002, **16**, 191–198.
90. P. A. Tyrakis, A. Palazon, D. Macias, K. L. Lee, A. T. Phan, P. Velica, J. You, G. S. Chia, J. Sim, A. Doedens, A. Abelanet, C. E. Evans, J. R. Griffiths, L. Poellinger, A. W. Goldrath and R. S. Johnson, *Nat*ure, 2016, **540**, 236–241.
91. P. Y. Lin, S. M. Chen, C. L. Hsieh, C. Y. Lin, Y. S. Huang, K. Hamase and J. A. Lee, *J. Pharm. Biomed. Anal.*, 2020, **188**, 113367.
92. Y. Tojo, K. Hamase, M. Nakata, A. Morikawa, M. Mita, Y. Ashida, W. Lindner and K. Zaitsu, *J. Chromatogr. B: Anal. Technol. Biomed. Life Sci.*, 2008, **875**, 174–179.
93. K. Hamase, Y. Nakauchi, Y. Miyoshi, R. Koga, N. Kusano, H. Onigahara, H. Naraoka, H. Mita, Y. Kadota, Y. Nishio, M. Mita and W. Lindner, *Chromatography*, 2014, **35**, 103–110.
94. A. Morikawa, K. Hamase, T. Inoue, R. Konno and K. Zaitsu, *Amino Acids*, 2007, **32**, 13–20.
95. H. Han, Y. Miyoshi, T. Oyama, R. Konishi, M. Mita and K. Hamase, *J. Sep. Sci.*, 2011, **34**, 2847–2853.
96. Y. Miyoshi, R. Konno, J. Sasabe, K. Ueno, Y. Tojo, M. Mita, S. Aiso and K. Hamase, *Amino Acids*, 2012, **43**, 1919–1931.
97. H. Han, Y. Miyoshi, R. Koga, M. Mita, R. Konno and K. Hamase, *J. Pharm. Biomed. Anal.*, 2015, **116**, 47–52.
98. M. Suzuki, Y. Gonda, M. Yamada, A. A. Vandebroek, M. Mita, K. Hamase, M. Yasui and J. Sasabe, *Sci. Rep.*, 2019, **9**, 1–6.
99. A. Hesaka, S. Sakai, K. Hamase, T. Ikeda, R. Matsui, M. Mita, M. Horio, Y. Isaka and T. Kimura, *Sci. Rep.*, 2019, **9**, 1–8.
100. F. Ianni, G. Saluti, R. Galarini, S. Fiorito, R. Sardella and B. Natalini, *Free Radicals Biol. Med.*, 2019, **144**, 35–54.
101. M. Gabbs, S. Leng, J. G. Devassy, M. Monirujjaman and H. M. Aukema, *Adv. Nutr.*, 2015, **6**, 513–540.
102. L. L. Mazaleuskaya, A. Salamatipour, D. Sarantopoulou, L. Weng, G. A. FitzGerald, I. A. Blair and C. Mesaros, *J. Lipid Res.*, 2018, **59**, 564–575.
103. F. Ianni, G. Saluti, R. Galarini, S. Fiorito, R. Sardella and B. Natalini, *Free Radicals Biol. Med.*, 2019, **144**, 35–54.
104. J. B. Ewaschuk, G. A. Zello, J. M. Naylor and D. R. Brocks, *J. Chromatogr. B: Anal. Technol. Biomed. Life Sci.*, 2002, **781**, 39–56.
105. G. R. Villani, G. Gallo, E. Scolamiero, F. Salvatore and M. Ruoppolo, *Clin. Exp. Med.*, 2017, **17**, 305–323.
106. O. A. Mamer and M. L. Reimer, *J. Biol. Chem.*, 1992, **267**, 22141–22147.
107. E. Haan, G. Brown, A. Bankier, D. Mitchell, S. Hunt, J. Blakey and G. Barnes, *Eur. J. Pediatr.*, 1985, **144**, 63–65.
108. J. Hagen, H. te Brinke, R. J. A. Wanders, A. C. Knegt, E. Oussoren, A. J. M. Hoogeboom, G. J. G. Ruijter, D. Becker, K. O. Schwab, I. Franke,

M. Duran, H. R. Waterham, J. O. Sass and S. M. Houten, *J. Inherited Metab. Dis.*, 2015, **38**, 873–879.
109. P. G. Barth, G. F. Hoffmann, J. Jaeken, W. Lehnert, F. Hanefeld, A. H. van Gennip, M. Duran, J. Valk, R. B. Schutgens and F. K. Trefz, *Ann. Neurol.*, 1992, **32**, 66–71.
110. R. A. Chalmers, A. M. Lawson, R. W. E. Watts, A. S. Tavill, J. P. Kamerling, E. Hey and D. Ogilvie, *J. Inherited Metab. Dis.*, 1980, **3**, 11–15.
111. M. Kranendijk, E. A. Struys, G. S. Salomons, M. S. Van der Knaap and C. Jakobs, *J. Inherited Metab. Dis.*, 2012, **35**, 571–587.
112. J. B. Brooks, J. V. Kasin, D. M. Fast and M. I. Daneshvar, *J. Clin. Microbiol.*, 1987, **25**, 445–448.
113. X. Li, Z. Xu, X. Lu, X. Yang, P. Yin, H. Kong, Y. Yu and G. Xu, *Anal. Chim. Acta*, 2009, **633**, 257–262.
114. W. Checkley, M. P. Deza, J. Klawitter, K. M. Romero, J. Klawitter, S. L. Pollard, R. A. Wise, U. Christians and N. N. Hansel, *Respir. Med.*, 2016, **121**, 59–66.
115. A. J. Buglass and S. H. Lee, *J. Chromatogr. Sci.*, 2001, **39**, 453–458.
116. M. S. Rashed, M. AlAmoudi and H. Y. Aboul-Enein, *Biomed. Chromatogr.*, 2000, **14**, 317–320.
117. M. S. Rashed, H. Y. Aboul-Enein, M. AlAmoudi, M. Jakob, L. Y. Al-Ahaideb, A. Abbad, S. Shabib and E. Al-Jishi, *Biomed. Chromatogr.*, 2002, **16**, 191–198.
118. D. Norton, B. Crow, M. Bishop, K. Kovalcik, J. George and J. A. Bralley, *J. Chromatogr. B: Anal. Technol. Biomed. Life Sci.*, 2007, **850**, 190–198.
119. P. Franco, T. Zhang, A. Gargano, M. Mahut, M. Lammerhofer, W. Lindner and K. E. Y. Points, *LC · GC Eur.*, 2012, **25**, 600–610.
120. F. Ianni, Z. Pataj, H. Gross, R. Sardella, B. Natalini, W. Lindner and M. Lämmerhofer, *J. Chromatogr. A*, 2014, **1363**, 101–108.
121. C. V. Hoffmann, R. Pell, M. Lämmerhofer and W. Lindner, *Anal. Chem.*, 2008, **80**, 8780–8789.
122. L. Deng, H. Nakano and Y. Iwasaki, *J. Chromatogr. A*, 2007, **1165**, 93–99.
123. L. Deng, H. Nakano and Y. Iwasaki, *J. Chromatogr. A*, 2008, **1198–1199**, 67–72.
124. P. García, P. Franco, R. Úlvarez and Ú. R. De Lera, *J. Sep. Sci.*, 2011, **34**, 999–1003.
125. W. Piyatheerawong, Y. Iwasaki and T. Yamane, *J. Chromatogr. A*, 2005, **1068**, 243–248.
126. T. Nagai, T. Kinoshita, E. Kasamatsu, K. Yoshinaga, H. Mizobe, A. Yoshida, Y. Itabashi and N. Gotoh, *J. Oleo Sci.*, 2019, **68**, 1019–1026.
127. M. Kalpio, J. D. Magnússon, H. G. Gudmundsson, K. M. Linderborg, H. Kallio, G. G. Haraldsson and B. Yang, *Chem. Phys. Lipids*, 2020, **231**, 104937.
128. M. Lísa and M. Holčapek, *Anal. Chem.*, 2013, **85**, 1852–1859.
129. T. Řezanka, I. Kolouchová, L. Nedbalová and K. Sigler, *J. Chromatogr. A*, 2018, **1557**, 9–19.

130. T. Řezanka, K. Pádrová and K. Sigler, *Anal. Biochem.*, 2017, **524**, 3–12.
131. R. Nagao, H. Tsutsui, T. Mochizuki, T. Takayama, T. Kuwabara, J. Z. Min, K. Inoue, K. Todoroki and T. Toyo'oka, *J. Chromatogr. A*, 2013, **1296**, 111–118.
132. T. Toyooka, *Anal. Sci.*, 2017, **33**, 555–564.
133. T. Takayama, T. Mochizuki, K. Todoroki, J. Z. Min, H. Mizuno, K. Inoue, H. Akatsu, I. Noge and T. Toyo'oka, *Anal. Chim. Acta*, 2015, **898**, 73–84.
134. T. Mochizuki, T. Takayama, K. Todoroki, K. Inoue, J. Z. Min and T. Toyo'oka, *Anal. Chim. Acta*, 2015, **875**, 73–82.
135. T. Takayama, T. Kuwabara, T. Maeda, I. Noge, Y. Kitagawa, K. Inoue, K. Todoroki, J. Z. Min and T. Toyo'oka, *Anal. Bioanal. Chem.*, 2015, **407**, 1003–1014.
136. I. Ilisz, R. Berkecz and A. Péter, *J. Pharm. Biomed. Anal.*, 2008, **47**, 1–15.
137. I. Ilisz, A. Aranyi, Z. Pataj and A. Péter, *J. Chromatogr. A*, 2012, **1269**, 94–121.
138. I. Ilisz, A. Aranyi and A. Péter, *J. Chromatogr. A*, 2013, **1296**, 119–139.
139. H. Brückner, T. Westhauser and H. Godel, *J. Chromatogr. A*, 1995, **711**, 201–215.
140. S. Sethi, J. Martens and R. Bhushan, *Biomed. Chromatogr.*, 2021, **35**, e4990.
141. R. Bhushan and H. Brückner, *J. Chromatogr. B: Anal. Technol. Biomed. Life Sci.*, 2011, **879**, 3148–3161.
142. M. Kobayashi, Y. Takano, H. Takishima, S. Sakaitani, M. Niitsu and T. Furuchi, *Chromatographia*, 2019, **82**, 705–708.
143. S. Hess, A Universal HPLC-MS Method to Determine the Stereochemistry of Common and Unusual Amino Acids, in *Amino Acid Analysis: Methods and Protocols*, 2019, ed. M. Alterman, vol. 2030, pp. 263–275, ISBN 978-1-4939-9639-1.
144. N. J. Ayon, A. D. Sharma and W. G. Gutheil, *J. Am. Soc. Mass Spectrom*, 2019, **30**, 448–458.
145. K. Moon, C. Lim, S. Kim and D. C. Oh, *J. Chromatogr. A*, 2013, **1272**, 141–144.
146. K. Fujii and K. I. Harada, *Anal. Chem.*, 1997, **69**, 5146–5151.
147. S. Hess, K. R. Gustafson, D. J. Milanowski, E. Alvira, M. A. Lipton and L. K. Pannell, *J. Chromatogr. A*, 2004, **1035**, 211–219.
148. R. C. Moldovan, E. Bodoki, A. C. Servais, J. Crommen, R. Oprean and M. Fillet, *J. Chromatogr. A*, 2017, **1513**, 1–17.
149. R.-C. C. Moldovan, E. Bodoki, A.-C. C. Servais, J. Crommen, R. Oprean and M. Fillet, *J. Chromatogr. A*, 2019, **1590**, 80–87.
150. R. Pérez-Míguez, B. Bruyneel, M. Castro-Puyana, M. L. Marina, G. W. Somsen and E. Domínguez-Vega, *Anal. Chem.*, 2019, **91**, 3277–3285.

151. S. Vijayasarathy, P. Prasad, L. J. Fremlin, R. Ratnayake, A. A. Salim, Z. Khalil and R. J. Capon, *J. Nat. Prod.*, 2016, **79**, 421–427.
152. R.-C. Moldovan, E. Bodoki, A.-C. Servais, J. Crommen, R. Oprean and M. Fillet, *J. Chromatogr. A*, 2019, **1590**, 80–87.
153. T. Toyo'oka, *Methods Mol. Biol.*, 2013, **970**, 233–248.
154. T. Toyo'oka, Chiral Metabolomics Using Triazine-Based Chiral Labeling Reagents by UPLC-ESI-MS/MS, in *Chiral Separations: Methods and Protocols, Methods in Molecular Biology*, ed. G. K. E. Scriba, vol. 1985, pp. 57–79, https://doi.org/10.1007/978-1-4939-9438-0_4, Springer Science + Business Media, LLC, part of Springer Nature, 2019.
155. B. Fransson and U. Ragnarsson, *J. Chromatogr. A*, 1998, **827**, 31–36.
156. S. Mason, C. J. Reinecke, W. Kulik, A. van Cruchten, R. Solomons and A. M. T. van Furth, Cerebrospinal fluid in tuberculous meningitis exhibits only the L-enantiomer of lactic acid, *BMC Infect. Dis.*, 2016, **16**, 251.
157. E.-H. Shim, C. B. Livi, D. Rakheja, J. Tan, D. Benson, V. Parekh, E.-Y. Kho, A. P. Ghosh, R. Kirkman, S. Velu, S. Dutta, B. Chenna, S. L. Rea, R. J. Mishur, Q. Li, T. L. Johnson-Pais, L. Guo, S. Bae, S. Wei, K. Block and S. Sudarshan, *Cancer Discovery*, 2014, **4**, 1290–1298.
158. J.-Y. Zheng, X. Jiang, J.-L. Zhou, Z.-Q. Shi, L.-F. Liu and G.-Z. Xin, *Anal. Chim. Acta*, 2019, **1077**, 174–182.
159. Q.-Y. Cheng, J. Xiong, W. Huang, Q. Ma, W. Ci, Y.-Q. Feng and B.-F. Yuan, *Sci. Rep.*, 2015, **5**, 15217.
160. J. Zheng, S. J. Zheng, C. B. Qi, D. M. Wu and Y. Q. Feng, *Anal. Chem.*, 2019, **91**, 11440–11446.
161. Y. Xie, G. M. Alexander, R. J. Schwartzman, N. Singh, M. C. Torjman, M. E. Goldberg, I. W. Wainer and R. Moaddel, *J. Pharm. Biomed. Anal.*, 2014, **89**, 1–5.
162. W. F. Visser, N. M. Verhoeven-Duif, R. Ophoff, S. Bakker, L. W. Klomp, R. Berger and T. J. De Koning, *J. Chromatogr. A*, 2011, **1218**, 7130–7136.
163. R. Bhushan and S. Dixit, *Amino Acids*, 2012, **42**, 1371–1378.
164. K. Takada, A. Ninomiya, M. Naruse, Y. Sun, M. Miyazaki, Y. Nogi, S. Okada and S. Matsunaga, *J. Org. Chem.*, 2013, **78**, 6746–6750.
165. C. Müller, J. R. Fonseca, T. M. Rock, S. Krauss-Etschmann and P. Schmitt-Kopplin, *J. Chromatogr. A*, 2014, **1324**, 109–114.
166. K. Inoue, D. Hosaka, N. Mochizuki, H. Akatsu, K. Tsutsumiuchi, Y. Hashizume, N. Matsukawa, T. Yamamoto and T. Toyo'Oka, *Anal. Chem.*, 2014, **86**, 797–804.
167. H. Iizuka, K. Ishii, Y. Hirasa, K. Kubo and T. Fukushima, *J. Chromatogr. B: Anal. Technol. Biomed. Life Sci.*, 2011, **879**, 3208–3213.
168. H. Iizuka, Y. Hirasa, K. Kubo, K. Ishii, T. Toyo'oka and T. Fukushima, *Biomed. Chromatogr.*, 2011, **25**, 743–747.
169. T. Mochizuki, S. Taniguchi, H. Tsutsui, J. Z. Min, K. Inoue, K. Todoroki and T. Toyo'oka, *Anal. Chim. Acta*, 2013, **773**, 76–82.

170. T. Takayama, H. Mizuno, T. Toyo'oka, H. Akatsu, K. Inoue and K. Todoroki, *Anal. Chem.*, 2019, **91**, 4396–4404.
171. S. Zhang, J. Shi, C. Shan, C. Huang, Y. Wu, R. Ding, Y. Xue, W. Liu, Q. Zhou, Y. Zhao, P. Xu and X. Gao, *Anal. Chim. Acta*, 2017, **978**, 24–34.
172. T. Sakamoto, S. Furukawa, T. Nishizawa, M. Fukuda, M. Sasaki, M. Onozato, S. Uekusa, H. Ichiba and T. Fukushima, *J. Chromatogr. A*, 2019, **1585**, 131–137.
173. S. Fukui, T. Takayama, T. Toyo'oka, H. Mizuno and K. Todoroki, *Talanta*, 2019, **205**, 120146.
174. M. H. El-Maghrabey, N. Kishikawa and N. Kuroda, *Biomed. Chromatogr.*, 2020, **34**, e4756.

Subject Index